Thirty-Second Hanford Symposium on Health and the Environment

Regional Impacts of Global Climate Change:

Assessing Change and Response at the Scales That Matter

October 19–21, 1993
Richland, Washington, U.S.A.

Edited by
SJ Ghan
WT Pennell
KL Peterson
E. Rykiel
MJ Scott
LW Vail

BATTELLE PRESS
Columbus • Richland

PACIFIC NORTHWEST NATIONAL LABORATORY
operated by
BATTELLE
UNITED STATES DEPARTMENT OF ENERGY
under Contract DE-AC06-76RLO 1830

Library of Congress Cataloging-in-Publication Data

Hanford Symposium on Health and the Environment (32nd : 1993 : Richland, Wash.)
Regional impacts of global climate change : assessing change and response at the scales that matter / edited by S. J. Ghan . . . [et al.].
p. cm.
"Thirty-second Hanford Symposium on Health and the Environment. October 19–21, 1993, Richland, Washington, U.S.A."
Includes bibliographical references and index.
ISBN 1-57477-017-9 (alk. paper)
1. Climatic changes—Mathematical models—Congresses. 2. Climatic changes—Simulation methods—Congresses. I. Title.
QC981.8.C5H253 1993 95-46745
363.73′87—dc20 CIP

Additional copies may be ordered through Battelle Press, 505 King Avenue, Columbus, Ohio, 43201, U.S.A., (614) 424–6393 or 1–800–451–3543.
Fax: (614) 424–3819

CONTENTS

PREFACE

The Thirty-Second Hanford Symposium on Health and the Environment examined the problem of predicting climate change at the scales relevant to human and natural systems and understanding the response of these systems to such change. The Symposium focused on the interface between the disciplines of effects assessment and climate change prediction. We examined how this interface might be bridged to improve our ability to understand the physical, economic, and social consequences of human-induced climate change and to evaluate the effectiveness of possible responses to it.

Physical scientists, ecologists, agronomists, social scientists, life scientists, and risk assessment practitioners discussed the problems and prospects for gauging the effects of global climate change on human and natural systems. The public and government policy implications of these climate change impacts are also discussed. The papers contained in this book are the first comprehensive collection of papers on the subject of regional, as contrasted to global, impacts of climate change. It represents a valuable resource for scientists, government administrators, and industrial and agricultural planners. The papers are written at a level that is understandable to a general scientific audience, but they also contain specific information that will be of value for researchers working in the areas of global warming and climate change modeling.

The editors would like to express appreciation to Ms. Dvara Lee Felton and Dr. Raymond W. Baalman for their dedicated attention to technical details in the editing of the symposium proceedings.

MODELING REGIONAL-SCALE CLIMATE

PERSPECTIVES FOR EARTH SYSTEM MODELING ON THE REGIONAL SCALE

F. Giorgi

National Center for Atmospheric Research, Boulder, Colorado

ABSTRACT

A number of experiments recently conducted at the National Center forAtmospheric Research (NCAR) and in other laboratories worldwide have shown the feasibility of applying limited-area regional climate models (RegCMs) to long-term climate simulation. Nested within general circulation models (GCMs), current RegCMs allow attainment of grid point spacings of up to a few tens of kilometers over selected regions. At such resolution the effects of complex topographical features, vegetation inhomogeneities, coastlines, and large inland bodies of water on the regional climate can be captured in a physically based way.

With the advent of the latest generation of computing systems (multiprocessor vectorial supercomputers, massively parallel machines, and large clusters of high-performance work stations), it may be feasible to have long-term (multiyear to multidecades), nested RegCM/GCM simulations at resolutions as fine as 10 km, or less, thereby filling a gap in the spatial-temporal scale in current climate modeling, i.e., long-term simulations at high spatial resolution. This would allow the coupling of RegCMs with regional Earth process models, such as hydrology, ecosystem, ocean-sea ice and chemistry models, at spatial and temporal resolutions most internally consistent from the viewpoint of available models and relevant processes. This paper reviews some of the latest experiments with an NCAR RegCM and discusses the opportunities that the nested modeling technique opens for the development of multipurpose, coupled, regional Earth process modeling systems.

INTRODUCTION

Following a series of initial experiments performed at the National Center for Atmospheric Research (NCAR; Dickinson et al., 1989; Giorgi, 1990; Marinucci and Giorgi, 1992; Giorgi et al., 1992; 1993a, 1994a), limited area models nested within general circulation models (GCM) have been increasingly recognized as useful tools to study climate issues on the regional scale. In fact, versions of several major mesoscale modeling systems have been modified and tested for climate application. Among them are the NCAR/Pennsylvania State University southeastern Australia (Pitman et al., 1991). The model has been run in climate mode at grid point spacings ranging from 20 to 80 km and for periods of time ranging from a few days (e.g., Dickinson et al., 1989) to monthly (Giorgi, 1990; Marinucci and Giorgi, 1992) to multiyear (Giorgi et al., 1993a, 1994a). In all these experiments the model was run continuously, i.e., it was initialized only once, then driven by large-scale fields provided at a model lateral buffer zone. Large-scale driving fields were obtained either from analyses of observations (e.g., Giorgi et al., 1993a) or from output of GCM simulations (e.g., Giorgi et al., 1994a). Finally, the model has been applied to simulations of present-day climate conditions (Giorgi and Bates, 1989; Giorgi, 1990; Marinucci and Giorgi, 1992), climate conditions under doubled carbon dioxide concentration (Giorgi et al., 1992, 1994a) and a paleoclimate (Hostetler et al., 1994).

An overall evaluation of the performance of the model, resulting from the various validation runs, indicates that the model reproduced realistic structures of synoptic events even after long integration times. It also captured the basic dynamical, hydrological, and thermal effects of the high-resolution forcings associated with the complex topography and coastline of the regions. Simulated regional temperature values were generally within a few degrees of those observed, with a reasonably good description of seasonal and diurnal cycles (e.g., Giorgi et al., 1993a). Simulated precipitation was generally within 10 to 50% of observed, with performance generally better during the cold season than the warm season.

For illustrative purposes, some results are briefly presented from two of the latest experiments performed with RegCM: those of Giorgi et al. (1994a) and Hostetler et al. (1994). Giorgi et al. (1994a) produced 2 × CO_2 regional scenarios over the conterminous United States by completing two continuous, 3.5-yr-long simulations: one for present-day conditions (control experiment), and one with doubled carbon dioxide concentration (2 × CO_2 experiment), with the RegCM driven by a version of the NCAR Community Climate Model (CCM) run at a resolution of about 4.5° latitude and 7.5° longitude. The RegCM domain encompassed the whole United States, southern Canada, northern Mexico, and adjacent ocean waters at a grid point spacing of 60 km, a resolution at which major features of United States topography (the Rocky Mountains, the Sierra Nevada, and the Appalachians) and coastlines (the

Baja and Florida peninsulas and the Great Lakes) are captured (see Giorgi et al., 1994a).

The results which best illustrate the performance of the nested regional model are shown in Figures 1A-E and 2A-B. Figures 1A-E show for observations and RegCM runs what Giorgi et al. (1994a) define as the large-scale and mesoscale cold season precipitation signal. The large-scale signal is obtained by performing spatial averaging over a scale comparable to the CCM resolution. The mesoscale signal is obtained by subtracting the large-scale signal from the original gridded data and is thus a measure of the effect of sub-GCM grid scale processes. The observed data were interpolated to the RegCM grid from the half-degree resolution dataset of Legates and Willmott (1990). For comparison, Figure 1E shows the cold season precipitation simulated by the driving CCM control run.

The result of interest here is that, although significantly overestimating precipitation, the CCM captures the main features of the large-scale precipitation patterns, with maxima over the eastern and western coasts and minima over the central plains. On the other hand, the CCM misses the mesoscale signal associated with the complex United States topography, i.e., relative maxima (positive signal) over the coastal ranges, Sierra Nevada, Rocky Mountains, and, though less pronounced, the Appalachians. The CCM also misses relative minima (negative signal), over the Columbia Basin, Great Basin, and Great Plains, produced by the precipitation shadowing effect of upwind mountain systems. Conversely, the RegCM captures this mesoscale signal quite well, both in location and magnitude.

The simulated 2 $\times$ CO_2-control changes in cold season precipitation (Figures 2A-B) show that the large-scale patterns in precipitation change produced by the driving and nested models are similar, with a precipitation increase over most of the continental United States except areas of the southwest and northeast. The details of the predicted changes, however, can be quite different. The nested model places most of the precipitation increase along a narrow band over the western coasts and simulates areas of negative change in the lee of the Rockies, where the CCM-simulated changes are positive. This is due to the presence in the RegCM of sharp topographical gradients which are not captured by the CCM topography. Temperature change scenarios are discussed in Giorgi et al. (1993b), while a detailed analysis of the simulated surface hydrology is reported in Giorgi et al. (1994b).

The examples above show that it is possible today to perform continuous, multiyear simulations of reasonably good quality for different climate conditions at resolutions of up to ~50 km with nested regional climate models. (As another example, two 5-yr-long simulations for present day and 2 × CO_2 conditions over eastern Asia have been recently completed at NCAR and are currently being analyzed). Recent developments in climate modeling indicate that the performance of nested RegCM/GCM regional modeling systems will improve substantially in the next few years. For example, the latest generation of the NCAR CCM, the CCM2, shows much improved simulation of present-day climate conditions compared with previous versions (Hack, 1994). Similar improvements have been produced by other GCM (Bengtsson, 1992). This indicates that the reliability of the driving large-scale GCM fields used in nested model simulations will likely be much improved in the near future. Parallel improvements are taking place in the development of regional models. As mentioned in the introduction, new versions of mesoscale modeling systems are being developed for climate application. Through the use of semilagrangian dynamic representations or multiple nesting capabilities, these models can reach regional resolution of a few tens of km or less.

The runs by Giorgi et al. (1994a) were completed with the first-generation RegCM and required about 180 hr of Cray Y-MP CPU hours per simulated year. The second generation model is over three times faster than the old one, and similar improved efficiency can be expected in other modeling systems. Therefore, with the advent of the next generation vector supercomputers, the use of models in multitasking configuration, and the use of massively parallel processing machines or clusters of high-performance workstations, it is the author's opinion that simulations at resolutions as high as 5 to 20 km for time periods of up to a few decades may become feasible in the near future. (For example, a series of month-long simulations over the Alpine region with the latest RegCM at 20-km grid point spacing, nested within a high-resolution version of the MPI GCM (ECHAM3) have already been completed on an NEC SX/3 supercomputer). The perspectives that this opens for regional ESP modeling are the focus of the next sections.

MODELING EARTH SYSTEM INTERACTIONS ON THE REGIONAL SCALE

As in Giorgi (1993), the discussion here is limited to three broad ESP components: vegetation/hydrology, ocean/sea-ice, and atmospheric chemosphere/aerosols. Giorgi (1993) discusses a number of issues whose treatment requires high regional spatial and temporal resolution. A few examples are useful.

One of the main issues in the study and modeling of atmosphere-biosphere interactions is the role of surface heterogeneities. Heterogeneous surface characteristics affect the energy and water cycles on a wide range of scales (Pielke and Avissar, 1990). Increasing attention is being devoted, however, to heterogeneities occurring on scales of a few tens of km, which can produce organized mesoscale circulations and modify the thermal and dynamical structure of overlying synoptic weather systems. This is because human activities related to agriculture (e.g., irrigation and deforestation) and urban development are beginning to modify the environment at those scales. The effects of such modifications may be important in transitional climatic settings (such as those in semiarid regions), and indications that extensive irrigation and deforestation affect the local climate are already quite strong (Barnston and Schickedanz, 1984; Henderson-Sellers and Gornitz, 1984). Variability in surface morphology (slope and soil properties) is also critical in regulating the surface-water budget (Wood et al., 1990), and extreme events associated with mesoscale phenomena can influence crop yields, ecosystem dynamics, and basin hydrology to a greater extent than "average" climate.

Studies of ocean-atmosphere interactions offer extremely interesting opportunities for coupling at the regional scale. The mutual effects of synoptic and mesoscale atmospheric systems and ocean mesoscale vortices and eddies are still poorly known. Mesoscale eddies of scales ranging from 50 to 300 km significantly contribute to horizontal energy transport in the oceans and can generate sharp horizontal gradients in the energy exchanges with the atmosphere (e.g., Washington and Parkinson, 1986). Another area where phenomena of mesoscale nature affect the ocean circulation is deep water formation, which often appears to be generated through very localized (scale of a few tens to a few hundreds km) and episodic events (Warren, 1981). Such events can be associated with intense brine release or strong cooling produced by cold and dry local circulations. In addition, strong and rapidly changing horizontal gradients can be found during formation and melting of pack and fast ice and in coastal ocean environments.

Finally, a description of high-resolution processes is necessary to study the behavior of short-lived atmospheric constituents. This is not only because the sources of such constituents may be highly variable in space and time, but also because removal and transport mechanisms can be dominated by the contribution of convective and mesoscale circulations (e.g., Crutzen, 1988).

As emphasized by this small, and by no means exhaustive, set of examples, coupling of regional ESP models can provide valuable insights for a multitude of problems. One aspect which makes this coupling especially appealing is that a wide range of ESP models are presently available which have been expressly designed for the regional scale. Examples of regional ecosystem models are CENTURY (Parton et al., 1990) and RHESS (Band et al., 1993). A number of hydrology models have been constructed for basins of sizes varying from 10^2 to 1000^2 km^2 (Dooge, 1986). Regional ocean and sea-ice models of complexity comparable to that of atmospheric models are available which can be run at resolutions of a few tens of km (Haidvogel and Bryan, 1992; Washington and Parkinson, 1986), and full chemistry-transport models, including several tracer constituents, have already been coupled to regional atmospheric models (NCAR ADMP, 1985).

So far, each class of ESP models has assumed climate as an external forcing input. One of the main reasons that have so far prevented interactive coupling with regional atmospheric models has been the need to perform long-term simulations. Before the advent of RegCM, limited-area models had been run for periods of a few days or less. Depending on the process under consideration, however, most regional ecosystem, hydrology, ocean and sea ice models operate on scales ranging from seasonal and multiyear to multidecadal and centennial. From examples already given, one can envision that nested RegCM can at least provide up to decade-long simulations, which would therefore allow coupling with other regional ESP components.

As an illustrative exercise, Figure 3 shows the design of a multipurpose coupled regional ESP modeling system. The entire system is driven by meteorological initial and lateral boundary conditions provided to a regional climate model either from a GCM or from global analyses of observations (e.g., from NMC or ECMWF). A key element in the coupling of the RegCM with the other model components is the Surface-Vegetation Atmosphere Transfer (SVAT) module, of which BATS is a typical example. The SVAT would provide physically based exchanges of momentum, sensible heat, water vapor, and flux of relevant tracers between the atmosphere and the land and ocean system.

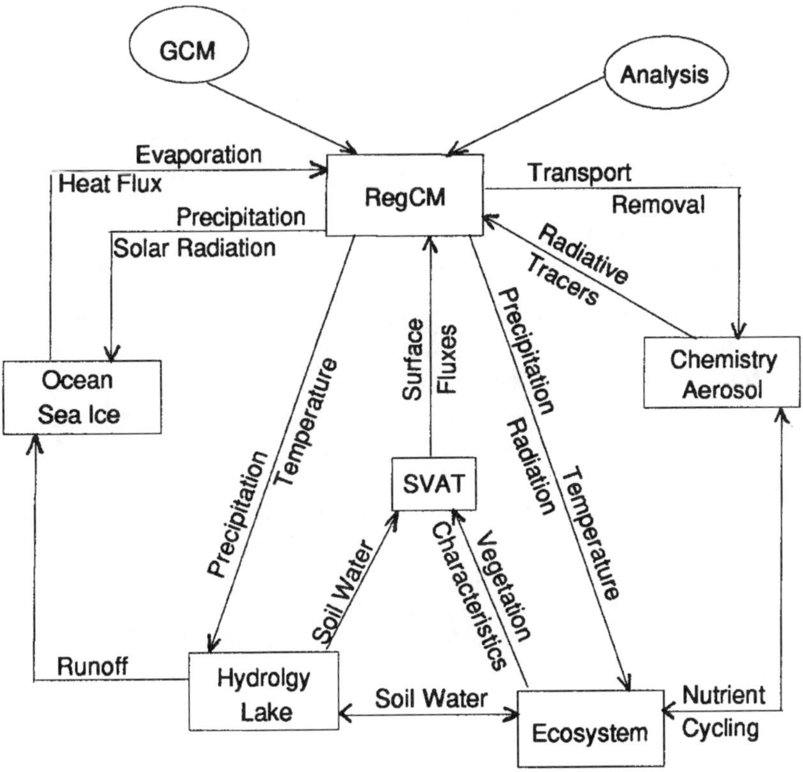

Figure 3. Possible design of regional Earth System model (from Giorgi, 1993).

Precipitation and radiative fluxes also enter most coupling arrows between the atmosphere and the other model components. Land hydrology interacts with basin ocean dynamics via the freshwater input associated with runoff. Hydrology and ecosystem models mostly interact through the distribution of soil moisture, which generally shows high spatial variability and regulates the effectiveness of evaporation and runoff as well as the ecosystem dynamics. The ecosystem model could be designed to provide vegetation characteristics (e.g. vegetation type, leaf area index) to the SVAT scheme, and the atmospheric model would provide wind and precipitation rates to the tracer transport removal models. Finally, interactions between the chemistry and ecosystem modules would mostly take place via the cycling of key nutrients such as carbon, nitrogen, and phosphorus.

The first attempts at coupling regional ESP models are under way. For example, plans are in place to couple CENTURY to RAMS (Kittel, personal communication). The NCAR RegCM has already been interactively coupled with a lake model (Hostetler et al., 1993), and it is planned to include in the model an improved representation of surface and hydrologic processes and a tracer chemistry component. RHESS (Band et al., 1993) is an example of interactive coupling between an ecosystem dynamics module and a spatially distributed hydrologic system of comparable conceptual complexity.

The discussion of Giorgi (1993) points out that a number of models of widely different characteristics are available. Coupling of such models is certainly technically possible, but care must be taken that the conceptual complexity and the scale of application of the coupled models are consistent. This may require suitable modifications of present modeling tools or development of new tools. Finally, an important aspect of regional modeling is the inclusion of human effects. Human activities, such as agriculture, deforestation and reforestation, grazing, urban development, and pollution production probably represent the strongest environmental forcings on the regional scale, and their interactions with other components of the system need to be suitably accounted for.

CONCLUSIONS

The primary thesis of this work has been that, although global models remain the fundamental tools for the study of the Earth System as a whole, coupling of regional models of ESP components can give useful information on the interactions between the atmosphere, the biosphere, the hydrosphere, the cryosphere, and the chemosphere. Several examples have been presented in support of this thesis. The main reason that coupling of ESP models on the regional scale is feasible is the recent development of high-resolution regional climate models, which fill a gap in the spatial-temporal scale present in current climate modeling, i.e., long-time integration in conjunction with fine spatial resolution.

Compared to global modeling, the main advantages of ESP modeling on the regional scale can be summarized as follows:

(1) Coupling between different model components is done at spatial scales which are more internally consistent, both in terms of the avail-

ability of present regional models and the scale of many relevant processes.

(2) The effects of high-resolution forcings and interactions can be more explicitly represented. This will help the study of the effects of, or the impacts on, human activities.

(3) Although still very demanding computationally, regional modeling systems are easier to develop and maintain (compared to GCM), once access to driving meteorological fields is achieved. This is mostly because only problems relative to a given region of interest need to be addressed and because computational resources are reduced. An important result is that coupled modeling can be accessed by a broader scientific community.

(4) It is easier to gather observational data, or design the field experiments necessary for model setup and or validation, than with GCM.

These considerations are not intended to place regional modeling in competition with global modeling. Global models are needed to provide the fields necessary to drive the regional system, so it is imperative that such large-scale fields be accurate and reliable. In fact, in several ways regional modeling can be complimentary to global modeling: by providing information on interactions at scales which global models might presumably reach in the future, and by helping to develop parameterizations of sub-GCM grid scale processes. A possible scenario for future ESP modeling is that a few large centers can develop global models and provide data for regional centers to use in their regional models to focus on specific issues of relevance for their area.

REFERENCES

Anthes, RA and TT Warner. **1978**. Development of hydrodynamical models suitable for air pollution and other mesometeorological studies. Mon. Weather Rev. 106, 1045-1078.

Anthes, RA, EY Hsie, and YH Kuo. **1987**. *Description of the Penn State/NCAR Mesoscale Model Version 4 (MM4)*, NCAR/TN-282+STR, pp. 66. National Center for Atmospheric Research, Boulder, CO.

Band, LE, P Patterson, R Nemani, and SW Running. **1993**. Forest ecosystem processes at the watershed scale: Incorporating hillslope hydrology. Agric. For. Meteorol. 63, 93-126.

Barnston, AG and PT Schickedanz. **1984**. The effect of irrigation on warm season precipitation in the Southern Great Plains. J. Climate Appl. Meteorol. 23, 865-888.

Bengtsson, LO. **1992**. Climate system modeling prospects, pp. 705-724. In: *Climate System Modeling*, K Trenberth (ed.). University Press, Cambridge, UK.

Bossert, JE, C-YJ Kao, JL Winterkamp, JO Roads, and SC Chen. **1992**. Regional scale simulation of the western U.S. Climate Proceedings, Third Symposium on Global Change Studies, Atlanta, GA. American Meteorol. Soc. 24-28.

Changnon, SA and DMA Jones. **1972**. Review of the influences of the Great Lakes on Weather. Water Resour. Res. 8, 360-371.

Crutzen, PJ. **1988**. Variability in atmospheric-chemical systems. In: *Scales and Global Change. Spatial and Temporal Variability in Biospheric and Geospheric Processes*, SCOPE 35. John Wiley, New York.

Dickinson, RE, PJ Kennedy, A Henderson-Sellers, and M Wilson. **1986**. *BiosphereAtmosphere Transfer Scheme (BATS) for the NCAR Community Climate Model,* NCAR/TN-275+STR. National Center for Atmospheric Research, Boulder, CO.

Dickinson, RE, RM Errico, F Giorgi, and GT Bates. **1989**. A regional climate model for the western U.S. Clim. Change 15, 383-422.

Dooge, JCI. **1986**. Looking for hydrological laws. Water Resour. Res. 22, 46S-58S.

Giorgi, F. **1990**. On the simulation of regional climate using a limited area model nested in a general circulation model. J. Climate 3, 94-963.

Giorgi, F. **1993**. Perspectives for regional Earth System modeling. Submitted to Global and Planetary Change.

Giorgi, F and GT Bates. **1989**. The climatological skill of a regional model over complex terrain. Mon. Weather Rev. 117, 2325-2347.

Giorgi, F and LO Mearns. **1991**. Approaches to the simulation of regional climate change: a review. Rev. Geophys. 29, 191-216.

Giorgi, F, MR Marinucci, and G Visconti. **1992**. A 2 × CO_2 climate change scenario over Europe generated using a Limited Area Model nested in a General Circulation Model. II: Climate change scenario. J. Geophys. Res. 97, 10,011-10,028.

Giorgi, F, GT Bates, and S Nieman. **1993a**. The multi-year surface climatology of a regional atmospheric model over the Western United States. J. Climate 6, 75-95.

Giorgi, F, MR Marinucci, and GT Bates. **1993b**. Development of a second generation regional climate model (RegCM2): Boundary layer and radiative transfer processes. Mon. Weather Rev. 121, 2794-2813.

Giorgi, F, MR Marinucci, G De Canio, and GT Bates. **1993c**. Development of a second generation regional climate model (RegCM2): Convective processes and assimilation of lateral boundary conditions. Mon. Weather Rev. 121, 2814-2832.

Giorgi, F, C Shields Brodeur, and GT Bates. **1994a**. Regional climate change scenarios over the United States produced with a nested regional climate model: Spatial and seasonal characteristics. J. Climate 7, 375-399.

Giorgi, F, SW Hostetler, and C S. Brodeur. **1994b**. An analysis of the surface hydrology in a regional climate model. Q. J. R. Meteorol. Soc. 120, 161-183.

Hack, J. **1994**. Climate statistics for the NCAR CCM2. J. Geophys. Res. 99, 20785.

Haidvogel, DB and F Bryan. **1992**. Ocean general circulation modeling, pp. 371-412. In: *Climate System Modeling*, K Trenberth (ed.). University Press, Cambridge, UK.

Henderson-Sellers, A and V Gornitz. **1984**. Possible climatic impacts of land cover transformations, with particular emphasis on tropical deforestation. Clim. Change 61 231-257.

Henderson-Sellers, A. **1990**. Predicting generalized ecosystem groups with the NCAR CCM: first steps towards an interactive biosphere. J. Clim. 3, 917-940.

Hostetler, SW, GT Bates, and F Giorgi. **1993**. Interactive coupling of a lake thermal model with a regional climate model. J. Geophys. Res. 8, 5045-5057.

Hostetler, SW, F Giorgi, GT Bates, and PJ Bartlein. **1994**. The role of lake-atmosphere feedbacks in sustaining paleolakes Bonneville and Lahontan 18,000 years ago. Science 263, 665-668.

Kida, H, T Koide, H Sasaki, and M Chiba. **1991**. A new approach to coupling a limited area model to a GCM for regional climate simulation. J. Meteorol. Soc. Japan 69, 723-728.

Legates, DR and CJ Willmott. **1990**. Mean seasonal and spatial variability in gauge-corrected global precipitation. Int. J. Climatol. 10, 111-127.

Leung, LR and SJ Ghan. Representing the influence of subgrid topography on hydrology. (This volume, in press).

Liu, Y, F Giorgi, and WM Washington. **1994**. Simulation of summer monsoon climate over east Asia with an NCAR Regional Climate Model. Mon. Weather Rev. 122, 2331-2348.

Marinucci, MR and F Giorgi. **1992**. A 2 × CO_2 climate change scenario over Europe generated using a Limited Area Model nested in a General Circulation Model. I: Present day simulation. J. Geophys. Res. 97, 9,989-10,009.

McGregor, JL and K Walsh. **1993**. Nested simulations of perpetual January climate over the Australian region. J. Geophys. Res. 98, 23283-23290.

Meehl, GA. **1992**. Global coupled models: atmosphere, ocean, sea ice, pp. 555-581. In: *Climate System Modeling*, K Trenberth (ed.), Cambridge, UK.

NCAR ADMP. **1985**. *The NCAR Eulerian Regional Acid Deposition Model*, p. 178, NCAR/TN-256+STR. National Center for Atmospheric Research, Boulder, CO.

Parton, WJ, DS Ojima, DS Schimel, and TGF Kittel. **1990**. Development of simplified ecosystem models for applications in Earth System studies: The Century experience, pp. 281-302. In: *Modeling the Earth System*, D. Ojima (ed.). UCAR/OIES, Boulder, CO.

Pielke, RA and R Avissar. **1990**. Influence of landscape structure on local and regional climate. Landscape Ecol. 4, 133-155.

Pitman, AJ, F Giorgi, and A Henderson-Sellers. **1991**. Southeast Australia wintertime precipitation: Sensitivity of climate prediction to model resolution. Aust. Meteorol. Mag. 39, 21-35.

Sasaki, H, H Kida, T Koide, and M Chiba. The performance of long term integration of a limited area model with the spectral boundary coupling method. J. Meteorol. Soc. Japan, in press.

Semazzi, F, YW Lin, and F Giorgi. **1994**. A nested model study of the Sahelian climate response to sea-surface temperature anomalies. Geophys. Res. Lett. 20:2897-2900.

Trenberth, K, (ed.). **1992**. *Climate System Modeling*, University Press, Cambridge, UK, 788 pp.

Warren, BA. **1981**. Deep circulation of the World Ocean, Chapter 1. In: Evolution of Physical Oceanography, BA Warren and C Wunsch, (eds.). MIT Press, Cambridge, MA.

Washington, WM and CL Parkinson. **1986**. An Introduction to Three-Dimensional Climate Modeling. University Science Books, Mill Valley, CA, 388 pp.

Wood, EF, M Sivapalan, and K Beven. **1990**. Similarity and scale in catchment storm response. Rev. Geophys. 28, 1-18.

STOCHASTIC CHARACTERIZATION OF REGIONAL CIRCULATION PATTERNS FOR DOWNSCALING OF CLIMATE MODEL PRECIPITATION SIMULATIONS

E. Zorita,[1] J. P. Hughes,[2] D. P. Lettenmaier,[3] and H. von Storch

[1]Max Planck Institut für Meteorologie, Hamburg, Germany

[2]Department of Statistics, University of Washington, Seattle, Washington

[3]Department of Civil Engineering, University of Washington, Seattle, Washington

ABSTRACT

Two statistical methods for linking large-scale atmospheric circulation patterns and daily local rainfall are described and applied to several General Circulation Model (GCM) simulations. Our ultimate objective is to simulate local precipitation associated with alternative climates. The index stations used to drive these statistical models are located near the North American west and east coasts. The first method is based on classification and regression trees (CART) analysis. It allows classification of observed daily SLP fields in weather types that are most strongly associated with the presence or absence of rainfall in a set of index stations. The best results were obtained for winter rainfall for the west coast, where a set of physically reasonable weather types could be identified, whereas for the east coast the rainfall process seemed to be spatially less coherent. The GCM simulations were validated against observations in terms of probability of occurrence and survival time of these weather states. Some discrepancies were found, but there was no systematic bias, indicating that this behavior depends on the particular dynamics of each model. This classification method was then used for the generation of daily rainfall time series from the daily SLP fields from historical observation and from the GCM simulations. Whereas the mean rainfall and probability distributions were rather well replicated, the simulated dry periods were, in all cases, shorter than in the rainfall observations.

The second rainfall generator, based on the analog method, uses information on the evolution of the SLP field in several previous days. It was found to perform reasonably well, although some downward bias in the simulated rainfall persistence was present.

Rainfall changes in a $2 \times CO_2$ climate were investigated by applying both methods to the output of a greenhouse-gas experiment. The simulated precipitation changes were small.

REGIONALIZATION OVER EUROPE OF GLOBAL MODEL SIMULATIONS USING THE HIRHAM MODEL

B. Machenhauer,[1] M. Botzet,[1] D. Jacob,[1] and J. H. Christensen[2]

[1]Max-Planck-Institut für Meteorologie, Hamburg, Germany

[2]Danish Meteorological Institute, Copenhagen, Denmark

Key word: *Climate, models HIRHAM*

ABSTRACT

Within the next few years, long-term (100 yr) climate change simulations with improved higher-resolution atmospheric/ocean general circulation models (AOGCM) are planned. It is expected that T42 resolutions, corresponding to about half the grid length used so far, i.e., about 250 km, can be used in these simulations. For various impact studies, however, climate change information is needed at much smaller scales.

A potentially useful technique to obtain higher-resolution information over a specific area of interest is to run a high-resolution, limited area model (LAM) with initial fields and boundary values taken from an AOGCM simulation.

In previous experiments with this technique, LAM covering relatively small areas have been used, which has limited the possibility of improving the large-scale flow in the high-resolution area. In addition the LAM have been driven by low resolution and, therefore, less accurate global simulations.

In a European project, coordinated by B. Machenhauer, the goal is to achieve improved regional-scale resolution by using improved LAM, larger areas, and higher resolution (T42) to drive global simulations.

Within this project a LAM, called HIRHAM, has been constructed by combining the dynamic part of the Danish Meteoroelogical Institute's high-resolution forecast model (HIRLAM) with the physical parameterization part of the Max-Planck-Institute für Meteorologie's global climate model ECHAM.

Preliminary verifications against observations over Europe of a 20-month-long HIRHAM integration driven by European Center for Medium-Range Weather Forecasting analyses ("perfect boundaries") are presented. In addition, results of verifications against climatological fields of a 2-yr HIRHAM integration driven by output from a T42 ECHAM present-day climate simulation are presented and compared with those obtained for the driving AGCM. In both HIRHAM integrations, a grid length of about 50 km is used.

With "perfect" boundaries, seemingly realistic regional details which could not have been resolved in a T42 resolution are simulated. This seems to be the case also in the simulations driven by the T42 ECHAM model. In addition, the verification indicates that with the relatively large area covered by HIRHAM, it is also capable of reducing large-scale systematic errors of the driving GCM.

INTRODUCTION

The Resolution Problem

Global coupled atmosphere-ocean models, usually referred to as atmosphere-ocean general circulation models (AOGCM), are the main tools available today for climate simulations. However, because of limited computer resources, very long integrations required in anthropogenic climate change simulations can be performed only at resolutions that are too coarse to adequately describe the regional climate details which are required for impact studies.

Over the next few years, improved higher-resolution AOGCM will be used in climate change experiments for the atmospheric component with a spectral resolution T42 or with corresponding grid point distances of about 250 km.

Although preliminary experiments with atmospheric T42 GCM indicate that significant improvements over previous lower-resolution (T21) simulations are obtained (Roeckner et al., 1992), a T42 resolution is still not sufficient for impact studies. In addition to insufficient representation of important surface fields for forcing at the lower boundary, the internal dynamics of weather systems are also adversely affected by the low resolution of GCM.

By the so-called "time-slice" technique, higher-resolution simulations can be obtained with affordable computer resources. The idea is to repeat a certain period of a long, low-resolution simulation with a higher-resolution GCM. One could, for instance, repeat the last 10 yr of a 100-yr T42 simulation with a T106 AGCM, using initial values and SST from the T42 simulation. This would cost about 13% of the CPU time required to run the whole 100-yr simulation with a T106 resolution. Preliminary experiments with this technique have been carried out and, during the next year, three AGCM T106 10-yr simulations with observed SST are planned at the European Center for Medium-Range Weather Forecasting (ECMWF), using the models of Météo-France, United Kingdom Meteorology Office (UKMO), and Max-Planck-Institut für Meteorologie (MPIM).

In the projects described here, alternative techniques are explored by which even higher resolutions can be obtained over a specific area of interest (e.g., Europe), with affordable costs of computer time.

Current and Planned European Regionalization Projects

In the climate modeling project "Climate of the 21st Century" (C21C), supported by the Commission of the European Communities (CEC) under the EPOCH program, one of the five subprojects is Group 5: "Regional climate simulations with nested models" (C21C,5). In this subproject, two institutions, the Danish Meteorological Institute (DMI) and Météo-France, are participating. The C21C project was started in August, 1990, and its original 3-yr period has been extended by 3 months, to the end of October, 1993.

The main objective in the proposal of the C21C,5 subproject was to develop and test dynamical regionalization techniques by which high-resolution climate model simulations could be obtained over Europe.

DMI would use the HIRLAM, a high-resolution limited area model developed for short-range forecasting by the Nordic, Dutch, and Irish meteorological institutes (Machenhauer, 1988). The idea was to explore a technique which had recently been applied in some pioneering experiments in the United States (Dickinson et al., 1989). By this technique, detailed simulations over Europe would be obtained by integrating HIRLAM for certain "time slices," taking the initial and boundary values needed from a long-period simulation made in advance with a global lower-resolution model (a GCM). The justification for using this "nesting" technique is, of course, that a high-resolution limited area model (LAM) is much less expensive in computer resources than a global model with the same resolution. The computation time and storage capacity are reduced by a factor approximately equal to the fractions of the earth covered by the LAM.

An alternative technique by which a high-resolution simulation may be obtained over an area of interest at a relatively low computational cost is to use a global model with variable horizontal resolution. The idea here is to gradually increase the resolution toward a point in a region of interest (to "zoom" at the region of interest) and to decrease it correspondingly on the opposite side of the earth in such a way that the mean resolution is unchanged. The ARPEGE/IFS global model code (Courtier and Geleyn, 1988; Courtier et al., 1991), which, at the start of the C21C, was under development for operational forecasting by Météo-France and ECMWF, offers this "zooming" possibility. Météo-France would, in the C21C,5 subproject, in cooperation with DMI, try it out in time-slice climate simulations focusing on Europe.

At the beginning of 1993 a new CEC-supported Regionalization project was started under the Environment program. It runs to the end of 1994 and includes six contractors with the following responsible scientists: Dr. B. Machenhauer, MPIM, Hamburg (Coordinator); Dr. L. Laursen, DMI, Copenhagen; Dr. A. Ruiz de Elvira, University of Alcalá de Henares (UAH) , Madrid; Dr. R. Sausen, Deutsche Forschungsanstalt für Luft-und-Raumfahrt (DLR) e.V., Oberpfaffenhofen, Germany; Prof. G. Visconti, Università Degli Stude-L'Aquila (UNIVAQ), Coppito, Italy; Dr. F. Nucciotti, Entre Regionale di Svieluppo Agnicolo, Servizio Meteorologico Regionale (ERSA/SMR), Bologna, Italy. The Regionalization project is, in effect, an extension of the C21C,5 subproject, with several additional participants and with essentially the same main objectives as C21C,5.

Also, UKMO (the Hadley Centre) has started up dynamical regionalization experiments, using the nesting technique. This work is supported untill the end of 1995 by CEC under the "Anthropogenic Climate Change" modeling project, which is another C21C follow-on project under the Environment program (coordinated by Dr. U. Cubasch, DKRZ, Hamburg).

The high-resolution LAM simulations that have been made or are planned in the above-mentioned projects are performed with a number of different LAM, using boundary conditions from different GCM simulations in spectral T42 or equivalent grid point resolution made at different centers, e.g., National Centre for Atmospheric Research (NCAR), UKMO, and MPIM. The approximate extent of the LAM areas to be used by the participants in the Regionalization project are indicated in Figure 1.

Covering relatively large areas (up to 1/8 of the earth), multiyear (3- to 5-yr) LAM simulations are being made by MPIM/DMI, UKMO and UNIVAQ with 40- to 60-km grid lengths. ERSA/SMR will make LAM simulations on a smaller area covering southern Europe but with a higher resolution, i.e., about 30-km grid length. Even higher-resolution LAM simulations, with a grid length of about 15 to 20 km, will be made by UAH over a smaller area around the Iberian Peninsula. Relatively high LAM resolution simulations (grid length about 30 km) will also be made by DLR over a small area of the Alps using a special statistical-dynamical technique.

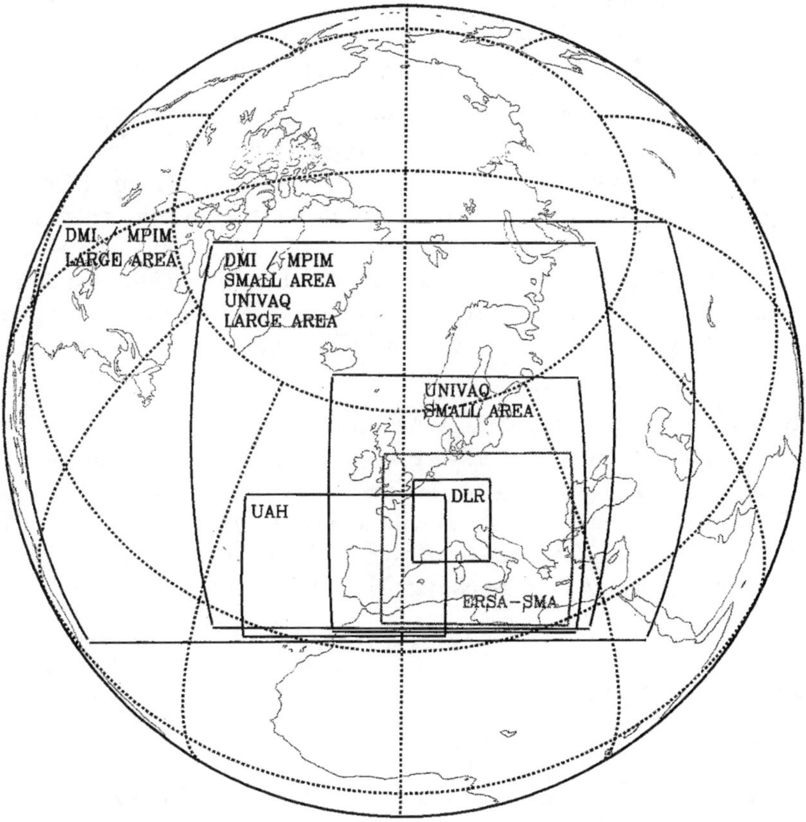

Figure 1. Approximate integration areas used by the participants of the Regionalization project.

Recently, a proposal for a further 1-yr extension, to the end of 1995, of the Regionalization project was made to CEC in the second call for proposals to the Environment program. In this extension, called "Regionalization of Anthropogenic Climate Change Simulations," we apply for support for the above-listed participants of the Regionalization project during 1995 and for UKMO and Météo-France during 1994 and 1995.

The C21C,5 and the Regionalization project concentrate on present-day climate simulations and their evaluation. If support is given by the CEC to extension of the Regionalization project, high-resolution anthropogenic climate change simulations and additional present-day climate simulations will be made. This will include new model set-ups, with ARPEGE zooming over southern Europe and a very high-resolution LAM (grid length about 20 km) covering northern Europe.

Two types of LAM simulations are being made in the C21C,5 and the Regionalization projects. In the first type, the initial fields and the time-dependent boundary conditions are determined from ECMWF analyses. These "perfect" boundary simulations are made in order to test the performance of the LAM to be used in the second type of simulations, in which the initial fields and boundary conditions are determined from the GCM simulations. For the "perfect" boundary simulations we have agreed to use the ECMWF analyses from the period January 1, 1990, to August 31, 1991.

A very important part of the Regionalization project is comparison with observations of the different high-resolution present-day climate simulations which are being produced. The degree of accuracy that can be obtained from these evaluations will constitute an upper limit of the accuracy that can be expected with the same techniques when applied to the anthropogenic climate change simulations which are planned in the extension project.

PRELIMINARY RESULTS

Regionalization in the C21C Project

As mentioned above, only two institutions, DMI and Météo-France, are participants in the C21C, 5 subproject, and from the start of the project in August, 1990, rather limited resources were devoted to it.

In cooperation with the HIRLAM project group, especially those at DMI, the HIRLAM forecast model was prepared for runs in climate mode, starting with simulations using "perfect" boundary conditions, i.e., ECMWF analyses. A series of test runs was then made with the HIRLAM model to test it in climate mode and, in particular, to test and subsequently modify the boundary relaxation scheme.

Similar start-up work was done at Météo-France. The ARPEGE model to be used in the "zooming" experiments was not expected to be ready until 1992. Instead, their operational LAM PERIDOT was adapted to run in climate mode, and month-long test runs were made with "perfect" boundary conditions.

Two coordinating meetings were arranged during the first 1½ yr of the project. Beside DMI and Météo-France, other groups interested in regional climate simulations for Europe participated in these meetings; in particular, representatives from NCAR (Dr. F. Giorgi), from UKMO (Dr. H. Cattle), and from the German Weather Service (DWD) (Dr. V. Renner). At these meetings the common period for which ECMWF analysis should be used to drive LAM, as well as standard outputs and verification schemes, were discussed.

DMI, Météo-France, and DWD presented their first month-long simulations and results of verifications against the ECMWF analysis that were used for boundary conditions. The results of these preliminary experiments indicated that relatively accurate simulations could be obtained with "perfect" boundary conditions, although systematic errors were identified. As expected, these errors were most evident in the simulation with the largest integration area, i.e., the HIRLAM simulations.

The NCAR group, with Dr. Giorgi as the leading scientist, had started their experiments with the nesting technique 1 to 2 yr before the start of C21C,5. They were using the Pennsylvania State University (PSU)/NCAR mesoscale model, version MM4, with a grid length of about 70 km. During the first 1 1/2 yr of the C21C,5 they published a series of papers, some presenting simulations from the western part of the United States and part of the Pacific Ocean, and others covering the European area marked "UNIVAQ, small area" in Figure 1 (Giorgi and Marinucci, 1991; Marinucci and Giorgi, 1992; Giorgi et al., 1990, 1992). For further planning of the C21C,5 and of its extension (the Regionalization project), which was started in the autumn of 1991, it was now possible to use the experience from these studies. The work on the NCAR high-resolution simulations over Europe was done in cooperation with Prof. G. Visconti at UNIVAQ. By including UNIVAQ as a participant in the Regionalization project, close cooperation was established with the NCAR/UNIVAQ group on further dynamical regionalizations over Europe. The NCAR group has made "perfect boundary" as well as GCM-driven LAM simulations. Giorgi and Marinucci (1991) report on month-long simulations with the MM4, using initial and boundary fields from January, 1979, and July, 1979, ECMWF analysis.

Meijgaard, 1992). A first version of the new combined model, which is now called HIRHAM, was ready in the beginning of this year, and after several shorter test runs, a long 1-yr simulation with "perfect" ECMWF boundary conditions from 1990 was made at DMI. This long run revealed further errors, and after correction, a new, 20-month "perfect" boundary simulation was made. In these first simulations the smallest integration area of the two marked DMI/MPIM in Figure 1 and a grid length of 0.51° were used.

As an illustration of the improvements obtained with the ECHAM physics, Figures 2 and 3 show verification results for area mean values of temperature and specific humidity at the model levels. The area mean values are computed for the inner domain, excluding the 10-point boundary relaxation zone. In Figures 2 and 3 the HIRLAM and HIRHAM simulations are verified. In each of the figures results are shown from 2-month-long integrations with ECMWF analyses from January, 1991, and July, 1990, used for boundary conditions. Shown are mean monthly biases and standard deviations computed from time series of six hourly values. In general, a substantial improvement is achieved with the ECHAM physics used in HIRHAM.

After further development of the ARPEGE model, the experiments with the "zooming" technique resumed at Météo-France. In stretching mode the resolution is defined by the spectral truncation, the location of the pole of stretching, and the factor of stretching. Initially, ARPEGE was integrated for 60 days with the pole over France in two resolutions: T63 with a stretching factor of 2, and T42 with a stretching factor of 3. In both cases, the maximal resolution over Europe corresponds to T126, approximately equivalent to a grid length of about 80 km in a grid point model. The time average of the last 30 days of the simulations (January means) was compared with two similar experiments with homogeneous resolutions T42 and T106. Encouraged by the results, the stretched and homogeneous T42 simulations were extended to include a full annual cycle. The results of these experiments indicate that climatology over Europe of the T42 and T63 stretched versions is similar to that of T106, and that both are more realistic than the homogeneous T42 simulation. This can be seen most conveniently in Figure 4, showing mean January surface air temperature fields over Europe. Here, the map marked "Clim EC Jan" shows the 10-yr mean January temperature field computed from ECMWF analyses from the period 1979-1988. "Jan" indicates first January and "Jan2" the second; "T42" denotes the homogeneous, and "T42SF3" the stretched model simulations.

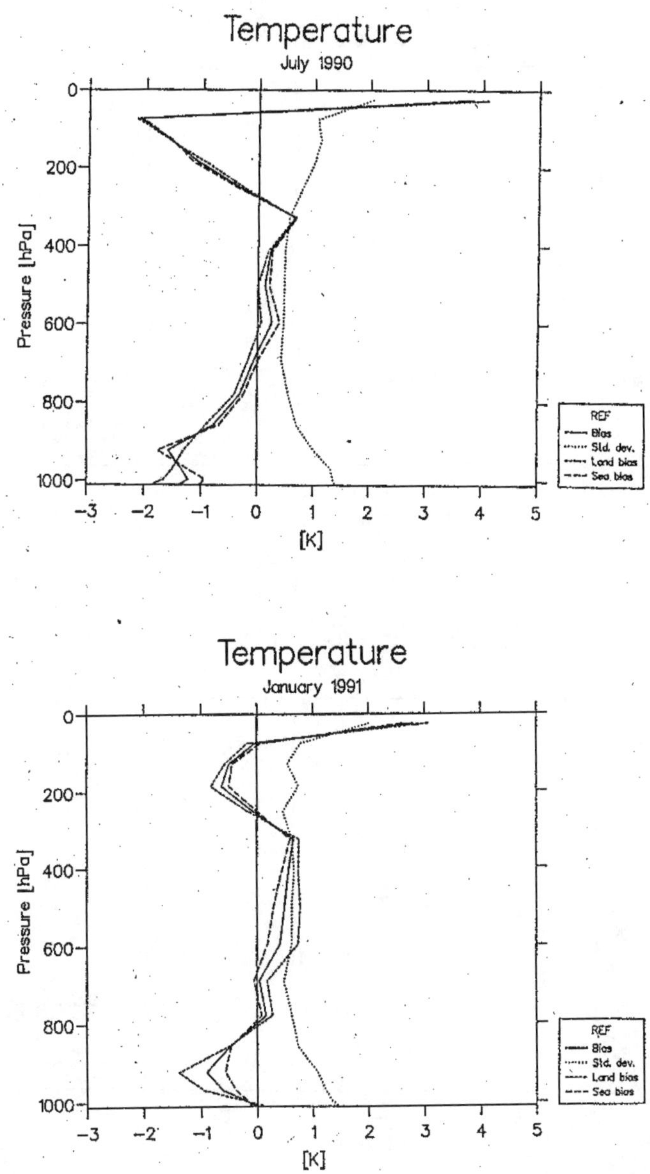

continued

Figure 2. Area averaged mean monthly biases (solid lines) and standard deviations (dotted lines) for 2-month-long simulations with the high-resolution limited area model (HIRLAM) model driven by "perfect" boundary conditions. Also shown are biases averaged over sea points only (fine dashed lines) and land points only (dashed lines).

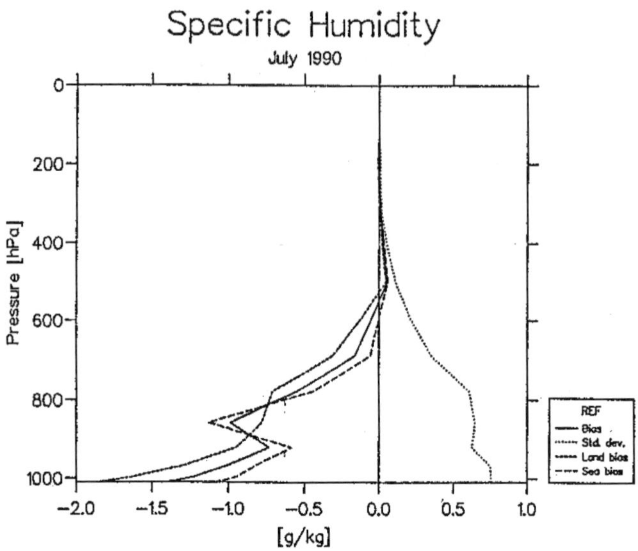

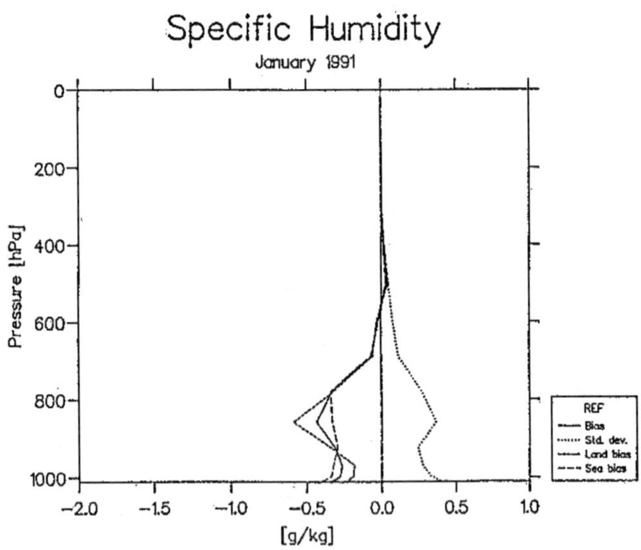

Figure 2. *Continued.*

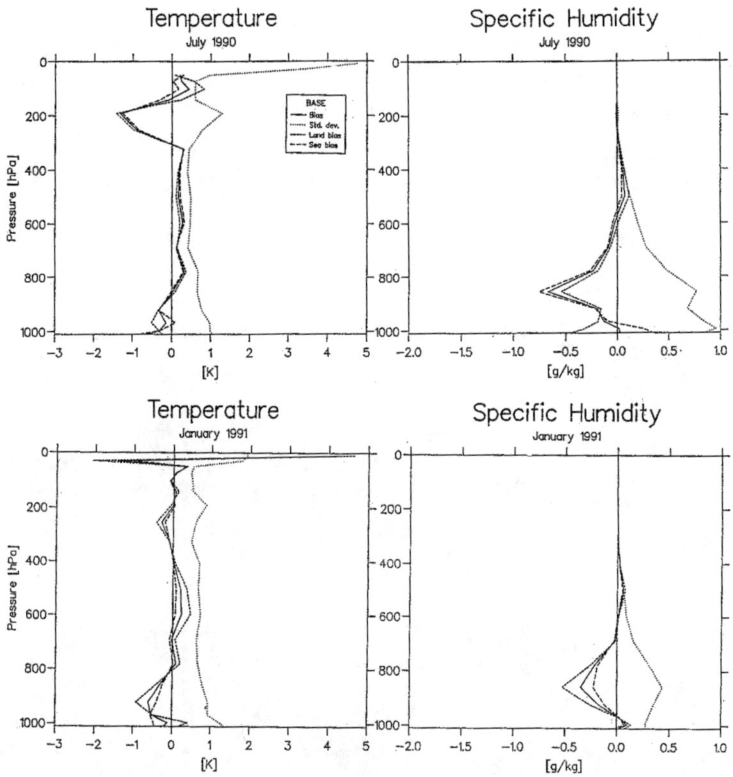

Figure 3. As in Figure 2, but for simulations with the HIRHAM model.

Besides the improvement of the temperature field over Europe shown in Figure 4 for January and for July, the use of the stretched coordinate has a positive impact on the precipitation field. Rainfall is increased in the mid-Atlantic and over southern Europe. The better representation of orography yields a more realistic simulation of the rainfall maxima on the Norwegian coast and northwestern Spain. The comparison of data for the first and the second January dates of the experiment does not reveal any strong drift which could have been induced by the discretization technique.

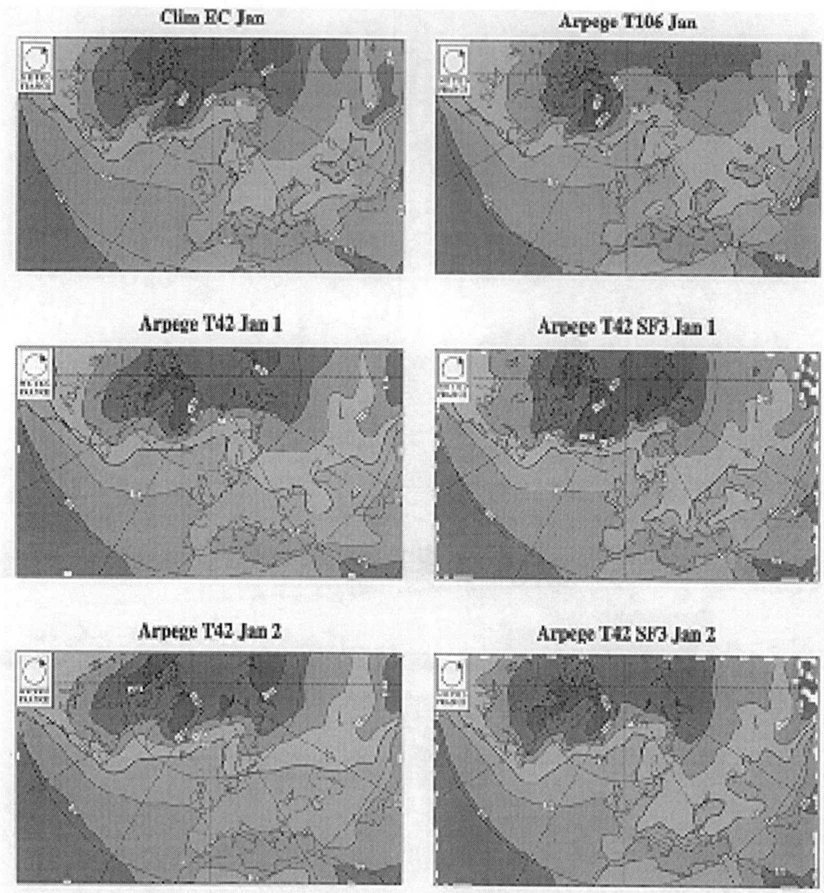

Figure 4. Maps of mean January surface air temperatures (contour interval 10°C). Upper left: 10-yr mean January temperature computed from ECMWF analyses for the period 1979-1988. The remaining maps are from simulations with the ARPEGE model in a homogeneous T106 resolution (upper right), in a homogeneous T42 resolution for first January (middle left) and second January (bottom left), and in a stretched T42 resolution (stretching factor = 3) for first January (middle right) and second January (bottom right).

The Regionalization Project

During the first half-year of the Regionalization project, mostly preparatory work, (i.e., model developments, developments of verification/analysis packages, and collection of data) were planned. This work has been completed according to the work plan schedule. Some experiments have been carried out; some of the results will be presented below.

Besides the above-mentioned 20 months "perfect" boundary HIRHAM simulation, similar 1-yr simulations were made by UNIVAQ, using a new, improved version of the MM4 model, called MM5. Two 1-yr MM5 simulations were made with grid lengths of 50 km and 200 km, respectively, both over a domain similar to the one used in the HIRHAM simulation and using ECMWF analyses from August 1, 1990, to July 31, 1991 (which is part of the period used in the 20-month HIRHAM simulation). These HIRHAM and MM5 simulations will be compared with land surface observations and ECMWF analyses from the periods in question.

Preliminary verification of the HIRHAM simulation shows that the area averaged deviations from the driving ECMWF analyses remain small during the whole 20-month period of integration. As an example, Figure 5 shows the area averaged mean monthly biases and standard deviations of temperature and specific humidity for January, 1990, and January, 1991, in the simulation.

So far, verifications of near surface fields have been made only for specific areas from which observations were readily available. These verifications seem to indicate that these fields are simulated realistically. Over Denmark, the simulated mean monthly area averaged wind speed, temperature, and precipitation amounts were found to match closely the observed values for all midseason months. Over northern Italy, an evaluation of precipitation in March, 1991, showed realistic values and patterns of monthly mean precipitation. Finally, the simulated precipitation over the Iberian peninsula has been compared with a detailed analysis of observed precipitation from January, 1991. We shall presently return to this comparison.

In the Regionalization project the UAH is working with its subcontractors, University Complutense (UC) and University of Lisboa. They use a very high-resolution atmospheric model, PROMES, developed by UC. It covers the area around the Iberian peninsula indicated in Figure 1 and has a grid length of 15 to 20 km and 31 vertical levels. After initial tests with perfect boundary conditions, the model will run with boundary conditions from 0.51° HIRHAM simulations, providing a further downscaling over the Iberian peninsula, with the aim of simulating the complex features produced by Iberian orography.

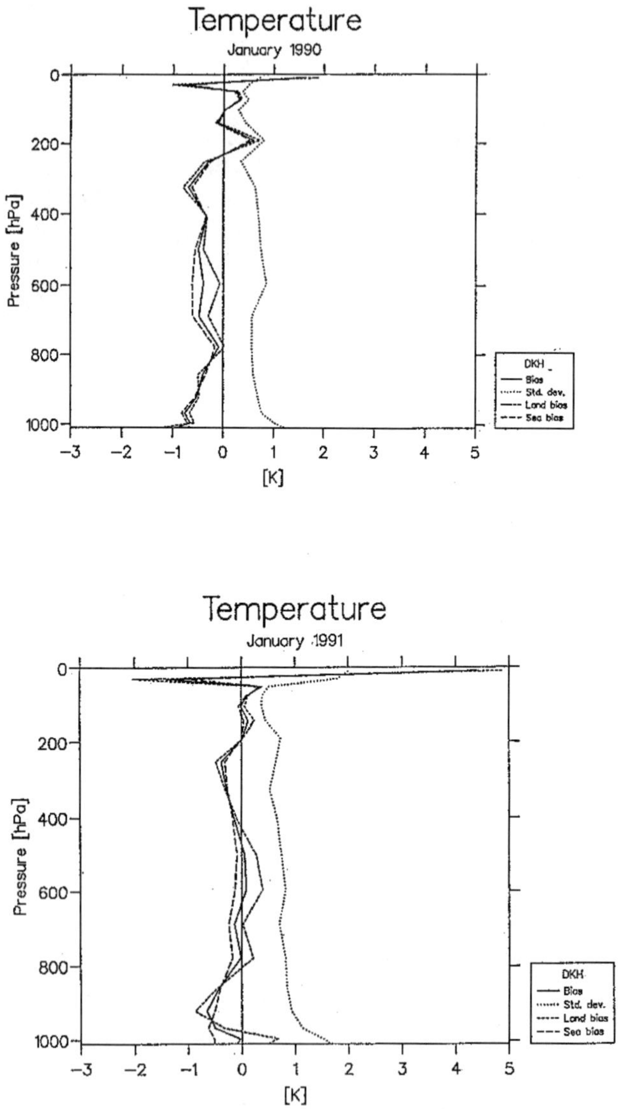

continued

Figure 5. As in Figure 2, but from a 20-month simulation with the HIRHAM model driven by "perfect" boundary conditions. January, 1990, is the first January, and January, 1991, is the second in the simulation period.

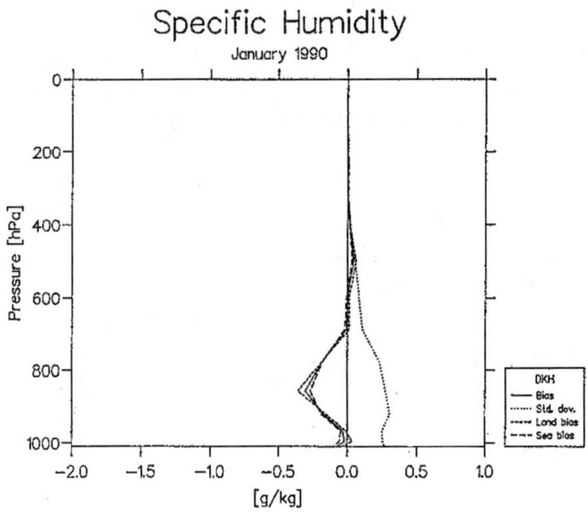

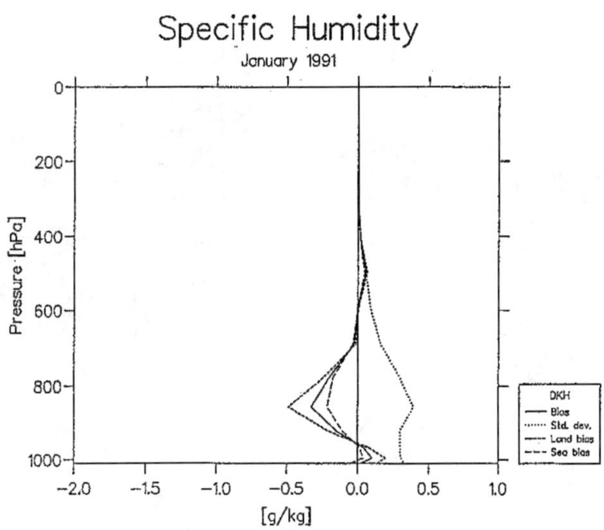

Figure 5. *Continued.*

A first test of PROMES was made in January, 1991, using "perfect" boundary conditions (ECMWF analyses). In Figure 6A the distribution of simulated and analyzed observed accumulated precipitation for the Iberian peninsula is shown. The model reproduces successfully the maxima observed in the northwestern part of the peninsula (simulated maximum: 548 mm, observed point maximum: 652 mm) and at the eastern coast (simulated maximum: 550 mm, observed point maximum: 364 mm). The rain in the northwestern part was associated with the passage of fronts from the Atlantic during the first half of the month, while the latter was caused by a shallow easterly flow, adverting moist air over the coastal Mediterranean hills. The model also reproduces successfully a broad band with low precipitation covering the peninsula from northwest to southwest, including the whole central plateau.

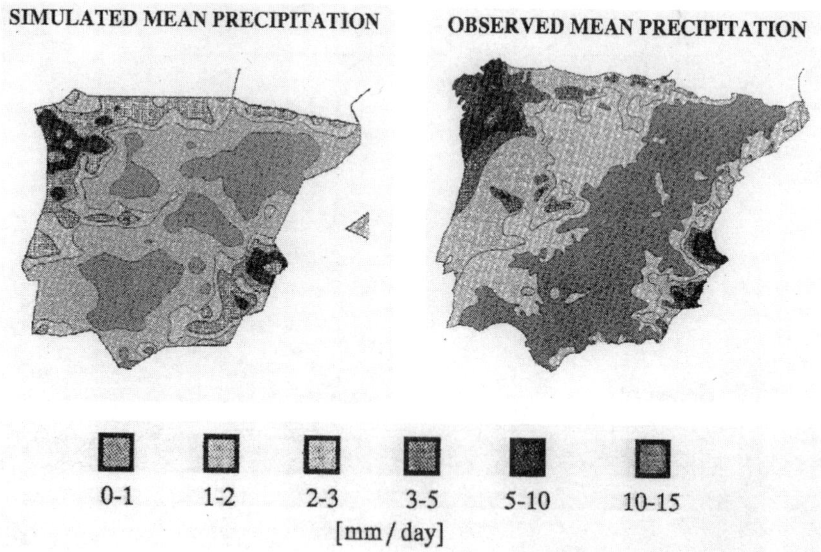

Figure 6A. Mean monthly precipitation pattern simulated with the PROMES model, using "perfect" boundary conditions from January, 1991 (left map), and the observed January, 1991, pattern (right map). The latter is based on a large number of climatological station reports.

Figure 6B shows the precipitation pattern simulated by HIRHAM for the same month, January, 1991, 1 yr into the 20-month simulation with perfect boundary conditions. In spite of the much larger area covered by HIRHAM and the fact that it has been forced with observed (analyzed) fields only at the sea surface (the SST) and at the lateral boundaries during the preceding year, the pattern of precipitation over the Iberian peninsula is very well simulated. Due the lower resolution of HIRHAM versus PROMES, the HIRHAM pattern is less detailed, and the peak values are slightly less than observed.

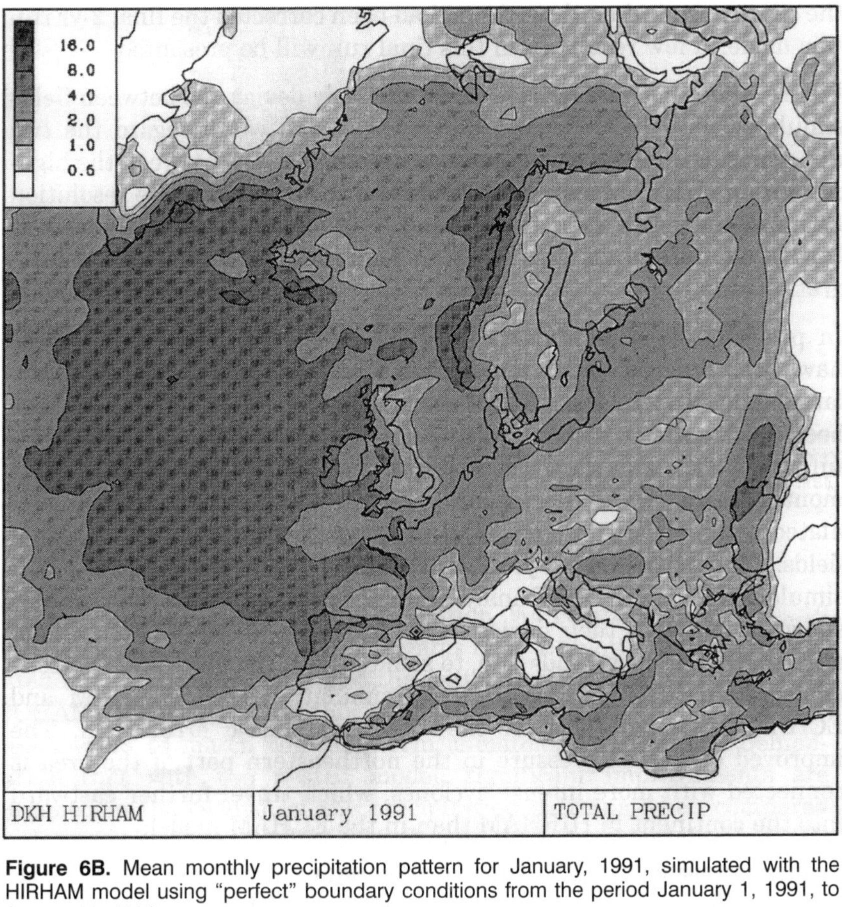

Figure 6B. Mean monthly precipitation pattern for January, 1991, simulated with the HIRHAM model using "perfect" boundary conditions from the period January 1, 1991, to August 31, 1991.

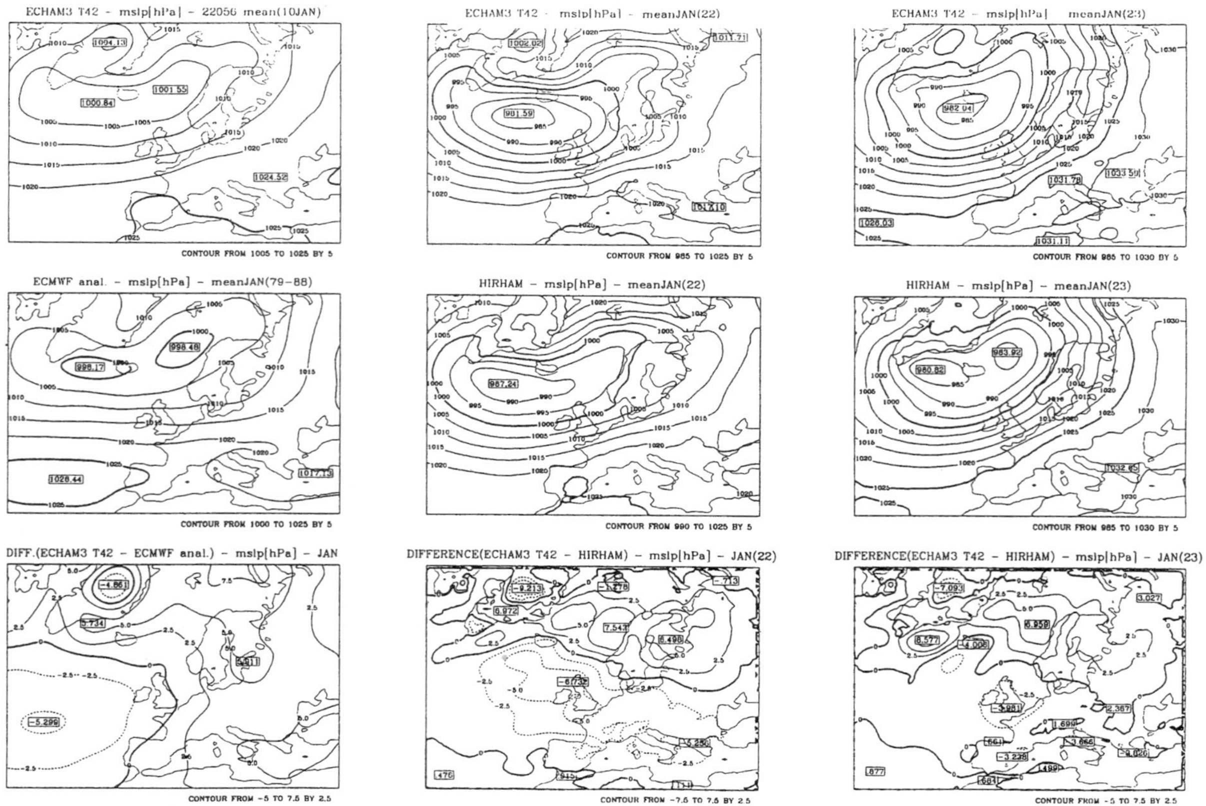

Figure 8. Mean sea-level pressure maps (contour interval, 5 hPa) and difference maps (contour interval, 2.5 hPa). In middle and righthand columns, mean January pressure fields from simulations with observed SST are shown: the HIRHAM field (middle), the driving ECHAM field (upper), and difference field (bottom). In the lefthand column, the 10-yr mean January pressure field from an ECHAM simulation with observed SST from 1979-1988 (upper), the 10-yr mean January pressure field from ECMWF analyses, 1979-1988 (middle), and the difference field (bottom) are shown.

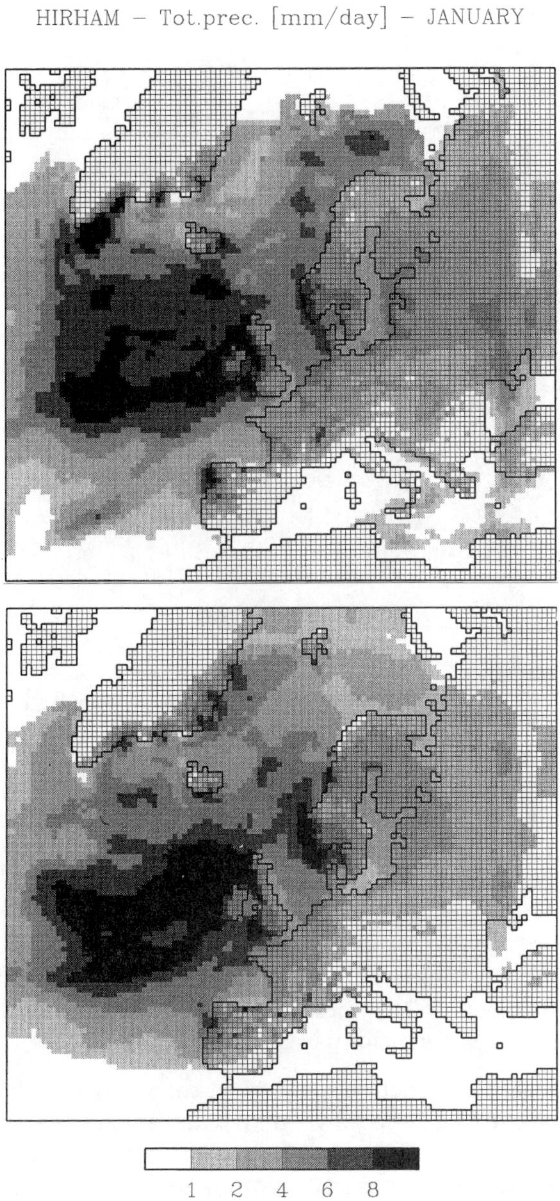

Figure 9. Pattern of mean January precipitation (mm/day) simulated by HIRHAM. From the 2-yr HIRHAM simulation with ECHAM fields and observed SST used for boundary conditions. Upper map: first January (1990). Bottom map: second January (1991).

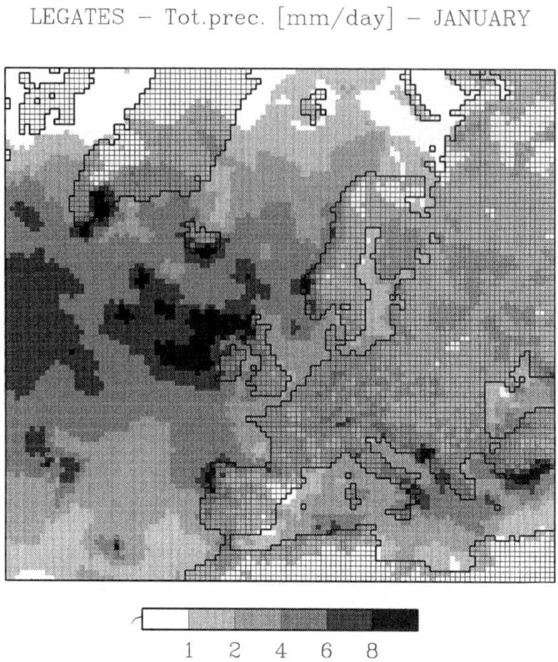

Figure 10. Pattern of mean January precipitation (mm/day) simulated by ECHAM with observed SST corresponding to those shown in Figure 9.

At UKMO, a 1-yr GCM-driven simulation with their high-resolution LAM has recently been completed. They are also using a relatively large area, and they came to the same conclusions as those made above from the 2-yr HIRHAM simulation: namely, that the high-resolution nested model is capable of improving the large-scale flow pattern within the model domain, and that significant improvements are obtained in the realism of the precipitation patterns when compared with the large-scale pattern simulated bv the GCM.

Definitive conclusions cannot be made from the results of these experiments. With the even larger area we plan to use in the near future and the longer, multiyear simulation periods we expect to establish more firmly the capability of the nesting technique to improve not only the small-scale but also the large-scale fields within the integration domain.

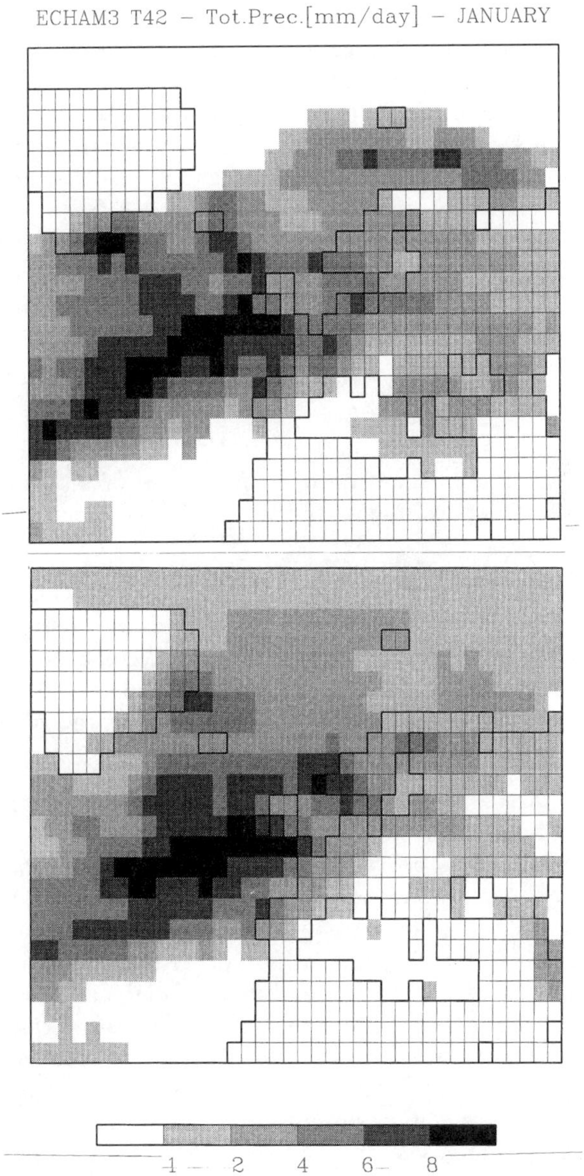

Figure 11. Observed mean January precipitation climatology (mm/day) from Legates and Willmott (1990b).

CONCLUDING REMARKS

As mentioned previously, verification of the high-resolution present-day climate simulations to be performed in the Regionalization project must be verified against observations in order to establish their accuracy. The resulting accuracy will constitute an upper limit of what can be expected in simulations of future climate scenarios that are planned using the same techniques in the extension project. These verifications are, therefore, a very important element of the Regionalization project. Since the start of the project, two workshops have been arranged for the project participants and other groups interested in regionalization. One of the main subjects of these workshops has been verification.

As most of the high-resolution simulations can be made for only relatively short time periods, e.g., 5 yr, the natural variability on longer time scales must be taken into account in the evaluations. In order to estimate such variabilities, we need a time series of monthly climate statistics for a sufficiently long period and with a sufficiently high spatial resolution.

After a survey of existing climatological data bases it was found that additional data would be desirable. We have therefore requested from the National Meteorological Services in Western Europe surface station data of monthly mean statistics for the period 1961 to 1990 with the highest density possible. We plan to use the data we receive as a supplement to the data bases already available.

On the basis of the results presented, which were obtained in the C21C and in the first half-year of the Regionalization project, it seems obvious that substantially improved regional climate simulations will be produced. For boundary conditions, using the best possible global present-day climate AOGCM simulations, the high-resolution simulations obtained by the nesting technique will be verified against observations. If support from CEC is given to extending the Regionalization project, the dynamical zooming technique and statistical techniques will be applied, and their results will be verified against observations. In the extension project, it is also planned to apply the different techniques to anthropogenic climate change simulations. The accuracy of the high-resolution climate changes computed by the different techniques could then be estimated on the basis of their intercomparison and the verification results of the present-day climate simulations.

REFERENCES

Christensen, J H and van E Meijgaard. **1992**. *On the Construction of a Regional Atmospheric Climate Model*, Technical Report No. 92-14. Danish Meteorological Institute, Copenhagen, Denmark.

Courtier, P and JF Geleyn. **1988**. A global numerical weather prediction model with variable resolution: application to the shallow water equation. Q. J. R. Meteorol. Soc. 114:1321-1346.

Courtier, P, C Freydier, JF Geleyn, F Rabier, and M Rochas. **1991**. The ARPEGE project at METEO-FRANCE, Vol. 2, pp. 193-231. In: *Proceedings of ECMWF workshop on Numerical Methods in Atmospheric Modeling*, Reading, UK.

Dickinson, RE, RM Errico, F Giorgi, and GT Bates. **1989**. A regional climate model for the Western United States. Clim. Change 15:383-422.

Giorgi, F and MR Marinucci. **1991**. Validation of a regional atmospheric model over Europe: sensitivity of wintertime and summertime simulations to selected physics parameterizations and lower boundary conditions. Q. J. R. Metdorol. Soc. 117:1171-1206.

Giorgi, F, MR Marinucci, and G Visconti. **1990**. Use of a limited area model nested in a general circulation model for a regional climate simulation over Europe. J. Geophys. Res. 95:18413-18431.

Giorgi, F, MR Marinucci, and G Visconti. **1992**. A 2 $\times$ CO_2 climate change scenario over Europe generated using a limited area model nested in a general circulation model. II: Climate change scenario. J. Geophys. Res. 97:10011-10028.

Legates, DR and CJ Willmott. **1990a**. Mean seasonal and spatial variability of global surface air temperature. Theoret. Appl. Climatol. 41:11-21.

Legates, DR and CJ Willmott. **1990b**. Mean seasonal and spatial variability in gauge-corrected global precipitation. Int. J. Climatol. 10:111-127.

Machenhauer, B (ed.). **1988**. HIRLAM *Final Report*, HIRLAM Technical Report No. 5. Danish Meteorological Institute, Copenhagen, Denmark.

Marinucci, MR and F Giorgi. **1992**. A 2 $\times$ CO_2 climate change scenario over Europe generated using a limited area model nested in a general circulation model. I: Present-day seasonal simulation. J. Geophys. Res. 97, D9:9989-10009.

Meehl, GA and WM Washington. **1990**. CO_2 climate sensitivity and snow sea-ice albedo parameterization in an atmospheric GCM coupled to a mixed-layer ocean model. Climate Change 16:283-306.

Oberhuber, JM. **1993**. Simulation of the Atlantic circulation with a coupled sea ice-mixed layer isopycnal general circulation model. Part 1: Model description. J. Phys. Oceanogr. 23:808-829.

Perlwitz, J. **1993**. Time-slice climate change experiments. Ann. Geophys. Suppl. II to Vol. 11, C176.

Roeckner, E, K Arpe, L Bengtsson, S Brinkop, L Dümenil, M Esch, E Kirk, F Lunkeit, M Ponater, B Rockel, R Sausen, U Schlese, S Schubert, and M Windelband. **1992**. *Simulation of the Present-Day Climate with the ECHAM Model: Impact of Model Physics and Resolution*, Report No. 93. Max-Planck-Institut für Meteorologie, Hamburg, Germany.

REPRESENTING THE INFLUENCE OF SUBGRID TOPOGRAPHY ON HYDROLOGY

L. R. Leung and S. J. Ghan

Pacific Northwest Laboratory, Richland, Washington

Key words: *Regional climate model, aggregation method, topography, cloud, precipitation, snow cover*

ABSTRACT

Estimates of the impact of global climate change on land surface hydrology require climate information on scales far smaller than those explicitly resolved by global climate models of today and the foreseeable future. To bridge the gap between what is required and what is resolved, we propose a subgrid-scale parameterization of the influence of topography on clouds, precipitation, and land surface hydrology. The parameterization represents subgrid variations in surface elevation in terms of discrete elevation classes. Separate cloud and surface processes are calculated for each elevation class. The simulated surface temperature, precipitation, snowpack, and soil moisture for each elevation class can then be distributed according to the spatial distribution of surface elevation within each grid cell. The scheme is being applied to the Pacific Northwest Laboratory's climate version of the Penn State/National Center for Atmospheric Research Mesoscale Model. Validation is being addressed by driving the model with observed lateral boundary conditions for the Pacific Northwest and comparing with surface observations. Preliminary results from the simulation will be presented.

INTRODUCTION

Human activities rely critically on the distribution of surface moisture and precipitation. These, together with water vapor and cloud in the atmosphere and snow cover on the ground surface, constitute the hydrological cycle that interacts intimately with the atmosphere-ocean system. An understanding of the impacts of global climate change on human society cannot neglect these complicated interactions. Yet it is this component that, at present, represents the most uncertainty in the modeling of the global environment.

The modeling of climate has evolved from simple one- or two-dimensional models that capture the energy balance of the climate system to complicated three-dimensional models that couple the atmosphere and ocean with rather detailed physical parameterizations. However, because of the diversity of spatial and temporal scales involved in hydrological processes of the atmosphere and the land surface, global climate models (horizontal resolution of a few hundred kilometers) are far from being able to model these processes at scales useful for impact assessment (less than 10 km).

One example is the modeling of precipitation. Evaluation of a number of General Circulation Models (GCM) performed by the Intergovernmental Panel on Climate Change (IPCC, 1991) indicated that there is an error of between 20 and 50% in the model-simulated seasonal precipitation when compared with observations averaged over regions of roughly 15° by 15°. Furthermore, sensitivity studies using a number of GCM in a doubled CO_2 concentration experiment reveal that, on regional scales, the differences in the projected changes in seasonal precipitation are of the same order of magnitude as the changes themselves (Grotch and MacCracken, 1991).

Undoubtedly, such disparity between model results and observations are due both to an insufficient understanding or representation of the physical processes and a lack of horizontal resolution to resolve either the atmospheric dynamics or characteristics of the lower boundary such as land surface and topography. In an effort to test the advantage of using higher horizontal resolution in climate models Giorgi and Bates (1989) and Giorgi et al. (1993) developed a regional climate model and used it to simulate climatology over the western United States. With more detailed parameterizations of cloud and much higher horizontal resolution (60 km) compared with GCM, their studies indeed show improved skill for simulating seasonal precipitation on the regional scale (several hundred kilometers). However, differences between the simulated and observed precipitation remain significant when comparisons are made on scales less than 100 km (Giorgi et al., 1992). This suggests that two areas of research are still needed to bring the model simulation of the hydrological cycle to closer agreement with observations on spatial scales useful for climate change studies. First, improvements in physical parameterizations of the processes in the atmosphere that govern the hydrological cycle are needed. Second, parameterizations that account for the subgrid variations due to unresolved dynamics and lower boundary are essential even in regional climate models. This paper represents an early effort targeting the latter:

to develop a parameterization of subgrid orographic precipitation for climate models.

Topography is a prominent feature in the western United States. An accurate simulation of precipitation in this area cannot be achieved without considering the effects of topography on precipitation. This is particularly true for explaining the spatial distribution of precipitation along coastal mountain ranges (e.g., Smith, 1987; Sumner, 1988). During winter, precipitation is often generated as a result of the lifting of the moisture-rich air by the elevated surface below. Condensation and subsequent evolution of clouds are controlled by more detailed air circulations and cloud microphysical processes. Past studies indicated an enhancement between 50 and 80% in precipitation by hills up to a few hundred meters high (Browning, 1980; Storebo, 1975; Hobbs et al., 1973). For mountains higher than 1-2 km, the enhancement can be more than 200% (Barros and Lettenmaier, 1994). The accumulation of snow on mountain peaks affects water resources over large geographical areas covered by the river drainage system. The capability of climate models to capture these climatological features is essential to an integrated modeling of the changing environment.

Barros and Lettenmaier (1994) reviewed and summarized a list of mechanisms identified from numerous previous studies (e.g., Hobbs et al., 1973; Fraser et al., 1973; Choularton and Perry, 1986) as important to the generation of orographic precipitation. Among those are the general dynamics of air flow over mountain barriers and cloud microphysical processes such as the seeder-feeder mechanisms (Bergeron, 1960; Browning et al., 1975). The former has more prevailing influence on long-term precipitation distributions. We have developed our parameterization scheme based on a simple air-flow model and a detailed mixed-phase cloud microphysics model to represent the two components mentioned above. To seek a balance between detailed physical representations and computational efficiency for climate applications, we have adopted an aggregation approach that represents the variations of surface elevation with a limited number of surface elevation classes. The model formulation will be described in the section that follows. The parameterization has been implemented into a regional climate model. Precipitation is simulated for all surface elevation classes within the grid cells of the model and distributed back to its geographical locations. Other aspects of the hydrological cycle (namely, soil moisture, snow accumulation, and surface evapotranspiration), are also accounted for at the subgrid scale in this framework. Model results and sensitivity tests will be discussed in a later section.

MODEL DESCRIPTION

Following the approach pioneered by Dickinson et al. (1989) and Giorgi and Bates (1989), a regional climate model has been developed at the Pacific Northwest Laboratory (PNL) that is based on the Penn State/NCAR Mesoscale Model (MM4) (Anthes et al., 1987). Several physical parameterizations important on climatic time scales have been included in the model. They are equivalent to those used in the PNL version of the NCAR Community Climate Model (CCM1; Williamson et al., 1987). The use of the same physical parameterizations in both regional and global models also facilitates their future coupling for climate change studies. These parameterizations include a two-stream delta-Eddington treatment for short-wave radiation (Taylor and Ghan, 1992) and an emissivity approach for long-wave radiation (Kiehl et al., 1987); a bulk cloud microphysics scheme that distinguishes liquid and ice phase (Cotton et al., 1986; Ghan and Easter, 1992); and a surface physics scheme (BATS1E). A split-explicit time integration scheme with efficient treatment of gravity waves (Madala, 1987) has been introduced in this version of MM4. It permits the use of a 6-minute time step for simulations with a horizontal resolution of 90 km. Grell's parameterization scheme (Grell, 1993) is used for treatment of cumulus convection. For the planetary boundary layer, the high-resolution model of Zhang and Anthes (1982) is used.

In mesoscale models, high horizontal resolutions are generally achieved through the use of nesting. As the horizontal resolution increases, shorter time steps and more spatial points in the simulation place a significant burden on computational requirements. Therefore, most models of orographic precipitation in the past were developed for the purpose of understanding the physical processes rather than for climate applications. Most of them have detailed air circulation as well as cloud microphysics schemes and require time steps of a few seconds (e.g., Fraser et al., 1973; Tripoli and Cotton, 1989). To explicitly resolve the often highly varying surface elevations in the continents would therefore not be feasible, particularly on climate time scales. We have adopted an alternative method based on aggregation of surface elevation. Subgrid variations in surface topography are represented by a limited number of surface elevation classes. The 12 categories used to define the surface elevation classes are shown in Table 1. From a surface topography dataset of 1.5-km resolution, information relating to the aggregation scheme is derived. This includes, for each grid cell of a model domain, the grid-cell mean surface elevation, the number of elevation classes it contains, the mean elevation of those areas belonging to the same elevation class, and the fraction of area covered by each elevation class. On average, 4.5 classes are carried by each 90-km grid cell over the continental United States.

Table 1. Classification of surface elevations used in subgrid parameterization.

Class Number	Surface Elevation (m)
1	0–100
2	100–200
3	200–300
4	300–400
5	400–500
6	500–700
7	700–1000
8	1000–1500
9	1500–2000
10	2000–3000
11	3000–4000
12	4000–5000

There are two main components in the subgrid parameterization of orographic precipitation: an air-flow scheme and a cloud microphysical model. As is implied by an aggregation approach, the model does not carry any information on the geographical locations of the surface elevation classes within a grid cell. An air-flow model that merely determines the degree of uplifting of an air parcel as it passes over a mountain barrier would be consistent with the simplicity of this subgrid surface topographical representation. Such a model would not be able to capture the sequence and details of air flow over a mountain but would still provide useful information for a stable orographic cloud where precipitation is mainly generated by the lifting and condensation of the moisture-rich air in the lower atmosphere (Cotton and Anthes, 1989). This approach would therefore be most useful for simulating wintertime precipitation along coastal mountain ranges.

Whether air flows over or around mountains depends on their height with respect to the grid-cell mean surface elevation, their horizontal dimension perpendicular and parallel to the upstream flow, the upstream wind speed, and the static stability. A fundamental parameter that controls upstream blocking is the Froude number, defined as

$$F_r = U/NH \tag{1}$$

where U is the upstream wind speed, N is the Brunt-Vaisala frequency, and H is the relative height of the mountain. The square of the Froude number can be interpreted as the ratio of the kinetic energy of an air parcel to the potential energy required to lift the parcel from the surface to the mountain top. Analytical theory (Sheppard, 1956; Smith, 1980, 1988, 1989) and high-resolution numerical modeling studies (Pierrehumbert and Wyman, 1985; Smolarkiewicz and Rotunno, 1989, 1990) have shown that air generally flows over mountain when the Froude number is large, but around it when the reverse is true. We have adopted a method that determines the height rise of an air parcel by its energy balance. An air parcel will be forced to lift over the mountain until all its kinetic energy is converted to potential energy, after which it simply flows around the mountain until it descends. Therefore, the maximum height rise attainable by an air parcel is given by $h_{max} = U/N$ (Figure 1). The actual height rise, h, would be limited by the height of the surface elevation where the air parcel originates relative to the elevation of the mountain it encounters. Hence,

$$h = min(z_s - \bar{z}_s, h_{max}), \tag{2}$$

where z_s is the height of the surface elevation class of which the air parcel is being considered, and $\bar{z}_s$ is the height of the grid-cell mean surface elevation.

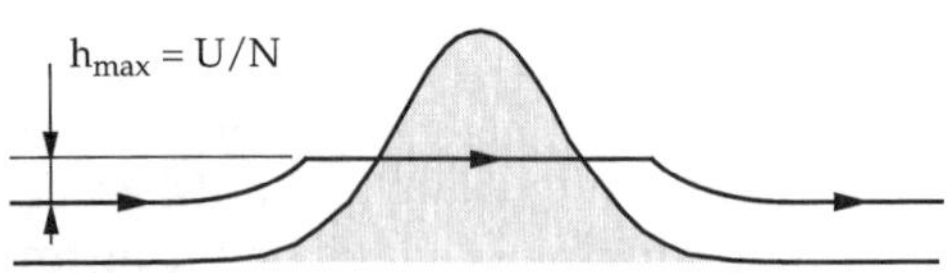

Figure 1. Schematic illustration of air flow over mountain. Maximum height rise attainable by an air parcel given by h_{max}. When lifting reaches maximum, air flows around mountain.

Once the height rise or descent of an air parcel is determined, values of the cloud variables in each elevation class can be diagnosed from profiles of those variables in the grid-cell mean. The thermodynamic and cloud variables θ_c and r_ω are used in this procedure because of their conservative properties during condensation as well as advection. They are defined as

$$\theta_c = \theta - (L/c_p)\,(r_{c/E}) \text{ and} \tag{3}$$

$$r_\omega = r_v + r_c, \text{ respectively,} \tag{4}$$

where $\theta = T/E$ is the potential temperature, $E = (p/p_0)^\kappa$ *with* $\kappa = R/c_p$ is the Exner function, and r_c and r_v are the mixing ratios of cloud and water vapor, respectively. Profiles of θ_c and r_ω are interpolated from the grid-cell mean to each elevation class according to the height rise/descent of the air parcel, determined by the air-flow scheme.

For cloud microphysical calculations, the temperature, T, cloud water, r_c, and water vapor, r_v, will be diagnosed from θ_c, r_w; and the pressure, p by assuming r_v never exceeds the water vapor saturation mixing ratio, $r_s\,(T,p)$. Here, the pressure, p, for each elevation class is estimated from its surface pressure, p_s, which can be inferred from the grid-cell mean surface pressure, $\bar{p}_s$, using the linearized hydrostatic relation

$$\mathrm{p_s} = \bar{p}_s\, e^{-(z_s - z_s)/\bar{\mathrm{H}}}, \tag{5}$$

where $\bar{\mathrm{H}}$ is the scale height. Therefore, elevation classes with higher surface elevations have lower surface pressures, which, from the definition of θ_c, yield colder temperatures and higher cloud water concentrations.

These procedures can be implemented either diagnostically or prognostically to account for subgrid cloud processes due to variations in surface elevations. In the former approach, cloud variables for each surface elevation class are diagnosed from the grid-cell mean values at the beginning of each time step, and cloud microphysical processes are calculated for each class accordingly. Because some microphysical processes such as precipitation operate on time scales much shorter than advection, θ_c and r_w may not be conserved in these procedures. Indeed, a preliminary testing of this approach resulted in orographic signatures of precipitation that were far too strong, violating the conservation of energy and water budgets. An approach that carries θ_c and r_w prognostically for each elevation class as well as the grid-cell mean avoids this problem. To implement such an approach, tendencies in θ_c and r_w due to air flow over or around mountains are calculated for each elevation class. The calculation is based on the difference between the vertical profiles of those variables for the grid-cell mean and those diagnosed for each surface elevation class from the air-flow scheme. The tendencies also depend on the time scale over which the lifting/descent completes. Physically, this time scale, τ, depends on the wind speed as well as the spatial distribution of the surface elevation classes within the domain grid cells. As a first test of the aggregation scheme, it is treated as an adjustable parameter in this study.

The parameterization described has been included in the PNL regional climate model. Advection and horizontal diffusion are calculated based on the grid-cell mean variables and applied to the prognostic equations of both the grid-cell means and the individual elevation classes. Physical processes such as radiation, planetary boundary layer and surface physics, cloud microphysics and cumulus convection are all performed for each elevation class. Tendencies from these processes are applied to prognostic equations for each class, then aggregated to the grid-cell mean variables.

Although the scheme described here offers some solutions to the problem of representing subgrid-scale orographic effects on clouds, it does not solve all of the difficulties. Most notably, the rainshadowing effect that results from dry-adiabatic descent of air on the lee side of mountains is not accounted for on the subgrid scale by this parameterization. The lee side of the mountains receives the same precipitation as the upwind side if it belongs to elevation classes within the same grid cell. This is not a problem for narrow ridges with advective time scales shorter than the time scales of precipitation processes, but it is potentially a serious limitation for broader orographic features not explicitly resolved by the regional model. In short, the grid resolution of the regional model should be chosen to be fine enough that the advective time scale is shorter than that of the formation of precipitation.

RESULTS

We have implemented the subgrid parameterization of orographic precipitation in the PNL regional climate model. Both the coarse and fine mesh used in this study are shown in Figure 2. The horizontal resolution of the coarse and the fine domains is 90 km and 30 km, respectively. The model has 23 vertical layers. A 1-yr simulation has been completed in the nested configuration without any aggregation for evaluating the general performance of the model. We will identify the model results for the coarse and fine domains in this simulation by CN and FN, respectively. For testing the aggregation scheme, a 1-month simulation has been performed over the same coarse domain shown in Figure 2. The aggregation of surface elevation classes is carried out over an area slightly larger than the nested domain indicated in the same figure. This simulation will be identified by CAGG. All simulations are performed with the same physical parameterizations and lateral boundary conditions from ECMWF analysis. Model results will be compared with observations made at more than 500 weather stations in the Pacific Northwest (Figure 3). Evaluation of the models will center on fields measured by these stations: daily maximum and minimum surface temperatures, and

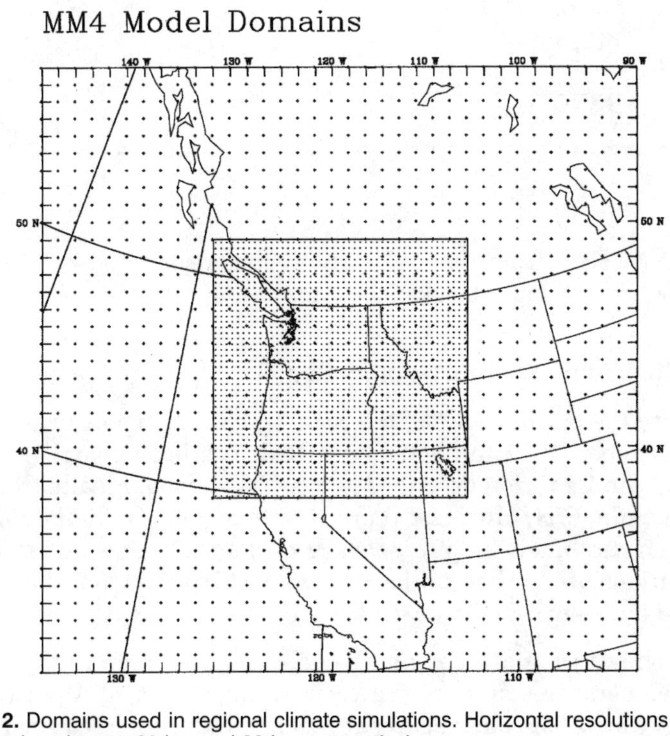

Figure 2. Domains used in regional climate simulations. Horizontal resolutions for coarse and fine domains are 90 km and 30 km, respectively.

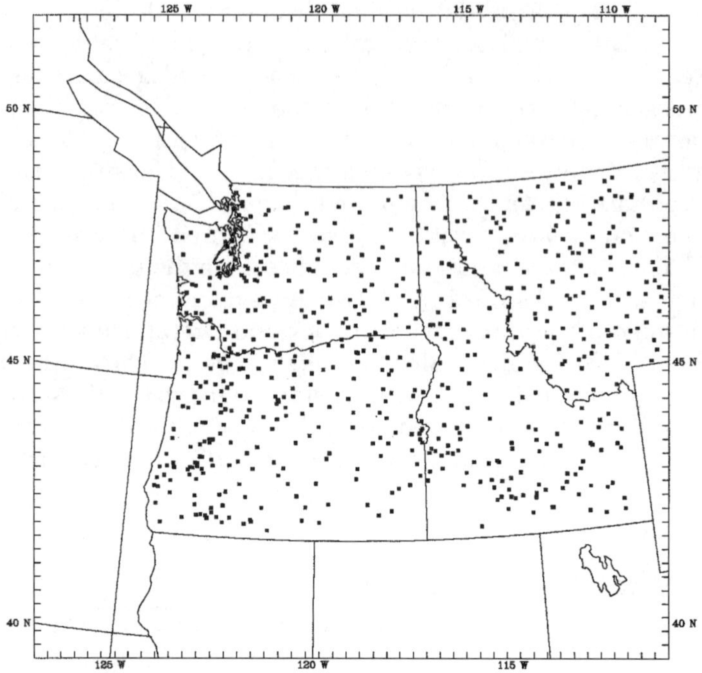

Figure 3. Locations of weather stations in Pacific Northwest used in model evaluations.

precipitation. Data from "snotel" stations that record snow cover over the same region will also be used for comparison.

Nested Simulation Without Aggregation

To evaluate the performance of the regional climate model, a 1-yr simulation (October, 1987 to September, 1988) has been completed with the nested configuration described above. Only model results from the nested domain (FN) are used here. Comparison of simulation and observations are made by interpolating model-simulated values to the exact locations of weather stations shown in Figure 3 from the closest four grid points surrounding each station.

Figure 4 shows the observed and model-simulated monthly averaged surface temperatures and precipitation averaged over all the weather station locations. The model is capable of simulating the seasonal cycle

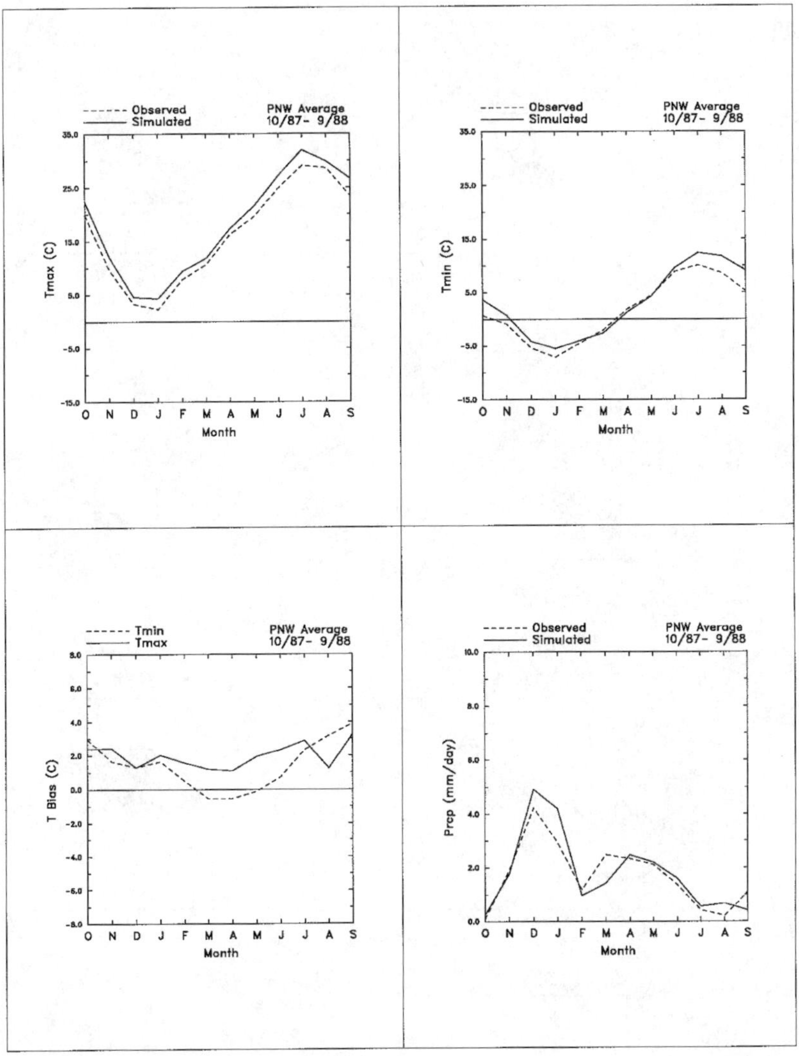

Figure 4. Simulated and observed monthly averaged daily maximum/minimum surface temperatures and precipitation averaged over all weather station locations. Simulation period, October, 1987 to September, 1988.

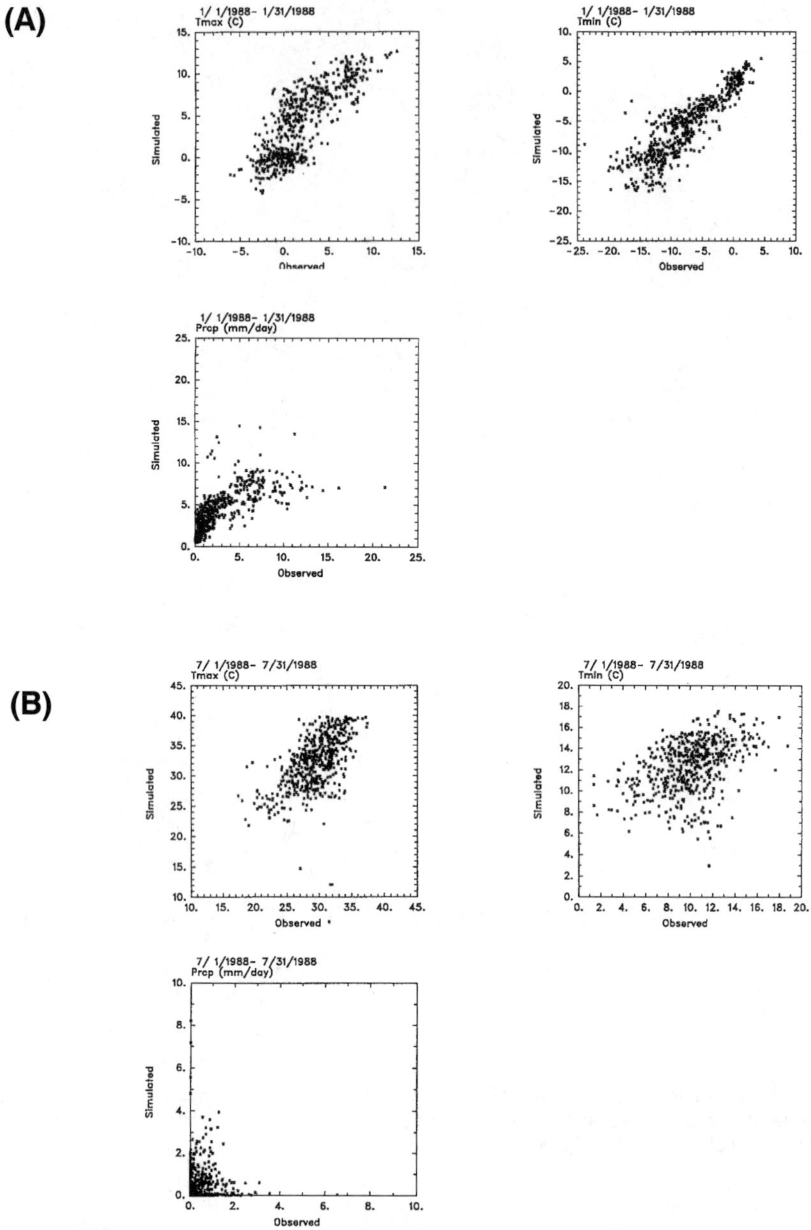

Figure 5. Scatter plots of monthly averaged simulated versus observed daily maximum/minimum surface temperatures and precipitation for (A) January and (B) July, 1988.

in the surface temperature with a bias of about 2°C. Maxima in precipitation during winter and early spring are also simulated well by the model. To evaluate the model performance beyond regional averages, Figure 5 shows scatter plots of model versus observed temperatures and precipitation for all stations during January and July of 1988. During winter, little bias is found in either the surface temperatures or the precipitation. The model explains between 50 and 75% of the spatial variance in the observations made over a wide variety of atmospheric conditions and surface boundaries. During summer, precipitation is generated more randomly at different locations by convective instability. Comparison of modeled and observed precipitation on a point-by-point basis indicates very little skill, consistent with the physical processes that are at play. With BATS included in the model, snow cover is also simulated and saved daily. A comparison of the simulated and observed snow cover (in inches of water equivalence) is shown in Figure 6 as a scatter plot for December, 1987. Not only is there little correlation between the observed and model-simulated snow cover, the bias in the mean values is also significant. It is clear that the

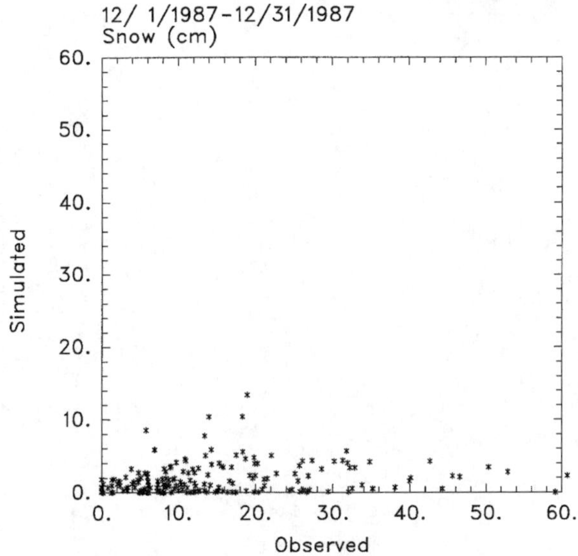

Figure 6. Scatter plot of monthly averaged simulated versus observed snow cover (December, 1987) in inches of water equivalence. Observations obtained from "snotel" stations across Cascades and Northern Rockies in Pacific Northwest.

simulation of snow cover has to be improved for credible modeling of surface hydrological processes.

Testing of Aggregation Scheme

Short simulations of 5 days each were performed to test the sensitivity of the aggregation scheme to the parameter τ , which controls the time scale for air flow over mountains. We tested a range of values for τ between 2 hours and 6 hours. Figure 7 shows an east-west cross section of three grid cells in western Washington state, covering the Olympic Mountains, Puget Sound, and the Cascade mountain range. Surface elevation varies from sea level to above 2000 m in the mountains. The model simulates precipitation for each elevation class within a grid cell. The precipitation is then distributed back to the geographical locations according to the 1.5-km, high-resolution surface elevation data from which the information of the aggregation scheme was derived. Interpolation has been performed for the simulated precipitation between

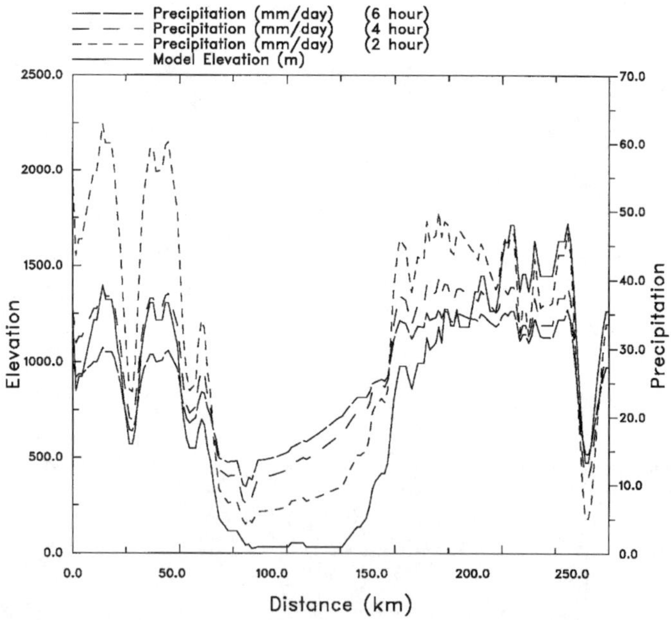

Figure 7. East-west cross-section of three model grid cells covering Olympic Mountains, Puget Sound, and western Cascade Mountains. Solid line represents surface elevations defined by elevation classes. Three dashed lines represent precipitation simulated by model using values of ranging from 2 hr to 6 hr.

elevation classes on adjacent grid cells. This ensures more continuous spatial distribution of precipitation in transition from one grid cell to another. Consistent with the physical interpretation of the parameter, the figure shows stronger orographic signatures for smaller values of τ .

The 1-month simulation with the aggregation scheme (CAGG) is based on τ = 6 hr. Figure 8 shows the same cross-section as in Figure 7 over the Puget Sound area for model-simulated rainfall, snowfall, and snow cover. Corresponding values for the CN simulation are shown for comparison. The orographic signatures in the CAGG simulation are very similar to what are usually observed in mountainous areas, with

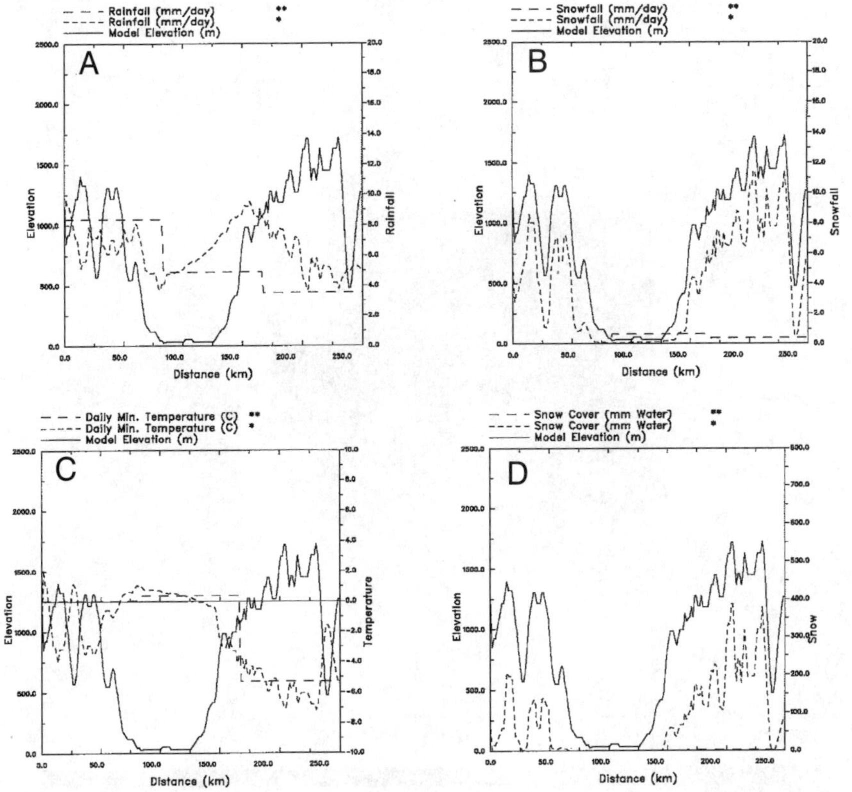

Figure 8. Same east-west cross-section as shown in Figure 7. Dashed lines represent model-simulated (A) rainfall, (B) snowfall, (C) daily minimum surface temperature, and (D) snow cover. Simulated values shown are monthly averages of December, 1987. The coarse domain (CN) and aggregation (CAGG) simulations are represented by ** and *, respectively.

increases in precipitation on the windward side and the rainshadowing effect on the lee side. Also, snow cover is now being simulated with reasonable magnitude. Figure 9 shows the spatial distribution of precipitation simulated with the aggregation scheme. Much more spatial detail is now captured by the model than would have been seen in the 90-km grid without aggregation. In this particular simulation, effects of surface topography on precipitation are strongest on the western Cascades.

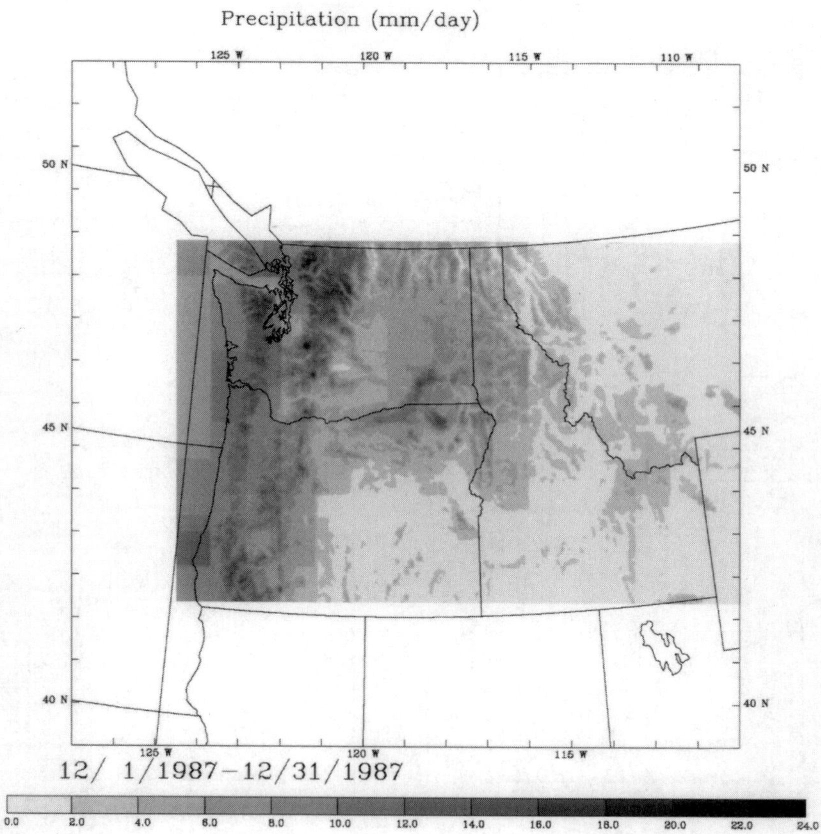

Figure 9. Spatial distribution of monthly-mean precipitation (mm/day) simulated by model with aggregation. Each tick mark represents 90-km horizontal resolution used in model.

Figure 10 shows scatter plots similar to Figure 5 for model simulations with the aggregation scheme (CAGG) and those of the coarse (CN) and fine domain (FN) without the aggregation scheme. With the aggregation scheme, the model is capable of simulating the stronger precipitation events that occur mostly on higher elevations. The correlation coefficients, r, are calculated for each scatter plot. More skill is demonstrated with the use of the aggregation scheme than even in the 30-km nested simulation. This is a significant improvement in representing subgrid orographic precipitation, not only in terms of improved skill, but also computational efficiency. A 1-day simulation for the model with aggregation takes 2.5 CPU hours on an IBM RS/6000-560, much less than the 8.5 CPU hours for the nested configuration. Finally, the

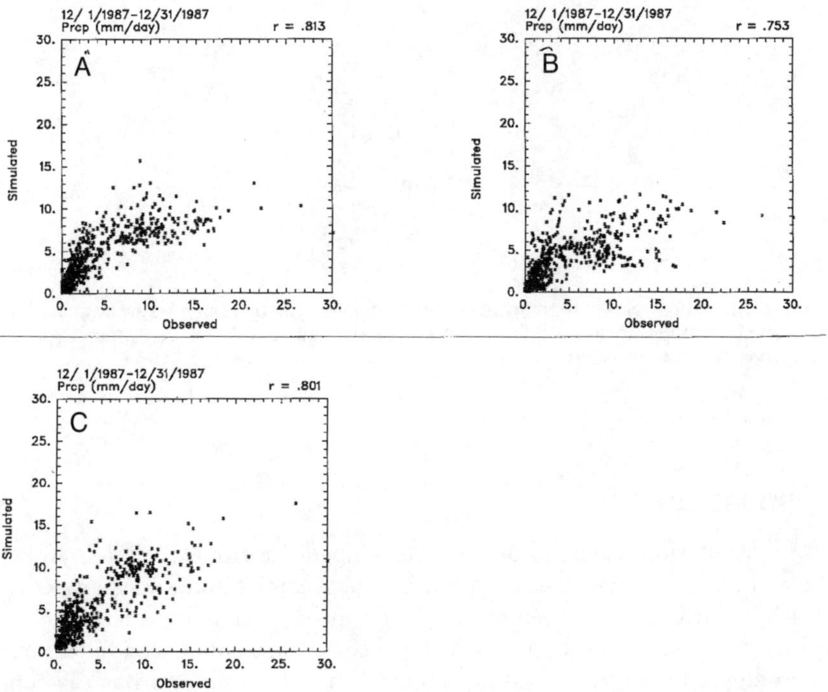

Figure 10. Scatter plots of monthly averaged simulated versus observed precipitation in three simulations: (A) 90-km grid with aggregation (CAGG), (B) 90-km grid without aggregation (CN), and (C) 30-km grid without aggregation (FN). Values for correlation coefficient, r, are shown on upper right corner of each scatter plot.

simulated snow cover for cases CAGG is shown in Figure 11. Although there is still a lot of scatter, the performance of the model with aggregation clearly surpasses that of the fine domain (FN) simulation shown in Figure 6.

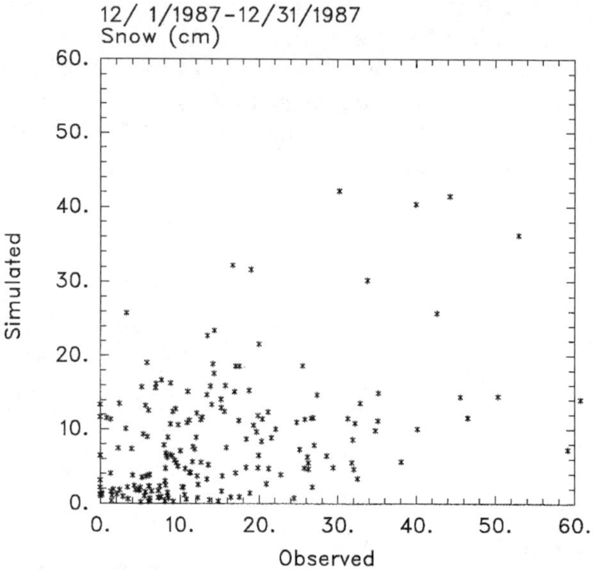

Figure 11. Scatter plot of monthly averaged simulated versus observed snow cover. Simulation performed with 90-km grid with aggregation. Contrast with Figure 6, 30-km grid simulation without aggregation.

DISCUSSION

An aggregation scheme has been developed for climate models to represent subgrid effects of topography on precipitation and snow cover. With a simple air-flow scheme and a detailed mixed-phase cloud microphysics scheme, cloud and surface processes are calculated for all subgrid elevation classes. Results indicate that such an approach is capable of capturing the enhancement of precipitation due to uplifting of air parcels by underlying mountain barriers. The simulated spatial distribution of precipitation and snow cover shows realistic climatological features typical of mountainous areas. Rainshadow effects are resolved on the scales of the model grid size, but not for mountain ranges within a grid cell. This, however, does not seem to have an adverse effect in this simulation, where most precipitation falls on mountain ranges along the coast. These mountains tend to be narrow enough that the time scale for advection is much shorter than that of the precipitation processes. More testing is needed to determine an optimal horizontal resolution for resolving rainshadow effects on broader mountain ranges, such as the Rockies.

Despite the preliminary success demonstrated in this study, the aggregation scheme now simulates, at best, the most general climatological features of orographic precipitation. There are inherent limitations in the aggregation approach because no information on the geographical orientation of the surface elevations is carried by the model. Physical processes related to the more detailed air circulation, such as lee waves and local thermally driven circulations, cannot be represented by such a method. Nevertheless, several aspects of the model can be further improved within the present framework for more realistic representations of the air circulation. These include a more detailed method for determining the rise of air parcels over mountain barriers, the use of a more physically based time-scale parameter, and, possibly, utilization of more information related to the aggregation of surface elevation. This study represents an important step toward the development of subgrid parameterizations that are essential even for regional climate models or global climate models of the future. The enhanced information produced from these parameterizations is valuable as climate input to models of surface hydrological processes.

ACKNOWLEDGMENTS

This work is supported by laboratory-directed research and development funding from the Pacific Northwest Laboratory. Pacific Northwest Laboratory is operated for the U.S. Department of Energy by Battelle Memorial Institute under Contract DE-AC06-76RLO 1830.

REFERENCES

Anthes, RA, EY Hsie, and YH Kuo. **1987**. *Description of the Penn State/NCAR Mesoscale Model Version 4 (MM4), NCAR Technical Note, NCAR/TN-282+STR*. NCAR, Boulder, CO.

Barros, AP and DP Lettenmaier. **1994**. Dynamic modeling of orographically-induced precipitation. Rev. Geophys. 32, 265-284.

Bergeron, T. **1960**. Problems and methods of rainfall investigation. Geophys. Monogr. 5:5-30.

Browning, KA. **1980**. Structure, mechanism and prediction of orographically-enhanced rain in Britain. Orographic Effects in Planetary Flows. GARP Pub. Ser. 23:88-108.

Browning, KA, CW Pardoe, and FF Hill. **1975**. The nature of orographic rain at wintertime cold fronts. Q. J. R. Meteorol. Soc. 101:333-352.

Choularton, TW and SJ Perry. **1986**. A model of the orographic enhancement of snowfall by the seeder-feeder mechanism. Q. J. R. Meteorol. Soc. 112:335-345.

Cotton, WR and RA Anthes. **1989**. *Storm and Cloud Dynamics*. Academic Press, New York.

Cotton, WR, GJ Tripoli, RM Rauber, and EA Mulvihill. **1986**. Numerical simulation of the effects of varying ice crystal nucleation rates and aggregation processes on orographic snowfall. J. Climate Appl. Meteorol. 25:1658-1680

Dickinson, RE, A Henderson-Sellers, PJ Kennedy, and MF Wilson. **1986**. *Biosphere-Atmosphere Transfer Scheme (BATS) for the NCAR Community Climate Model, NCAR Technical Note, NCAR/TN-275+STR*. NCAR, Boulder, CO.

Dickinson, RE, RM Errico, F Giorgi, and GT Bates. **1989**. A regional climate model for the western U.S. Climate Change. 15:383-422.

Fraser, AB, RC Easter, and PV Hobbs. **1973**. A theoretical study of the flow of air and fallout of solid precipitation over mountainous terrain: Part I. Airflow Model. J. Atmos. Sci. 30:801-812.

Ghan, SJ and RC Easter. **1992**. Computationally efficient approximations to stratiform cloud parameterization. Mon. Wea. Rev. 120:1572-1582.

Giorgi, F, and GT Bates. **1989**. On the climatological skill of a regional model over complex terrain. Mon. Wea. Rev. 119:2870-2888.

Giorgi, F, GT Bates, and SJ Nieman. **1992**. Simulation of the arid climate of the southern Great Basin using a regional climate model. Bull. Am. Meterol. Soc. 73:1807-1822.

Giorgi, F, GT Bates, and SJ Nieman. **1993**. The multiyear surface climatology of a regional atmospheric model over the western United States. J. Climate 6:75-95.

Grell, G. **1993**. Prognostic evaluation of assumptions used by cumulus parameterizations. Mon. Wea. Rev. 121:764-787.

Grotch, SL and MC MacCracken. **1991**. The use of general circulation models to predict regional climatic change. J. Climate 4:286-303.

Hobbs, PV, RC Easter, and AB Fraser. **1973**. A theoretical study of the flow of air and fallout of solid precipitation over mountainous terrain: Part II. Microphysics. J. Atmos. Sci. 30:813-823.

IPCC. **1991**. *Climate Change: The IPCC Scientific Assessment*. IPCC, Cambridge, UK.

Kiehl, JT, RJ Wolski, BP Briegleb, and V Ramanathan. **1987**. *Documentation of Radiation and Cloud Routines in the NCAR Community Climate Model (CCM1), NCAR Technical Note, NCAR/TN-288+IA*. IPCC, Boulder, CO.

Madala, RV. **1987**. Efficient Time Integration Schemes for Atmosphere and Ocean Models. *Finite-Difference Techniques for Vectorized Fluid Dynamics Calculations*, DL Book (ed.). Springer-Verlag, New York.

Pierrehumbert, RT and B Wyman. **1985**. Upstream effects of mesoscale mountains. J. Atmos. Sci. 42:977-1003.

Sheppard, PS. **1956**. Airflow over mountains. Q. J. R. Meteorol. Soc. 82:528-529.

Smith, RB. **1980**. Linear theory of stratified hydrostatic flow past an isolated mountain. Tellus 32:348-364.

Smith, RB. **1987**. The Role of Moist Processes in Mountain Effects. Seminar/Workshop 1986 on Observation. Theory and Modelling of Orographic Effects. European Center for Medium Range Weather Forecasts 1:77-111.

Smith, RB. **1988**. Linear theory of hydrostatic flow over an isolated mountain in isoteric coordinates. J. Atmos. Sci. 45:3889-3896.

Smith, RB. **1989**. Mountain induced stagnation points in hydrostatic flows. Tellus 41A: 270-274.

Smolarkiewicz, PK and R Rotunno. **1989**. Low Froude number flow past three dimensional obstacles. Part I: Baroclinically generated lee vortices. J. Atmos. Sci. 46:1154-1164.

Smolarkiewicz, PK and R Rotunno. **1990**. Low Froude number flow past three dimensional obstacles. Part II: Upwind flow reversal zone. J. Atmos. Sci. 47:1498-1511.

Storebo, PB. **1975**. Small scale topographical influences on precipitation. Tellus 28:45-59.

Sumner, G. **1988**. *Precipitation Process and Analysis*. John Wiley & Sons, New York.

Taylor, KE and SJ Ghan. **1992**. An analysis of cloud liquid water feedback and global climate sensitivity in a general circulation model. J. Climate 5:907-919.

Tripoli, GJ and WR Cotton. **1989**. Numerical study of an observed orogenic convective system. Part I: Simulated genesis and comparison with observations. Mon. Wea. Rev. 117:273-304.

Tripoli, GJ and WR Cotton. **1986**. An intense, quasi-steady thunderstorm over mountainous terrain. Part IV: Three Dimensional Numerical Simulation. J. Atmos. Sci. 43:894-912.

Williamson, DL, JT Kiehl, V Ramanathan, RE Dickinson, and JJ Hack. **1987**. *Description of NCAR Community Climate Model (CCM1), NCAR Technical Note, NCAR/TN-285 +STR*. NCAR, Boulder, CO.

Zhang, DL and RA Anthes. **1982**. A high-resolution model of the planetary boundary layer-sensitivity tests and comparisons with SESAME-79 Data. J. Appl. Meteorol. 21:1594-1609.

SIMULATION OF A PRECIPITATION EVENT IN THE WESTERN UNITED STATES

J. Kim[1] and S.-T. Soong[2]

[1]Lawrence Livermore National Laboratory, Livermore, California

[2]University of Calilfornia—Davis, California

ABSTRACT

Results from a simulation of an episode of 12-day precipitation in California and the surrounding states are presented. The simulation is performed with a moderately fine-resolution, three-dimensional, primitive-equation, limited-area model which is developed for long-term integrations over regional domains such as the western United States. The model employs an explicit treatment of cloud microphysical processes to simulate precipitation and clouds. A two-layer, soil-canopy-snow model is used to compute heat fluxes and hydrology at land surfaces.

The calculations indicate that there is a close correlation between the observed and simulated total precipitation. The orography of the region plays an important role in the geographical distribution of the precipitation, as exhibited by the heavy precipitation over the upslope of major mountain ranges. The model predicted a snowline location of 1.8 km above sea level, which led to a significant accumulation of snow over the Sierra Nevada Range during the simulation period. Local topography causes the prevailing low-level westerly wind to be deflected northwestward along the western slope of the Sierra Nevada Range, supplying additional moisture to the mountains in the northern part of the Central Valley during the period of heavy precipitation.

The simulation suggests that the important features of precipitation over mountainous terrains can be predicted by incorporating an explicit treatment of cloud microphysics within fine-resolution hydrostatic models. The explicit treatment of cloud microphysical processes enables the model to distinguish rain and snow from total precipitation. In addition, the properties of clouds that are important for computing radiative heat transfer are directly predicted from such a formulation.

INTRODUCTION

Wintertime precipitation is the major water resource in the western United States. Thus, correct assessment of wintertime precipitation, especially snowfall, is important in planning for the summertime water supply and the development of future water resources. Forecasts of local precipitation and soil hydrology are also important for early warnings of floods.

Regional precipitation depends chiefly on the large-scale flow and local topography. Large-scale moisture influx determines the amount of moisture available for precipitation. Topography affects two major factors in local precipitation: low-level vertical motion and local water-vapor transport. Topography also affects partitioning the total precipitation into rain and snow, since the snowline is usually lower than the peaks of major mountain ranges during winter.

Modeling the features of regional precipitation has been intensely studied, especially in conjunction with regional climate (Giorgi, 1991; Dickinson et al., 1989). Previous studies show that limited-area models nested within large-scale models or analysis can capture realistic features of storms. It was also found that the simulated structure of individual storms and precipitation vary significantly, depending on the parameterized precipitation processes (Zhang et al., 1988). As limited-area models usually have much finer spatial resolutions than large-scale models, parameterized precipitation processes developed for large-scale models may not be adequate for limited-area models (Frisch and Chappel, 1980).

We present a simulation of 12-day precipitation over California from February 11 to February 23, 1986. We focus on features of regional precipitation, such as local distribution, total amount, and occurrence of snowfall and rainfall. The simulation was carried out using a primitive-equation, limited-area model developed jointly by University of California at Davis and Lawrence Livermore National Laboratory.

MODEL

The model predicts the horizontal wind, surface pressure, potential temperature, and the mixing ratios of the water vapor, cloud ice, cloud water, snow, rain, and graupel. The governing equations are the flux-form of the primitive equations written in the σ-coordinates. The governing equations are discretized on the c-grid, using the finite-difference formulation by Takacs (1985), later modified by Hsu and Arakawa (1990). This finite-difference scheme ensures the advection equation to be third-order accurate in space and time. This finite-difference formulation yields no phase error and causes minimal oscillation in space and time. It is an important advantage for this modeling study, since steep orography in the region tends to generate spurious oscillations in the simulated flow field. The model has 14 layers between the ground surface and the 50-mb level (Figure 1).

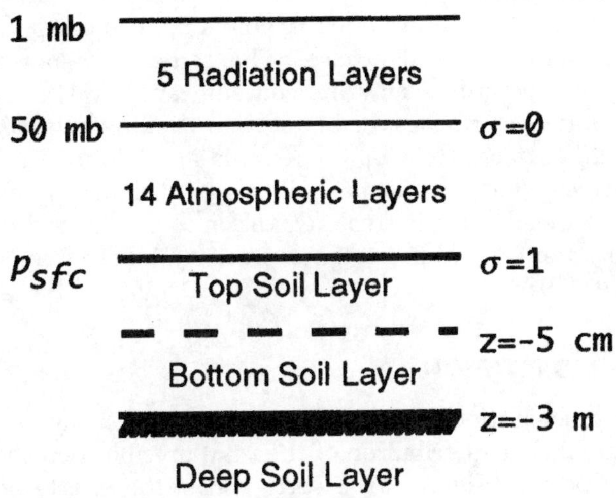

Figure 1. Vertical structure of the atmosphere-soil model.

The source/sink of hydrometeors (heating/cooling due to the phase change of water substances) and precipitation are computed by combining a Kuo-type cumulus scheme with parameterized cloud microphysics by Cho and Iribame (1987). This cloud microphysics model includes ice-phase processes as well as warm rain processes. As is discussed later, inclusion of explicit cloud microphysical processes for large-scale condensation appears to be crucial for the precipitation simulation in this study. In addition, inclusion of the cloud microphysical processes eliminates the need for separate estimation of cloud to compute the radiative transfer.

The solar and terrestrial radiative transfer processes are computed using the parameterizations by Harshvardhan and Corsetti (1984). The effects of clouds are computed separately for water and ice particles using the parameterizations by Stephens (1978) and Starr and Cox (1985), respectively. Vertical profiles of the temperature, ozone, and water vapor mixing ratio in the layer between the top of the model domain (50 mb) and the top of the radiation domain (1 mb) are prescribed from the midlatitude winter profiles by Ellingson et al. (1991).

A soil-snow-canopy model developed at Oregon State University (Mahrt and Pan, 1984; Ek and Mahrt, 1991) is used to compute heat fluxes and hydrology over land surfaces. This soil-canopy-snow model predicts the soil temperature, soil water content, canopy water content, and snow depth. It also provides the temperature and water vapor mixing ratio at land surfaces to compute sensible and latent heat fluxes from the surface. The two-layer soil model (Figure 1) uses a single-species, big-leaf model to treat vegetation cover. The soil texture and initial soil water content are obtain from 1°-by-1° resolution climatology by Zobler (1986).

OBSERVED PRECIPITATION

The observed precipitation in California during the 12-day period is shown in Figure 2. The distribution of the total precipitation accumulated over the period (Figure 2A) clearly shows the effects of local topography, as heavy precipitation is concentrated at the upslopes of major mountain ranges. The heaviest precipitation occurred over the western slope of the Sierra Nevada and the Coastal Range. Precipitation was light in the Central Valley and decreased toward southern California.

The precipitation maximum at the north of the Central Valley (near Redding) is of particular interest. The low-level wind during the period of heavy precipitation is southwesterly in general. Therefore, this region is located at the lee side of the Coastal Range, where weak precipitation is expected. The cause of this local maximum appears to be the low-level wind perturbation due to the Sierra Nevada. It will be discussed further in the next section.

The daily precipitation (Figure 2B) shows that the local precipitation was strongly controlled by the large-scale environment. Figure 2B shows that the daily precipitation was positively correlated with the large-scale water vapor flux convergence, as heavy precipitation events occurred when the large-scale water vapor flux convergence in the region was strong.

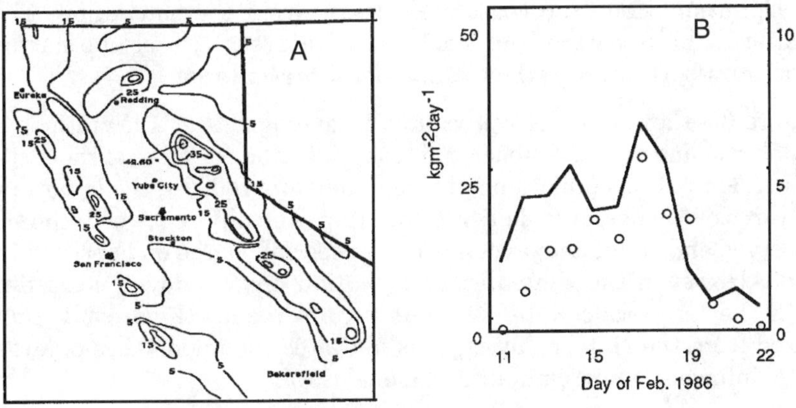

Figure 2. A, Total precipitation from February 11 to 22, 1986. Contours start from 5″ with 10″ interval. B, Daily precipitation averaged over 165 stations in California (circles) and vertically integrated water vapor flux convergence in the model domain (solid line).

SIMULATED PRECIPITATION

The initial and time-dependent lateral boundary conditions for the simulation were obtained from the twice-daily NMC analysis during the period from 0000Z, February 11 to 0000Z, February 22, 1986. The computational domain covers an 1140-km × 1260-km-wide region which contains much of the states of California, Nevada, and southern Oregon. This domain is divided with A 20-km × 20-km grid mesh.

The topography in the region is characterized by two major mountain ranges, the Coastal Range and the Sierra Nevada, with the Central Valley between these mountain ranges. At the northern end of the Central Valley, the Coastal Range joins the Sierra Nevada, forming a mild upslope toward the north. Such features of topography in the region are not only favorable for the orographic lifting of the low-level westerlies but also play an important role in the local transport of water vapor in the lower atmosphere.

Simulated precipitation (Figure 3A) agrees well with observations. The location and magnitude of the observed precipitation maxima at the western slope of the northern Sierra Nevada are well reproduced in the simulation. Despite good agreement, fine structures in the observed precipitation along the coast are missing from the simulation. This indicates the importance of small-scale features, such as the narrow mountain barrier along the coast, in local precipitation.

Figure 3B shows the daily precipitation averaged over 165 stations in California and over the model grids containing one or more of those stations. Again, the simulation appears to have reproduced the important features of the observed daily precipitation. The difference between the observed and simulated precipitation is especially large on February 14, 1986. However, the simulated precipitation on this day is consistent with the large-scale water vapor flux convergence (Figure 2B) computed from the NMC analysis. This indicates the importance of large-scale information in simulating regional flow.

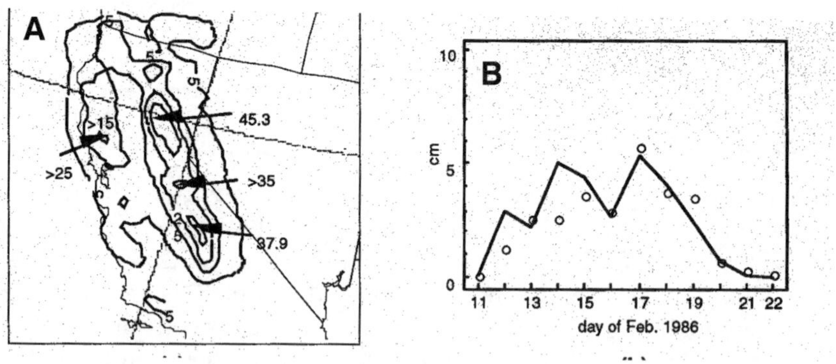

Figure 3. A, Simulated total precipitation during 12-day period. Contour interval same as in Figure 2A. B, Observed (circles) and simulated (solid lines) daily mean precipitation at 165 California stations.

The sea-level pressure perturbation associated with the western slope of the Sierra Nevada appears to cause the low-level wind in the Central Valley to turn northeastward, which is approximately parallel to the Sierra Nevada Range (not shown). This perturbation in the low-level wind appears to provide a significant amount of water vapor to the northern Sierra Nevada and to northern California, where orographic lifting produces precipitation. The isolated precipitation maximum near Redding, discussed in the previous section, appears both in the observations and in the simulation. It is thought to be caused by local transport of water vapor.

Explicit treatment of cloud microphysical processes in the model appears to be crucial in modeling precipitation in this study. Much of the precipitation in the valley and over the eastern slope of the Sierra Nevada appears to be due to the precipitating and cloud particles formed over the Coastal Range and the Sierra Nevada, respectively, and was transported downstream by the upper-level wind. The simulation with the cumulus scheme alone yields much less precipitation than observed. It also failed to simulate the local distribution of precipitation: the simulation with only cumulus parameterization is confined within narrow bands over the western slopes of the Sierra Nevada and Coastal Range. On the other hand, a simulation with only cloud microphysics (but without cumulus parameterization) produced almost the same results as reported here. The cloud microphysics parameterization includes the ice-phase process as well as the warm rain process. This feature enables the model to predict rainfall and snowfall separately without any further assumptions. More importantly, the simulated precipitation depends strongly on the ice-phase parameterization. Without ice-phase processes, the precipitation over the Coastal Range is significantly underestimated. The vertical distribution of rain (Figure 4A) and snow (Figure 4B) along the section crossing the Bay area is presented in Figures 4A and B. Most of the rainfall in the Sierra Nevada and the Central Valley appears to be due to melting snow, while the warm-rain process seems to be the major source of rainfall over the Coastal Range. The snowline is located 1.8 to 2 km above sea level during the simulated period (Figure 4B). As a result, the snowfall is confined to the high Sierra Nevada and northern California (Figure 5A). The precipitation in the valley and over the Coastal Range is almost entirely rain (Figure 5B).

ACCUMULATED SNOW

Snowpack over high country is especially important for the summertime water supply in the western United States. In warm springs, the runoff from the melting mountain snowpack can cause flooding. Accurate assessment and prediction of snow cover at the end of the wet season is therefore important for planning the short- and long-term water supply for industries, the environment, and rural and urban areas.

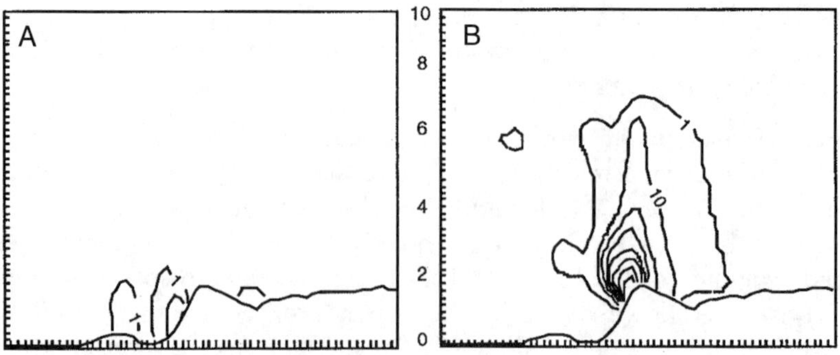

Figure 4. Vertical distribution of simulated precipitation along section crossing Bay area. A, rain; B, snow. Contour starts at 0.01 g/kg with an interval of 0.1 g/kg. Ordinate is height above sea level (km).

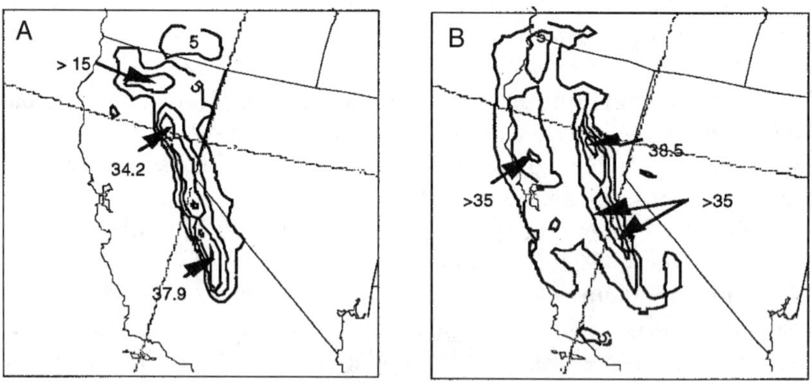

Figure 5. Simulated 12-day total (A) snowfall and (B) rainfall. Contour interval is as in Figure 2A.

The soil-canopy-snow model also separately computes accumulated snow over land surfaces as the difference between the fresh snowfall and snow melt. Snowfall is computed by the atmospheric model, and snow melt is computed as a part of the surface energy balance on the snow-covered land. Since the amount of snow cover at the beginning of the simulation is not available, only the net accumulation during the simulated period is computed. The results show that most of the fresh snow on the high regions of the Sierra Nevada is added to the snowpack at the end of the simulated period. On the other hand, much of the fresh snow in northern California has melted.

SIMULATED CLOUDS AND SURFACE RADIATION

The effects of the cloud cover on the solar and terrestrial radiative transfer is important in the regional and global energy budget and in climate study. The simulated cloud and precipitating particles are shown in Figure 6A, together with the insolation (Figure 6B) and downward long-wave radiation at the surface (Figure 6C) at 0000Z February 17, 1986, along a section crossing the Bay area. The effects of cloud cover on the surface insolation is clearly seen in Figures 6A and B. Over the Coastal Range and the Sierra Nevada, the surface insolation is reduced by more than 150 Wm^{-2} from the clear-sky value of 200 Wm^{-2}. Notice the slight increase of insolation in the valley where the cloud is relatively thin. The thin ice cloud at the lee side of the peak of the Sierra Nevada slightly reduced the surface insolation. The variation of the downward long-wave radiation (Figure 6C) is largely due to the topography, as the air temperature is colder over higher ground. Note that the horizontal variation of the downward long-wave radiation is almost the mirror image of the surface topography. On a constant-z level near the ground (not shown), the thick cloud cover over the upslopes of major mountain ranges enhances the downward long-wave radiation by a few tens of Wm^{-2}.

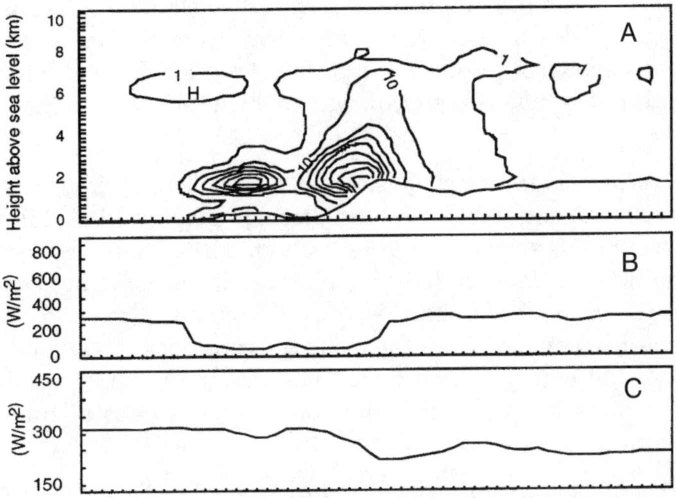

Figure 6. A, Vertical cross-section of the concentration of total hydrometeors (cloud water, cloud ice, rain, snow, and graupel) along section crossing Bay area. B, Variation of surface insolation along same section. C, Same as B, but downward long-wave radiation at ground surface.

The effects of different phases of water substances appear to be the most significant for solar radiation. Due to great optical thickness of water clouds, the surface insolation appears to be much smaller below an all-water cloud than below an all-ice cloud. On the other hand, the effects of both ice and water clouds on long-wave radiative transfer were small.

CONCLUSIONS

A 12-day episode of heavy wintertime precipitation in the western United States was simulated using a limited-area model nested within the NMC operational analysis. Combination of a Kuo-type cumulus scheme with cloud microphysics parameterization successfully reproduced important features of the observed precipitation.

The observed and simulated precipitation show that the local topography and large-scale water vapor flux are the two major factors in local precipitation. The low-level southwesterly from the Pacific Ocean turns southeasterly over the Central Valley due to the low-level pressure perturbation caused by the Sierra Nevada. The water vapor transport associated with this low-level wind perturbation appears to be an important source of water vapor for precipitation in the northern Sierra Nevada and in northern California. The cloud droplets and precipitating particles transported downstream from the upslope regions of the major mountain ranges caused most of the light precipitation on the lee side of the major mountain ranges, including that in the Central Valley.

Parameterized cloud microphysics, or an explicit moisture scheme, appears to be essential for simulating local precipitation in this case. A Kuo-type cumulus scheme alone severely underestimated the observed precipitation since most of the precipitation, in the simulated case was from stratiform clouds or shallow convective clouds. Rain and snow can be separately computed from the cloud microphysics. The results show heavy snowfall over the high Sierra Nevada and northern California. Precipitation in the Central Valley and over the Coastal Range was mostly rain.

The effects of clouds on the solar and terrestrial radiative transfer can be computed from the explicit treatment of cloud microphysics and the radiative transfer model. The short-wave transfer was particularly sensitive to water droplets, as the optical thickness of water clouds is much larger than that of ice clouds. The water and ice clouds did not cause much difference in the long-wave radiative transfer.

ACKNOWLEDGMENT

The authors express great appreciation to Professor John Roads of Scripps Institute of Oceanography for providing the station precipitation data over California and for useful discussions. This work is supported by National Institute for Climate Change (NIGEC) under the grant W/GEC0017, Institutional Collaborative Research (INCoR) Program of the University of California, and Laboratory Directed Research and Development (LDRD) Program of Lawrence Livermore National Laboratory. Work performed under the auspices of the U.S. Department of Energy by the Lawrence Livermore National Laboratory under contract No. W-7405-Eng-48.

REFERENCES

Cho, H-R and J Iribame. **1987**. The role of clouds and precipitation in long-range transport and acid rain in Canada. Phase II: Programming, testing and refining of cloud dynamics, microphysics and chemistry models, Vol. 2, Appendix 1: Cloud Microphysics., Report for the Canadian Electrical Association, Research and Development, Toronto, Canada.

Davies, R. *Documentation of the Solar Radiation Parameterization in the OLAS Climate Model*. NASA Tech. Memo 83961, Greenbelt, MD.

Dickinson, R, R Errico, F Giorgi, and G Bates. **1989**. A regional climate model for western United States. Climate Change 15: 383-422.

Ek, K and L Mahrt. **1991**. *OSU 1-D PBL Model User's Guide*, 118 pp. Department of Atmospheric Science, Oregon State University, Corvallis, OR.

Ellingson, RG, J Ellis, and S Fels. **1991**. The inter-comparison of radiation codes used in climate models: Long wave results. J. Geophys. Res. 96: 8929-8953.

Frisch, J and C Chappel. **1980**. Numerical prediction of convectively driven mesoscale pressure systems. Part I: Convective parameterization. J. Atmos. Sci. 37: 1722-1733.

Giorgi, F. **1991**. Sensitivity of simulated summertime precipitation over the western United States to different physics parameterizations. Mon. Weath. Rev. 119: 2870-2888.

Harshvardhan and T Corsetti. **1984**. *Longwave Radiation Parameterization for the UCLA/GLAS GCM*, NASA Tech. Memo 86072. NASA, Houston, TX.

Hsu, Y and A Arakawa. **1990**. Energy conserving and potential-enstrophy dissipating schemes for the shallow water equations. Mon. Weath. Rev. 118: 1960-1969.

Mahrt, L and H-L Pan. **1984**. A two-layer model of soil hydrology. Boundary-Layer Meteorol. 29: 1-20.

Starr, D and S Cox. **1985**. Cirrus clouds. Part I: A cirrus cloud model. J. Atmos. Sci. 42: 2663-2681.

Stephens, G. **1978**. Radiation profiles in extended water clouds. II: Parameterization schemes. J. Atmos. Sci. 35: 2123-2132.

Takacs, LL. **1985**. A two-step scheme for the advection equation with minimized dissipation and dispersion errors. Mon. Weath. Rev. 113: 1050-1065.

Zhang, D-L, E-Y Hsie, and M Moncrieff. **1988**. A comparison of explicit and implicit predictions of convective and stratiform precipitating weather systems with a meso—scale numerical model. Q. J. R. Meteorol. Soc. 114: 31-60.

Zobler, L. **1986**. *A World Soil File for Global Climate Modeling*, NASA Tech. Memo 87802. NASA, Houston, TX.

POLICY ISSUES OF GLOBAL CLIMATE CHANGE

CLIMATE CHANGE: RISKY BUSINESS ON THE POLICY FRONT

J. T. Ratchford

George Mason University, International Institute,
Arlington, Virginia 22203

Key words: *Climate change, policy, uncertainty, research needs*

INTRODUCTION

For those of you who saw the preliminary program for this symposium, you may recall that my presentation had a different title. I had originally conceived of this policy session as one that addressed the risks for all those involved in the global change business—risks of all kinds: Risks that data are wrong, or at least incomplete or inconclusive. Risks that the intellectual and mathematical frameworks for treating data are flawed. Risks that the international public policy process (i.e., politicians, scientists, economists, businesspersons, and others) would not act on valid scientific conclusions, or would act wrongly, or (perhaps worst of all) would act on incorrect scientific results. And there are myriad other risks, including risks of reducted research funds, for all involved in the search for truth in the field of climate change.

But when I saw the preliminary program I was heartened. "Risky" had been transmuted into "rocky"; like lead into gold. "Rocky Business on the Policy Front" is probably a more relevant descriptor of the climate change policy front than my original title.

I toyed with several approaches. One was contrasting rock-solid scientists with rock-tossing politicians. Another involved the rocky shoals the United States and international climate change policy ships have encountered in the past few years. As for rock music and rock musicians, the opportunities seemed almost endless. With them, you don't care what the climate is. And since both scientists and policy works are always thinking of the next job in the future, there could have been Rocky II the next year, and Rocky III the following year, and. . . .

But after much soul-searching, I decided to decline the helping hand of fate, manifesting itself in this case as a typographical error, and return to the risk theme. But, as I am the first to admit, that decision is itself risky.

Right here at the beginning I also want to put in a disclaimer. It's not the usual one that my university may not necessarily hold the same views as those presented in this paper. Everyone knows that universities, as institutions, rarely care what their professors say or write, as long as the footnotes are consistent with the *Chicago Manual of Style*. This is a welcome change from the White House, where the social mores do not encourage individual innovation in policy pronouncements.

My disclaimer is more fundamental: this is not a scientific paper. The science of global change, be it physical, life, or social science, is not where I have a comparative advantage. This concept of comparative advantage is common among economists. And to quote a *Nature* book review that appeared earlier this year, ". . . the best way to think of economics is not as a science, but rather as a combination of political philosophy and applied mathematics . . ." The same might be said of much of the literature in the field of climate change—some of that on the science side of the ledger as well as some on the policy side.

There is one other comment that I should like to make to insure that you are in the right mind-set to absorb this risky paper. It relates to the contribution of scholarship and reasoned thinking to the public policy process.

Many years ago I spent a year as a Congressional Fellow of the American Political Science Association. I was the only "hard" scientist in the program, and I soon regretted that I had never taken a course in political science as an undergraduate. In any case I had a ball, working for a relatively liberal Democrat in the House of Representatives and a conservative Republican in the Senate. It was in the Senate that I received a useful lesson in the relevance of scholarship to the policy process. It seems the Senator had never had a Ph.D. physical scientist in his office before and, as a lawyer, had little use for them professionally. In spite of this, we hit it off pretty well, and he soon came to refer to me around the office as his "token intellectual." I thought this was great, until one day he shared with me his working definition of an intellectual: ". . . someone educated beyond his or her ability."

It turned out that after my fellowship year I returned to the Hill for several years to work for what was then called the House Committee on Science and Technology. Never, in the years I spent on the Hill, did I admit voluntarily to having a Ph.D.

I shall now, in a more serious vein, talk about the fascinating topic of policy formulation in the climate change arena. I shall review briefly the scientific facts as currently understood, the uncertainties that exist, and how all this relates to the bubbling cauldron we call international policy development. I shall close with comments on what our responsibilities are as scientists.

THE HISTORICAL FACTS

Voltaire, in 1756, is reported to have said, "Three things influence human thought: climate, politics and religion." Notice that science did not even make the short list. In spite of this, I shall try to summarize some of the scientific work on climate change that is relevant to policy formulation.

Let's start when the earth was young, delving for a moment into palaeoclimatology. It would be nice to know how the earth got to its present climatic regime (temperature) without being diverted like Venus (boiling) or Mars (freezing) into clearly undesirable states of affairs. But even here we run into problems. Models of stellar (in our case solar) evolution indicate that solar luminosity when the earth was young should have been 25% to 30% less than it is today. This means that, if all else remained the same, the earth would have stayed frozen. But there is evidence of liquid water on earth 3.8 billion yr ago. Hence the "faint young sun paradox."

We do know that the natural greenhouse effect is very important. Greenhouse gases, such as carbon dioxide, methane, chlorofluorocarbons, and nitrous oxide, "trap" heat in the atmosphere and raise the mean surface temperature of the earth above what it would be otherwise. The Intergovernmental Panel on Climate Change (IPCC), the international arbiter of climate change "facts," concluded in 1990 that this natural greenhouse effect keeps the earth warmer by about 33°C than it would otherwise be (IPCC, 1990a, b). Further, changes in the earth's surface temperature over long periods of time have been correlated with changes in the concentration of greenhouse gases.

Natural variations in the earth's climate have been quite large. Alternating ice ages and interglacials are one indication of this. For the last 10,000 yr or so the current (holocene) interglacial has been a period of steady, dependable climate. Before that, according to results from studies of ice cores taken from central Greenland (GRIP, 1993), climate was more fickle. These cores cover the last 200,000 yr and include the most recent Ice Age, which lasted about 100,000 yr, and the preceding interglacial (Eemian), which lasted from approximately 135,000 to 115,000 yr ago. These studies indicate that very rapid shifts in both greenhouse gas concentrations and in temperature are possible in interglacial periods.

During the Eemian interglacial temperatures were, on average, 2°C higher than at present. More startling is the evidence that changes of up to 10°C occurred over periods of a couple of decades or less. These shifts might last for thousands of years or for only a few decades. We don't know why, nor do we know whether the single state, stable holocene or the multistate, "flickering" Eemian interglacial is the norm.

Thus, over the last 200,000 yr we now think that the global temperature has been spiky for both the Ice Ages and the interglacial that occurred over the period from 115,000 to 135,000 yr ago. Only the last 10,000 yr, which constitute the present interglacial, have been relatively steady as far as temperature is concerned. As the July 15, 1993, *Nature* noted: "We humans have built a remarkable socio-economic system during perhaps the only time when it could be built, when climate was stable enough to let us develop the agricultural infrastructure required to maintain an advanced society . . . If the Earth had an operating manual, the chapter on climate might begin with a caveat that the system has been adjusted at the factory for optimum comfort, so don't touch the dials." (Dansgaard et al., 1993)

MODERN CONCERN OVER WARMING

The July 17, 1991, *New York Times* carried the obituary of one of the greatest scientists of this century, Roger Revelle. As the *Times* noted, it was he who noticed in the 1950s that carbon dioxide was increasing in air samples taken from atop the Hawaiian volcano Mauna Loa and analyzed at Revelle's Scripps Institution of Oceanography at La Jolla, California. Roger Revelle (all 6 ft 4 in.) was an intellectual giant and a patient and inspiring teacher. He cared mightily about people everywhere, and pioneered efforts to apply science and technology to the solution of problems in many of the developing nations of the world. When President Bush presented him the National Medal of Science in 1990 at the White House, he was the star of the show.

I remember well his patient explanation given over and over again to scientists and policy-makers alike—of what we know and what we don't know about climate change and global warming. "About all we know for sure," he would begin, "is that carbon dioxide has been increasing over Hawaii since 1958." He didn't stop there, however, and his own pioneering work was carried on by many others. Models began to take shape that permitted a view of the future as well as the past. It is the very large and complex general circulation models (GCM) upon which predictions of future climate change are made. Unfortunately, these are very crude.

The IPCC concluded in 1990 that a global warming of between 1.5 and 4.5°C is likely for a doubling of carbon dioxide in the atmosphere. The majority of GCM projections fall within this range. Various studies indicate that carbon dioxide in the atmosphere has increased about 25% during the last 250 yr, since the advent of the industrial era.

RISKS AND UNCERTAINTIES

Since prediction of future climate change is based mostly on GCM, and since they are, as noted above, crude, we have problems on the policy front. Some of the biggest technical problems are the effects of clouds and how their effects change in a world with changing temperature; the role of the oceans; and how natural discontinuities such as volcanos and large meteorites affect climate, and for how long. There is also the case of the missing carbon: we can't find about one-fourth of the anthropogenic carbon, although scientists have looked carefully in the air, the oceans, and under rocks.

It's also hard to find out what's happening around the world. There are great disagreements as to what is happening to tropical forests, one of the giant sinks for carbon dioxide. This is important, because cutting or burning forests is at least as bad for the carbon balance as burning coal. There is also a lot to be learned about huge natural carbon dioxide sources, such as seaweed.

I shall not attempt to make a complete list of the risks and uncertainties facing the scientific and policy communities. Suffice it to say they are very, very large and diverse.

THE POLICY DILEMMA

Science has always dealt with insufficient data and imperfect models in developing explanations of the universe. In fact, the beauty of science is that it is a method, not a collection of facts and theories. As scientists, we are never sure of anything, so if today's model is not right, eventually we will develop one that is—or is at least more so.

Policy-makers, politicians in particular, don't have that luxury. If they are not at least perceived as being right they tend to lose elections. So how should they deal with global warming? To some extent, this depends whether the politician is in office or is just running for office. Those in the latter category have more flexibility with both their policy options and their facts.

In an attempt to be helpful, the scientific community, in 1990, through the IPCC, made some projections of what the mean world temperature rise would be if carbon dioxide levels in the atmosphere doubled. Their projection was between 1.5 and 4.5°C.

We are now 250 yr into the era of anthropogenic carbon, and a quarter of the way to doubling. Is there warming? Surface measurements seemed to indicate so in the late 1980s. But experimental design problems, such as urbanization of the sites at which much of the time-series data were taken, cast doubt on these results. And measurements by satellites of global air temperature variation over the last 15 yr show no sign of warming. *The Washington Post*, in a story in July of this year, characterized these satellite vs. surface measurement differences as "two points of view on global warming."

Going back to the Greenland ice cores, I find the abrupt discontinuities (as if there could be any other kind) to be instructive. There are plausible causes of these in nature, as indicated by the effects of El Niño and Mount Pinatubo. But is the threat from carbon dioxide different?

The answer to this question is "maybe." Recent work by Manabe and Stouffer (1993) addresses the resilience of the North Atlantic "conveyor belt" for heat transport with increases in carbon dioxide. They found that a doubling of CO_2 left the belt weak but intact; a quadrupling, however, might be catastrophic. These discontinuities are what we all fear.

One reaction to all this is to try to develop policies that are prudent. This means going part way to assuage your concerns without spending too much money or losing the next election because of the economic consequences of taking too much action. During the Bush administration this was known as a "no regrets" strategy. It meant planting trees and taking many small steps toward energy conservation and increased efficiency, but not imposing a $3.00 per gallon gasoline tax or pursuing an aggressive nuclear power strategy. It is also the stance reflected in the book *Buying Greenhouse Insurance*, by Manne and Richels (1992). We shall see the Clinton administration's strategy shortly.

WHAT SHOULD THE RESEARCH COMMUNITY DO?

This is a scientific symposium; the message here is directed toward that community. So what are we supposed to do as individuals and institutions that reflects our interests as good scientists searching for the truth, and as good citizens searching for what is good for our descendants many generations into the future?

First, we need bright ideas for answering the many important questions outstanding. For example, we still don't know whether or not the global mean temperature is increasing. Detecting the "greenhouse signal" by ocean acoustic thermometry is the kind of bright idea that we need many more of. That effort also demonstrates the unpredictable obstacles that can appear, seemingly out of nowhere. Similar experimental or conceptual breakthroughs are needed in paleoclimatology and almost every other climate-related field. And, of course, superior models are needed.

But bright, new ideas are not the whole story. Hard work, and staying the course on small bits and pieces of the puzzle are just as important—perhaps more so. In our country we are always looking for the magic bullet. We lament the fact that students in other countries do so much better than ours in mathematics and science education. We search for secrets in their curriculum or teaching methods. But upon closer examination, we find their students do better because they work longer and harder to accomplish the more rigorous tasks they are assigned. In climate research, slow and steady is just as important as innovation. And this holds not only for the bench scientist, but for the laboratory director and, even more importantly, the funding agency.

The bottom line is that we in the scientific community have a responsibility to make the policy debate informed. I am reminded of an instance many years ago, when I was on the staff of what is now the

House Committee on Science, Space and Technology. At that time I was one of the few staffers on the committee (perhaps the only one) with a Ph.D. in the "hard" sciences, although I was careful to not admit this fact, as noted earlier. In any case, there was some minor bill that was going to the floor that was controversial. Since I was the only one with intimate knowledge of the issues involved, I ended up writing all of the floor statements, on both sides of the issue.

After the bill was either passed or defeated (I don't remember now which, nor do I remember the subject of the legislation), I was chided by one of my colleagues as "an intellectual prostitute" for providing ammunition to both sides. My response was that it is in the public interest that debates on legislation be informed rather than uninformed.

The scientific community of which we are members has that same responsibility. Certainly, there is risk that our results will be misused. There is also risk that our results will be wrong, and many of them are certain to be. There are risks that investments are in second-best research because funding agencies are not aware of the best. And in this day of congressional earmarking of research funds, there are substantial risks that research teams or institutions that are less than the best will be funded to attack a particular problem.

But the greatest of all risks is of having an uninformed debate. By doing our part, each and every one of us, that won't happen.

REFERENCES

Manabe, S and RJ Stouffer. **1993**. Century-scale effects of increased atmospheric CO_2 on the ocean—atmosphere system. Nature 364:215-217.

GRIP (Greenland Ice-core Project Members). **1993**. Climate instability during the last interglacial period recorded in the GRIP ice core. Nature 364:203-207.

Dansgaard, W, SJ Johnsen, HB Clausen, D Dahl-Jensen, NS Gundestrup, CU Hammer, CS Hvidberg, JP Steffensen, AE Sveinbjörnsdottir, J Jouzel, and G Bond. **1993**. Evidence for general instability of past climate from a 250-kyr ice-core record. Nature 364:218-220.

IPCC. **1990a**. Climate Change: The IPCC Scientific Assessment. Intergovernmental Panel on Climate Change. World Meteorological Organization/United Nations Environment Program, Island Press.

IPCC. **1990b**. Climate Change: The IPCC Response Strategies. Intergovernmental Panel on Climate Change. World Meteorological Organization/United Nations Environment Program, Island Press.

Manne, AS and RG Richels. **1992**. Overview: The Greenhouse Debate, *Buying Greenhouse Insurance: The Economic Costs of CO_2 Emission Limits*, pp. 1-22. MIT Press, Cambridge, MA.

WHAT DECISION MAKERS WANT TO KNOW FROM RESEARCHERS ABOUT GLOBAL CLIMATE CHANGE

J. C. Bernabo

Science and Policy Associates, Inc., Washington, DC 20005

Key words: *Climate change, policy, research needs, assessment*

ABSTRACT

Climate research in the United States has focused on improving scientific understanding of the Earth's climate system and not on providing information for the more immediate needs of decision-makers. During the Bush administration, the primary objective was to determine if there was a problem, rather than to develop information on what to do about the issue. The Clinton administration is trying to focus research to more effectively assist decision-makers.

The *Joint Climate Project to Address Decision Makers' Uncertainties* is sponsored by EPRI, DOE, DOI, EPA, NOAA, USDA, and USFS to assess what information is most relevant for decision-making and how research could address those needs. The results of this multiple needs assessment is being used to help reorient the federal program. Initial findings indicate that many of the research areas germane to policy have not yet been addressed. Uncertainties in our understanding that are significant to decision-making must be extracted from the broader realm of overall scientific uncertainties. Many curiosity-driven research questions have only marginal value for helping to develop specific policies.

Decision-makers' priorities include greater emphasis on impacts and responses, not just on better understanding of earth system dynamics. More work is required in the social, economic, and ecological sciences to provide information for making policy. Researchers need periodically to provide interim information and policy-relevant assessments while continuing to improve long-term scientific understanding.

Some of the key questions raised by decision-makers could be answered in 2 to 5 years. Although more complete scientific understanding is decades away, much urgently needed input can be gleaned from analysis and interpretation of existing data. The Joint Climate Project is beginning a second phase of refining decision-makers' questions and identifying specific research opportunities for assisting decision-makers in the United States and internationally.

INTRODUCTION

Climate research efforts over the last few decades have been driven primarily by a desire to understand and reduce the scientific uncertainties related to predicting global climate change. This basic, long-term research is critical for developing a thorough understanding of the climate system and a foundation for more detailed and reliable predictions. However, the fundamental questions of interest to scientists are not necessarily the questions of most relevance to decision makers. Specifically, it is clear that the actual or potential impacts of climate change are the foremost concern for the latter.

Climate research in the United States has focused on long-term understanding of earth systems, rather than emphasizing the more immediate needs of decision makers. During the Bush Administration, the objective was to determine if there was a problem, rather than developing information on what to do about the issue. The Clinton Administration is trying to focus on the research that better assists decision makers.

Defining the research most relevant to policy is not simply a technical question that can be answered by scientists. Decision makers need and value information not necessarily sought by discipline-driven scientists. Therefore, decision makers must first define their needs so that scientists then can determine how—and by when—these could be addressed by research. This vital dialogue between both communities typically has been more ad hoc than systematic. The complexity of the global climate change issue and its urgency for decision makers require frequent two-way communication between scientists and decision makers on the information needed for policy development and what research can provide.

There has been a need for a broad assessment of what questions and uncertainties are most useful to address from the perspective of decision makers. A systematic and iterative dialogue between researchers and decision makers is required to define what scientific information can be provided over various time frames in order to address pending policy questions on global climate change.

The Joint Climate Project To Address Decision Makers' Uncertainties was designed to achieve more explicit communication between decision makers and scientists on the informational requirements for policy development and the capabilities of research to meet them. The project's findings will help to manage decision makers' expectations about the kinds of assistance they can anticipate from research, and also will assist research planners in identifying policy-relevant priorities. The project was a federal-private sector partnership (Table 1).

Table 1. Groups that participate in identifying decision makers' concerns.

Government	Non-Government
Congress	Coal
Office of Technology Assessment	Gas
Environmental Protection Agency	Oil
Department of Energy	Chemical
Department of Agriculture	Manufacturing
Department of the Interior	Forest Products
Department of State	Transportation
Office of Management and Budget	Electric Utility
Council of Economic Advisors	Environmental
Council on Environmental Quality	Groups
National Governors' Association	

UNIQUE APPROACH

The Joint Climate Project identified policy-relevant research using two distinct phases: U.S. decision makers first defined their information needs; then, scientists determined the research required to meet those policy-relevant needs and the time frame for providing better information. This two-step process avoided assuming implicitly that scientists somehow know what is relevant to policy or that policy makers should know what is good science. By iterating the process, the input from both groups can shape the outcome to what is technically sound and policy-relevant (Figure 1).

In the first phase of the project, dozens of U.S. government and private sector officials convened to define their key questions about global climate change. During the second phase of the project, leading experts in climate-related disciplines assembled to discuss the decision makers' questions that were developed during the preceding months. The workshop participants examined (1) the research needed to address the questions and (2) the expectations for providing better information in the future. In addition, the project examined the needs and opportunities for improving the dialogue between decision makers and researchers.

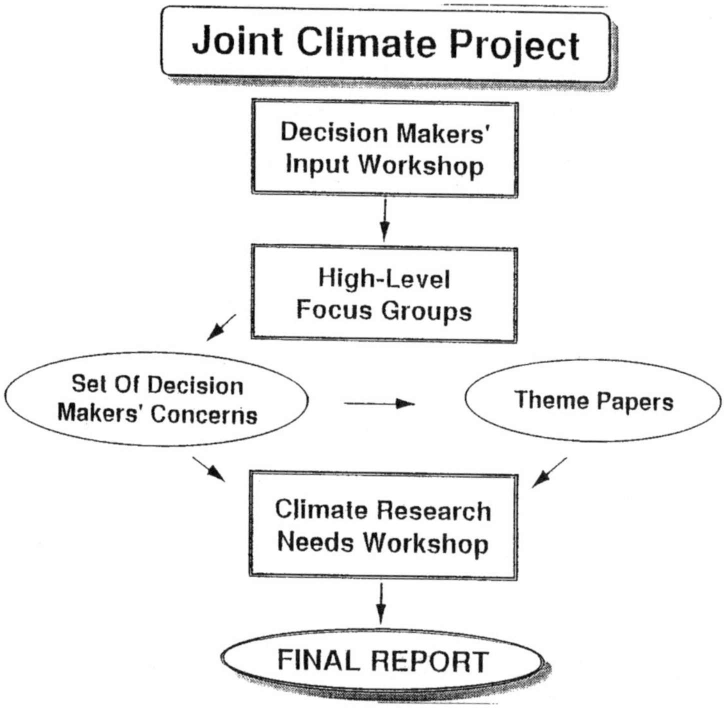

Figure 1. Flow chart, Joint Climate Project.

FINDINGS

The consensus-identifying approach of this project yielded several key findings that reflect the general concerns of decision makers and the responses of the research community. In discussions with these two communities, several common themes also emerged for enhancing communication and increasing the value of research results.

The Concerns Of Decision Makers

This section describes several of the guiding principles, suggested by the participants, from which the policy-relevant questions were derived. In particular, they stressed the importance of augmenting atmospheric research with the economic, social, and ecological sciences in order to inform policy decisions.

Climate Change Impacts and Human Responses-Key to Decision Making—Decision making is driven by concerns about the potential impacts of changing climate at the regional level, rather than predictions of changing global mean values of climate variables. Any response to the threat of climate change must be measured against what is at stake. Therefore, more information is needed on the ecosystems, regions, and human populations that are most at risk from potential climate changes—even if atmospheric research is still unable to provide reliable predictions of the specific changes that will drive effects.

Interim Information Needed—Decision makers recognize that, in many cases, researchers may be unable to resolve uncertainties or to furnish complete answers in time for policy development. The need for interim information is driven by the decisions made during relatively short time periods, depending on cycles in politics, budgets, and public concern.

Implications Of Uncertainties Need Clarification—There is a need to define better which uncertainties are most important for policy development and resource management, and the practical implications of these uncertainties for decision makers.

Certainty Not a Prerequisite for Action—Resolution of all scientific uncertainties is not a prerequisite for policy action. Decision makers will apply their constituents' values to determine how much certainty they judge is enough to take a given policy action. Decision makers regularly deal with uncertainties in complex issues and do not often demand scientific proof as a prerequisite for action.

The Response Of Researchers

In the next phase of the project, a diverse group of U.S. experts in climate-related fields was convened to examine how research could best address these questions posed by decision makers. Specifically, the scientists examined what types of research are needed to reduce the uncertainties in these policy-relevant questions and estimated the time frames for possible results.

Timely Results—Although more complete scientific understanding of climate change may be decades away, much of the information needed to begin addressing decision makers' questions can be provided within 2 to 5 yr. This could include a comprehensive evaluation of indicators of global climate change, a preliminary vulnerability analysis for systems and regions most sensitive to climate change, and an assessment of the sources and levels of greenhouse gas emissions for use in identifying potential mitigation and adaptation options.

Parallel Approach To Climate, Impacts, And Human Responses Research—It is neither necessary nor practical for research to progress sequentially from the climate system to the impacts and then to the potential human responses in order to provide useful results for decision makers. Much can be done to improve the understanding of impacts without waiting for accurate regional climate predictions.

Greater Emphasis Needed On Impacts And Human Responses Research—Information on climate change impacts and response strategies has the greatest potential for assisting decision makers, yet these fields are the least researched. Many of the key questions identified by decision makers involve a significant amount of new socioeconomic, behavioral, and ecological research. Social science and economic research, in particular, receive a small percentage of federal funding but are critical for making decisions about climate change.

Global View—An international perspective for research is essential, given the global dimensions of the issue. Even though decision makers may be most concerned with regional and local consequences, developing world issues (such as population and economic development, as well as the pace, quality, and sustainability of development) will be critical.

Integrated Assessments And Case Studies—Integrated assessments of the causal linkages from emissions through impacts and human responses would help structure information for effective use in decision making. A coordinated examination of case studies of regional climate variability is needed—one based on historically documented events that show how societies have responded to past climatic variations. This information would provide valuable insights on how to treat future events.

Expect The Unexpected—Multidisciplinary research on potential surprises is also important, given their potentially serious implications for decision making (i.e., climate change could be much worse than anticipated, or it could be insignificant). Decision makers and scientists should frequently re-examine research on potential surprises, given that scientific progress is incremental, and new information may become available. Based on this information, contingency plans could be developed to prepare for unforeseen events.

International Perspective—Because of the global dimensions of the issue, an international perspective for research is essential. Assessing the ability of the international community to implement mitigation and adaptation measures is important for evaluating the effectiveness of response strategies on the climate system.

Lessons in Communication

Not an Either/Or Decision—Researchers need to provide a broad array of information to address the complex and interacting decisions on global climate change. Decision makers, for their part, need to recognize the long time scales involved in research and, thus, the importance of continuity of funding and program goals.

Global Climate Change In A Relative Risk Context—Prediction of changes in mean global temperatures does not give an adequate picture of the societal risk that can be related to everyday experiences. The risk of global climate change needs to be compared to the risks of other economic, social, and environmental issues. Because the public tends to respond to perceived crises, assigning relative risk would assist decision makers to distinguish between verifiable serious threats and possibly misplaced public concern. Given that risk is a function of both the probability and the magnitude of the expected consequences, better data on possible impacts are critical to better estimates of societal risk.

Urgent Need For Education—A concerted effort is needed to educate decision makers on the facts and uncertainties of global climate change. Considering that public concern is often the impetus for formulating policy, scientists need to communicate technical information to the public more effectively as well as more frequently. In addition, scientists need to learn more about the decision-making process and the types of information most useful for policy. Frequent, two-way communication between decision makers and researchers is fundamental if research is to play an effective role in the decision-making process.

Research Does Not Always Provide The Answer—Decision makers need to realize that additional research actually could increase the amount of uncertainty in some areas. Researchers should inquire about how much certainty decision makers are requiring to take a specific action. To this end, uncertainties that do not matter for decision making should be identified early in the process. Decision makers and researchers should also seek ways to manage continuing uncertainties. For example, building resilient institutions would provide a flexible response to any future changes in climate, albeit at potentially significant costs. Contingency plans could allow decision makers to prepare for possible climate outcomes through R&D on response technologies without needing to deploy them.

Develop An Ongoing Assessment Process For Research—To improve communication and better inform decision makers, research efforts should include an iterative assessment process. These assessments would not only help to identify the relevant questions but would also serve to structure the research results and, thus, facilitate clearer communication between the two communities. Furthermore, the assessment process would provide valuable input to the planning of policy-relevant research.

CONTINUING THE PROCESS

The Joint Climate Project represents a first step in determining how researchers can assist U.S. decision makers over the coming years and decades, thereby helping to bridge the communication gap between these two communities. A more frequent and systematic two-way dialogue will be needed between decision makers and researchers in order for research to effectively inform the decision-making process.

Discussions with decision makers and researchers during the project revealed that both communities are very interested in developing and sustaining a productive dialogue. Both would like to increase the effectiveness of the research community in the decision-making process. Decision makers suggested meetings with principal investigators and their collaborators during the course of a project to discuss results, methodology, and data, as well as with the researchers upon the completion of a project. Researchers expressed their desire for more guidance and for frequent contact with the petitioners and ultimate consumers of the information their projects produce.

Given the debate over how to address potential climate change, representatives from industry, environmental organizations, and other groups also said that they would welcome the chance to participate in a consensus-building effort on areas of agreement for the policy options. Researchers can then focus on areas of disagreement to attempt to enlighten the debate. Similarly, scientists could pursue more frequent communication among and within their various disciplines—especially between those concerned with climate change prediction and those focused on impacts and human responses. In this way, researchers can identify areas of agreement and divergence, in order to communicate more effectively with decision makers.

The project's goal was to identify national-level decision makers' broad questions and to determine the general response of the research community. This project is an initial step in the ongoing, iterative exchange needed between science and policy. Further efforts will be required to update these initial results as both science and policy continue to evolve. These national-level findings need to be supplemented with similar efforts on the local, regional and international levels. Similarly, the Joint Climate Project approach could be applied to individual sectors, such as agriculture and energy, to yield significant details on the research needed to address their specific concerns. In any case, the process used by this project serves as a model for studying how to improve the process whereby research and policy are integrated and accommodated in research programs addressing societal needs.

INTEGRATED ASSESSMENT AND INTERDISCIPLINARY RESEARCH

J. Lancaster

Environmental Science and Policy Institute, Lexington, Massachusetts

Key words: *Integrated assessment, impact analysis global climate change, policy*

ABSTRACT

Assessment is used ambiguously in the context of global change policy. Integrated assessment and impact assessment are distinguished here, and a mental model for deriving environmental policy from environmental science is discussed. Reasons why the U.S. Global Change Research Program has made slow progress on impact assessment are given, and specific recommendations are offered to attend to this critical portion of integrated assessment.

INTRODUCTION

"Integrated assessment" has emerged as the newest "buzz phrase" in applied environmental science and policy research. I will relate it here to the similarly famous phrase, "global change," which emerged in the 1980s. Federal science funding agencies are now announcing a major commitment of funds from the U.S. Global Change Research Program (USGCRP) to support research in integrated assessment. What does the phrase "integrated assessment" mean? And will this funding commitment help to develop methodology for "impact assessment" as a contributing piece of integrated assessment?

"Assessment" is an evolving concept that means different things to different people. In environmental policy, it usually represents a critical transition, or bridge, between environmental science and policy formulation. To assess a situation, to assess the value of a resource, or to assess a level of risk, the assessor must integrate multiple factors affecting value or risk. Building an integrated view, then, is at the heart of any assessment. Similarly, many types of assessment are contained within another emerging concept called . . . integrated environmental management. Thus, the emergence of the term integrated assessment has led to confusion. What are we assessing? Are we integrating certain basic studies to make one assessment, or are we integrating a series of assessments to provide another, overarching assessment?

Some define integrated assessment simply as the combining of physical and social science information. Others characterize integrated assessment as an analysis that surrounds governmental, stakeholder, and citizen categories. Often, people characterize integrated assessment in a way that suggests impact assessment. However, although "impacts" are seen as a fundamental question within both the global change science and policy research communities, usually a next level of analysis, "policy-driven" assessment, is more closely linked to integrated assessment. Before the federal government spends too much money on integrated assessment, it would be useful to distinguish more clearly the following terms: integrated assessment, science assessment, integrated impact assessment and policy-driven assessment.

Integrated assessment has become an all-encompassing phrase that includes at least the following activities: (a) assessing the physical science; (b) assessing the potential impacts of environmental change on ecology, public health, and economics; (c) assessing potential societal responses for effectiveness and potential economic impact; and (d) assessing political feasibility in terms of the myriad of public policy interests (see Figure 1).

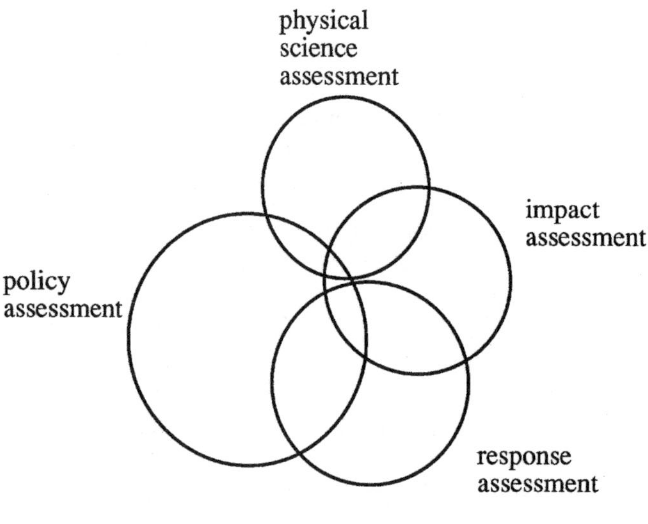

Figure 1. Integrated assessment.

Some of these activities are evolving areas of social science, and some are new mergers of physical and social science disciplines. Assessing the physical science, which has been one of the jobs of the Intergovernmental Panel on Climate Change (IPCC), is not itself a new branch of scientific research. However, new interdisciplinary projects are growing around this task, such as the Global Analysis, Interpretation and Modeling (GAIM) task force within the International Geosphere Biosphere Program (IGBP). Developing methodology for global change impact assessment is a new interdisciplinary research endeavor, and assessing potential societal responses is an evolving branch of research within the social sciences. The need for a comprehensive impact assessment to include an analysis of potential responses adds to the complexity.

As illustrated in Figure 2, the all-encompassing definition of integrated assessment suggests "vertical integration" because it begins by looking at the basic causes of the problem and moves up through physical impact, societal impacts, potential responses, and management issues. In contrast, we might describe as "horizontal integration" research such as that on impact assessment, combines many physical and socioeconomic parameters that pertain to a geographical region.

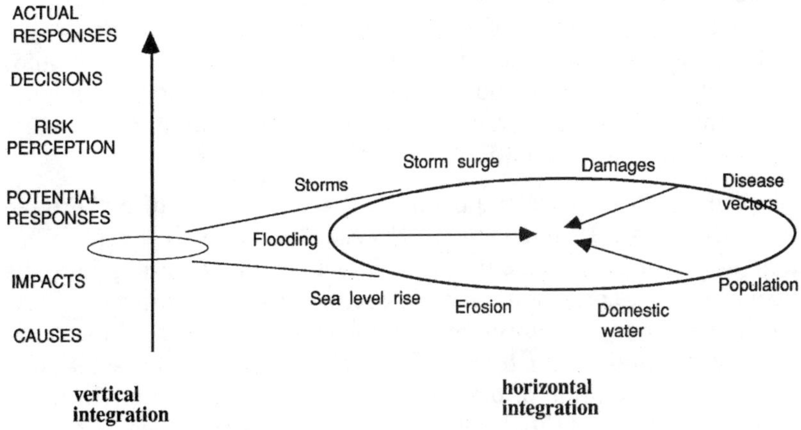

Figure 2. Differing dimensions and tasks of integrated assessment.

A Mental Model About Science Informing Policy Making

Science policy at the highest level has become confused by not recognizing those aspects of assessment that are themselves developing fields of scientific research. The view that assessment is mainly a policy function carried out by analysts, who could do a better job if only they could get the needed information from the scientific research community, needs to evolve. In this view, or mental model, the problem is seen as depicted in Figure 3A, where the scientific community, S, passes information to the policy-driven assessment, A_{pol}, which then allows choice of policy, P. Many current recommendations essentially aim to improve the assessment process, A, and the information flow i, both directly and by improving the policy-oriented feedback, f_1, to guide that communication.

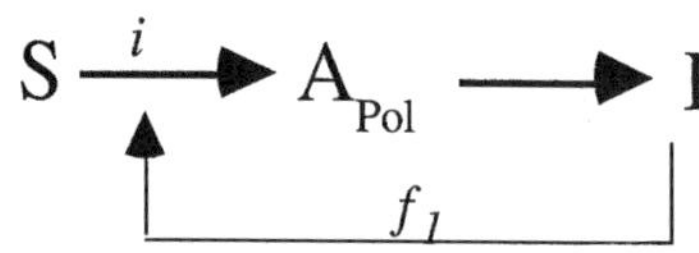

Figure 3A. Information Flow.

Such recommendations do not focus strongly enough on developing new scientific methodology for assessment or on the information needs to support this new methodology. It might be useful to add a little more detail to the mental model of how global change science connects to policy. Being clearer about the differing nuances of assessment will allow researchers to make specific recommendations to build new science, and to aim other recommendations toward the domain of policy analysis and transfer of information.

In Figure 3B, three parts of a potential "research community" are separated: the physical sciences, S_{phys}; the social sciences, S_{soc}; and the sciences of impact assessment, S_{imp}. The task of assessing potential responses, A_{resp}, because it relates closely to impact assessment, is set apart from other aspects of the policy assessment, A_{pol}. This model invites us to think about how feedback from policy-driven assessment can improve evaluation of potential societal responses to environmental change (f_2) and to the developing science of impact assessment (f_3). In turn, the urgent need for impact assessment research to direct new social science investigations is illustrated by an additional feedback loop (f_4). The policy-driven feedback to improve information flow from one step to another (f_1) is still an important recommendation, although distinct from f_2, f_3, and f_4.

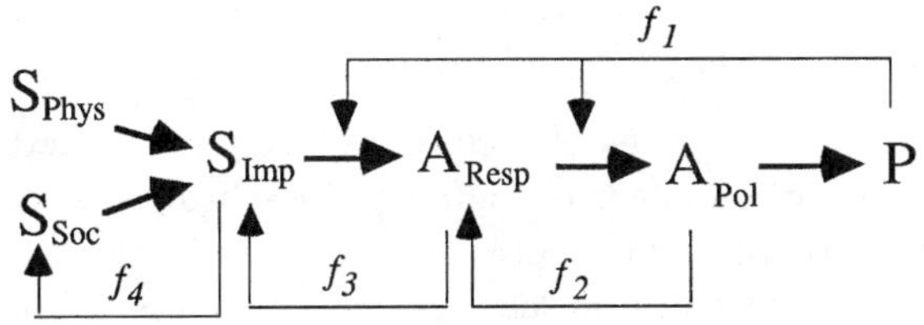

Figure 3B. Information Flow.

Failure of the U.S. Global Change Research Program to Address Impact Assessment

Impact assessment is half of the global environmental change problem. With many important planning and developmental decisions being made at local and regional levels, demonstrations of how environmental changes actually affect people, based on sound research at these local scales, can generate a deeper and genuine deliberation. Decision makers want and need evidence concerning the resiliency of their regions. A showing of extreme resiliency or vulnerability could lead governments to direct research money more appropriately to serve public health.

The USGCRP has virtually ignored this crucial impacts research. The existing research program has asked, *"Is the global environment changing?"* and *"How fast do we expect global environment to change?"* and *"What might be potential impacts on non-human ecosystems?"* and even, *"What would be the impact on our economy if we were to respond to potential climate change by reducing emissions?"* But we have not yet asked, nor begun to answer, the threshold question: "How vulnerable are people and their communities to global environmental change?"

Most often, scientists describe the potential impacts of increasing greenhouse gases as potential physical changes in climate, such as a global average temperature rise over many decades. However, they do not describe well the actual impacts on humans, such as deaths from extreme weather events, loss of fresh water, higher food costs owing to drought, effects on disease vectors, forced migrations, and loss of property to sea-level rise.

What information on each of these potential impacts actually exists? As yet, there has not been a comprehensive assessment at any level of society. There have been only localized case studies of specific potential threats, such as accelerated sea-level rise, or drought. Integrated analyses of combined stresses have not been carried out. There has been insufficient funding to even develop the needed methodology and tools.

At least six reasons why critical impact assessment research has been neglected can be identified:

(1) The global change research program was launched by earth scientists with neither the expertise nor the passion for pursuing impact assessment research. Naturally, the research program grew around the various scientific aspects of global warming and ozone depletion, particularly atmospheric chemistry, oceanography, and climatology. To many in this group the word "assessments" has meant assessments of the earth science.

(2) Some scientists involved with setting the USGCRP agenda have asserted incorrectly that impact and risk assessments could move ahead only after credible predictions of regional environmental change were achieved. They blur the distinction between "likely impacts" and "vulnerability." Thus, they have not mapped out a path for developing methodology for integrated assessment based upon (a) historical societal impacts from environmental change, combined with (b) a reasonable suite of hypothetical scenarios for future environmental change.

(3) The development of impact assessment methodology has occurred mainly within the corporate environmental consulting world. This group of researchers is not focused on global environmental problems, nor will it be unless a regulatory framework requires global impact statements before permitting some industrial activity. This group of researchers is not connected to the USGCRP mission, nor is much of this work published in a form useful to the academic community.

(4) Funding through disciplinary divisions within the National Science Foundation (NSF), the National Oceanic and Atmospheric Administration (NOAA) and the Department of Energy (DOE) has let impact assessment research fall through the cracks. Program managers are almost always specialists in some avenue of research, and they tend to fund proposals that advance their specialty. This is understandable and forgivable; a program manager would be hard-pressed to take funding from an associate with whom the manager has had a long-established relationship, only to shift it over to a mixed bag of new researchers from other disciplines, whose collaborative proposals seem quite foreign to the manager.

(5) No methodology has been developed because funding has been so inadequate and discontinuous. A few researchers have made starts on the problem. The major U.S. Environmental Protection Agency (EPA) study in the late 1980s attempted to assess potential impacts based on guesses, with no foundation in research. The best effort thus far has been the MINK study (Missouri, Indiana, Nebraska, and Kansas project, conducted in the late 1980s by Resources for the Future under auspices of DOE and the Pacific Northwest Laboratory). However, DOE stopped funding too early for that study to develop tools and approaches that could be exported to other regions. As part of a collaborative effort under DOE/NIGEC, our group of five teams worked 18 months to develop methodology before DOE recently cut off our funding (Lancaster et al. 1993). Part of this funding problem may have resulted from a "turf battle" over policy-oriented research internal to DOE. Also, DOE has only very recently begun to raise the priority on the impact assessment problem. At the international level, the IPCC Assessments Report is a summary of piecemeal studies, with an acknowledgment that we have not yet done the needed work.

(6) The databases required to assess potential impacts in various regions have yet to be identified and are not easily accessible to researchers. Our collaborative effort over the last 2 yr under NE-NIGEC started to address this need, but it was painfully slow going with such inadequate funding. The integrated assessment research at Carnegie-Mellon, Lawrence Livermore, and the Massachusetts Institute of Technology will move things ahead, but none of these groups is addressing the impact assessment topic very methodically. We hope that the Consortium for International Earth Science Information Network (CIESIN) will make inroads on this problem. The impediment is a serious one because multiple databases must be assembled before the

systemic analysis can proceed. Thus, useful results will follow by years a decision finally to commit serious funds to this research direction.

Recommendations to Improve Global Change Impact Assessment Research: How Can the Needed Information be Developed?

(1) Given the nature of the impediments, "top-down" guidance and motivation are needed. Impact assessment should be given top billing within the USGCRP by installing one or more researchers in this area at the Federal Coordinating Committee on Science, Engineering, and Technology (FCCSET) level. Given that we have a Vice President with fairly sophisticated understanding of global environmental issues, an appeal should be made directly to his office to make this change. Fully 50% of the USGCRP budget should be directed toward research on impact assessment.

(2) NSF, EPA, DOE, and NOAA should each have an integrated impact assessment program staffed by impact assessment experts, offering significant support to university researchers. Impact assessment research should not be funded through program managers devoted to specific disciplines of the physical or social sciences. These agencies could still subdivide the topic along their traditional lines (e.g., NSF more theoretical and tool oriented, EPA and DOE stressing public health, and NOAA stressing coastal zone), but linkages between groups and projects should be soundly supported.

(3) Close attention should be paid to developing databases that can be tapped for regional analyses at scales determined by the impact assessment researcher. Tools should allow a variety of geographic information systems to interact with multiple databases accessed on-line. Development of observations should aim to coincide with development of impact assessment methodology 5 yr out.

(4) Methodological development should be encouraged now through case studies where data sources are currently rich, and long time series exist.

(5) Methodology and tools should be developed so that they will be exportable from one case study region to the next, i.e., so that when the databases are developed for various START [Strategic Analysis, Research and Training (a program under IGBP and the Human Dimensions of Global Change Programme)] regions, the impact assessment capability will be immediately available.

(6) Attention should be paid to developing methodology for assessing the impacts in a region of combining stresses, i.e., systemic or synergistic impacts, where single stresses may not initially appear noteworthy.

(7) Attention should be paid to developing methodology for aggregating potential impacts over multiple geographical regions within certain political boundaries, from state to global scales.

REFERENCE

Lancaster, J, A Shlyakhter, and R Wilson (eds.). **1993**. *Global Warming Risk Analysis and Integrated Impact Assessment: A 5-Team Collaboration to Develop Methodology and Tools*. Report of DOE/NIGEC Research 1991-1993. Harvard University Physics Department, Cambridge, MA.

TRACE GAS INDICES USING AN OPTIMAL CONTROL FRAMEWORK

M. Kandlikar and A. Patwardhan

Department of Engineering and Public Policy,

Carnegie Mellon University, Pittsburgh, Pennsylvania 15213

Keywords: *Abatement, greenhouse index, damages, climate change*

ABSTRACT

A number of trace gas indices have been proposed to quantify the relative importance of different greenhouse gases for climate change, with much debate about their advantages and shortcomings. In this paper, the problem of determining an appropriate index is placed within an optimal control framework. A general form for a trace gas index is determined using cost-benefit and cost-effectiveness framings. Other proposed indices are shown to be special cases under very restricted and unrealistic assumptions. The role of scientific and economic uncertainty in trace gas index calculations is examined.

INTRODUCTION

A key element of possible policy responses to global climate change is the abatement of emissions of radiatively active gases. There has been considerable recent interest in viewing the emissions abatement problem as one involving multiple gases and not CO_2 alone. Comprehensive abatement strategies were proposed by the U.S. government (USDJ, 1991), based on the rationale that all greenhouse gases, some more than others, contribute to climate change. Under a comprehensive plan, greenhouse gas abatement must be carried out where . . . "it costs least to do so" (Stewart, 1992). In such an approach, a minimum-cost solution would require the control of multiple gases.

It has been argued that a comprehensive abatement strategy requires the formulation of a greenhouse gas index that would allow for an evaluation of tradeoffs between greenhouse gases for a variety of possible abatement policies, including licensing and tradable permits, taxes, and proportional control. Such a greenhouse gas index needs a dynamic representation of the greenhouse forcings, as well as the global cycles of individual gases. The most widely discussed metric is the Global Warming Potential (GWP) (Lashof and Ahuja, 1990; Houghton, 1990). A number of problems have been identified with the formulation and use of GWP. These include uncertainties in greenhouse lifetimes and the requirement of an arbitrary time horizon for integration (Enting and Rodhe, 1991; Hammond et al., 1991; Handel, 1991). Attempts have been made to provide alternatives (Hammond et al., 1990) and modifications (Harvey, 1993) to GWP.

Additionally, as stated by Eckaus (1992), "GWPs do not account for the time variation in the economic opportunity costs of an increment of radiative forcing." Consequently, they do not provide any insights for abatement policy formulation. Damages linked to climate change and abatement costs associated with emissions reductions are both time-varying quantities and need to be explicitly included in the calculation of a metric that determines gas-by-gas greenhouse responsibility. Reilly and Richards (1993) have used an economy-climate model to estimate greenhouse gas indices based on the relative economic impact of greenhouse gas emissions, assuming specific damage functions.

Not only are greenhouse gas indices required for comprehensive abatement, they are also required for use in international negotiations concerning abatement policies across nations. It is likely that climate change damages and costs of abatement will differ by nations. Developed and developing countries have historically differing contributions to climate forcing and different abilities to pay for abatement. Politically mediated equity criteria may have to be applied in order to allocate greenhouse responsibility on a national basis. The issue of allocating national responsibility has been extensively discussed (Fujii and Grubler, 1991; Agarwal and Narain, 1991; Mitchell, 1992; Epstein and Gupta, 1990; Smith, 1993), and some recent work attempts to reconcile efficiency and equity concerns of greenhouse indices (Smith et al., 1993). Although no attempt is made in this paper to address this issue, equity concerns lie at the heart of negotiations toward a climate treaty.

This paper attempts to integrate the complex scientific and economic aspects of the comprehensive abatement problem, using an optimal control framework. Optimal control and dynamic optimization models have been extensively used in natural resource economics (e.g., Kamien and Schwartz, 1981). Optimal control models have also been used to study greenhouse gas abatement policies (Nordhaus, 1992; Peck and Tiesberg, 1993; Falk and Mendelsohn, 1993) and for trace gas index calculations (Reilly and Richards, 1993). We use an optimal control framework to analytically examine the effects of problem formulation and damage functions on trace gas index calculations. Because our knowledge of dynamic damage functions and abatement costs is very limited, an attempt is made to quantify trace gas indices only in the simplest cases; the focus of this paper is to reveal the complexity inherent in such a calculation, not the calculation of a single set of values for the indices. Recognizing that, in practice, the specification of a dynamic damage function is a daunting task, we also formulate a cost-effectiveness framework, where climate change and damage information is incorporated through constraints, and we provide some remarks on the correspondence between the two frameworks.

A cost-benefit (C-B) analysis formulation is set up in the next section, and trace gas indices, defined as the cost of abating the next unit of a non-CO_2 trace gas (relative to CO_2), are determined using a general form of the damage function. Simplifications using special cases are provided. In the following section, a similar analysis is carried out using a cost-effectiveness (C-E) framing. A later section provides policy implications of this research, and conclusions are provided in the last section.

TRACE GAS INDICES USING A COST-BENEFIT FRAMING

We begin with the assumption that a single-actor model is valid for the global abatement problem; i.e., a single damage function and a set of abatement costs can be used to represent the globe. (For a more detailed discussion of this issue, see Dowlatabadi and Lave, 1993.) We also restrict our analysis to two gases, CO_2 (referred to as gas 1), and a non-CO_2 gas (referred to as gas 2), since the analysis can easily be generalized to more than two gases. In a cost-benefit framing, we would like to minimize the total costs, i.e., the sum of costs of abatement and the damages due to climate change. (Benefits due to CO_2 fertilization, and costs of ozone depletion due to CFCs are not considered.) The optimal control problem can be stated as:

$$\min \int_0 \{A_1(a_1(t)) + A_2(a_2(t)) + D(c_1(t), c_2(t))\} e^{-rt} dt \tag{1}$$

$$\text{s.t} \quad \frac{dc_1}{dt} = -\gamma_1 c_1(t) + \beta\,(s_1(t) - a_1(t)) \tag{2}$$

$$\frac{dc_2}{dt} = -\gamma_2 c_2(t) + (s_2(t) - a_2(t))\,. \tag{3}$$

Equation (1) is the objective function, and equations (2) and (3) are constraints corresponding to trace gas mass balance equations gases 1 and 2, respectively. (Equation (2) is the simplest form of a linear system approximation for the carbon cycle. In general, the atmospheric transfer function for CO_2 can be written as a sum of exponentials. The form in equation (2) assumes a single exponential transfer function. Details of linear system approximations to the carbon cycle are provided in Enting and Newsam, 1990.) $A_{1,2}$, $a_{1,2}$, $c_{1,2}$, $s_{1,2}$, $_{1,2}$ are the costs of abatement, levels of abatement, atmospheric concentration, business-as-usual emissions, and inverse of atmospheric lifetimes of gases 1 and 2, respectively. is the atmospheric airborne fraction of CO_2, r is the discount rate, and $D(c_1,c_2)$ is the damage function due to climate change. The Hamiltonian H is given by

$$H = \{A_1(a_1(t)) + A_2(a_2(t)) + + D(c_1, c_2)\} e^{-rt} + \lambda_1(t)[-\gamma_1 c_1(t) + \beta(s_1(t) - a_1(t))] + \lambda_2(t)[-\gamma_2 c_2(t) + (s_2(t) - a_2(t))] . \quad (4)$$

The first-order necessary conditions are:

$$\frac{\partial H}{\partial a_i} = 0 \quad (5)$$

$$\frac{\partial H}{\partial c_i} = -\dot{\lambda}_i \quad (6)$$

where the dot (•) superscript denotes the derivative with respect to t. Therefore,

$$\lambda_1 = \frac{1}{\beta} * \frac{\partial A}{\partial a_1} e^{-rt} \quad \text{and} \quad \lambda_2 = \frac{\partial A_2}{\partial a_2} e^{-rt} \quad (7)$$

$$\dot{\lambda}_1 = \gamma_1 \lambda_1 - \left[\frac{\partial D(c_1,c_2)}{\partial c_1}\right] e^{-rt} \quad \text{and} \quad \dot{\lambda}_2 = \gamma_2 \lambda_2 - \left[\frac{\partial D(c_1,c_2)}{\partial c_2}\right] e^{-rt} . \quad (8)$$

Equation (7) is the static optimality condition that relates the shadow price of the emissions constraint to the cost of abatement. Equations (7) and (8) in conjunction with equations (2) and (3) and initial conditions on c_i (e.g., the current value of atmospheric concentrations) and final conditions on λ_i [λ_i (t_{final}) = 0] represent a two-point boundary value problem that can be solved numerically.

The quantities $\frac{\partial A_1}{\partial a_1}$ and $\frac{\partial A_2}{\partial a_2}$ are the marginal costs of abatement for gases 1 and 2, heretofore denoted as M_1 and M_2. Assuming that the airborne fraction for CO_2 is a constant (for details on the limitations of this assumption see Enting and Newsam, 1990) we can eliminate γ_1 and γ_2 and $\gamma^{\bullet}_1$ and $\gamma^{\bullet}_2$ from (7) and (8) and obtain

$$\dot{M}_1 = (\gamma_1 + r)M_1 - \beta\left[\frac{\partial D(c_1,c_2)}{\partial c_1}\right] \text{ and } \dot{M}_2 = (\gamma_2 + r)M_2 - \left[\frac{\partial D(c_1,c_2)}{\partial c_2}\right] \tag{9}$$

Equation (9) is a key equation that relates the marginal costs of abatement of each gas to the marginal damages from climate warming, and contains the greenhouse gas attributes we need to capture: the gas lifetimes ($1/\gamma_1$, $1/\gamma_2$) , marginal costs of abatement (M_1, M_2)., marginal climate damages $\left(\frac{\partial D}{\partial c_1}, \frac{\partial D}{\partial c_2}\right)$ and discount rate, r. We now need to consider the specification of physically realistic damage functions.

Typically, damages due to climate change will be a function of particular climate variables. In the simplest case, the damage function may be assumed to be a function of the mean global temperature T(t). (We limit ourselves to damage functions that are dependent on the instantaneous value of temperature. In reality, damages will depend on both the instantaneous value and the temperature trajectory, precipitation, and extreme events. For a more detailed discussion of damage functions, see Peck and Tiesberg, 1993.) Thus we can write:

$$\frac{\partial D(c_1,c_2)}{\partial c_{i\,(i=1,2)}} = \frac{\partial D}{\partial c_i} = \frac{\partial D}{\partial T} * \frac{\partial T}{\partial c_i} . \tag{10}$$

Equation (9) can then be rearranged to give the following expression:

$$\frac{\dot{M}_1 - (\gamma_1 + r)M_1}{\beta\left(\frac{\partial T}{\partial c_1}\right)} = \frac{\dot{M}_2 - (\gamma_1 + r)M_2}{\left(\frac{\partial T}{\partial c_2}\right)} . \tag{11}$$

The terms $\frac{\partial T}{\partial c_1}$ and $\frac{\partial T}{\partial c_2}$ depend on the temperature trajectory, T(t), which is the response of the climate system to radiative forcing, $R[c_1(t), c_2(t)]$. A simple model to evaluate T(t) which treats the climate system as linear is the convolution integral:

$$T = \int_0^t R(\tau)H(t-\tau)d\tau \tag{12}$$

where H(t) is the impulse response of the climate system. Models with linear system approximations of the climate system have been formulated by Dickinson (1981) and Schneider and Thompson (1981) and used for policy analysis (Houghton, 1990; Nordhaus, 1992). We further assume that the radiative forcing, $R[c_1(t), c_2(t)]$ is linear separable, i.e., $R(c_1(t),c_2(t))=R_1(c_1(t))+R_2(c_2(t))$. While this assumption does not hold for gases with significant overlap, such as CH_4 and N_20, it is reasonable in most cases. In order to evaluate $\frac{\partial T}{\partial c_1}$ and $\frac{\partial T}{\partial c_2}$, we apply Leibnitz's rule for the differentiation of integrals and obtain

$$\frac{\partial T(t)}{\partial c_i} = \int_0^t \frac{\partial R(C_i(t))}{\partial C_i(t)} * H(t-\tau)\, d\tau \quad . \tag{13}$$

We define the dynamic trace gas index I as the ratio of the marginal costs of abatement of two gases, i.e.,

$$I = \frac{M_2}{M_1} \tag{14}$$

or the cost of abating an additional unit of gas 2 relative to the cost of abating an additional unit of CO_2. From equations (9),(10), and (13) the marginal cost of abating a greenhouse gas i is given by:

$$M_i = \int_0^{T_h} \frac{\partial D(T(t))}{\partial T(t)} * \left\{ \int_0^t \frac{\partial R(C_i(t))}{\partial C_i(t)} * H(t-\tau)\, d\tau \right\} e^{-(r+\gamma_i)t} dt \,. \tag{15}$$

For CO_2 the marginal cost expression is similar with an additional multiplier, β .

In general, the dynamic trace gas index I (= M_2/M_1) will have to be numerically evaluated; however, under some assumptions regarding the form of abatement functions, closed-form analytical solutions can be calculated. In particular, if abatement $A_i[a_i(t)]$ is a linear function of the level of abatement, $a_i(t)$, and exponential in time, i.e,

$$A_i(t) = (\beta_{1i}a_i + \beta_{2i})e^{p_i t}, \tag{16}$$

expressions for closed-form analytical solutions can be determined. Abatement costs for gas i, $A_i(t)$, grow at a rate p_i, depending on the nature of technological change and resource constraints. As pointed out by Falk and Mendelsohn (1993), if $p_i < 0$, the costs of abatement decrease with time, i.e., technological change could, over time, reduce the costs of abatement. On the other hand, under resource constraints, p_i could be greater than zero, and costs of abatement may increase with time. Substituting for equation (16) in equation (11), the analytical form for the index I can now be derived for this form of abatement costs as:

$$I = \left[\frac{\left(\int_0^t \frac{\partial R(C_i(t))}{\partial C_i(t)} * H(t-\tau)d\tau\right)_{i=1} * (p_1 - \gamma_1 + r)}{\beta\left(\int_0^t \frac{\partial R(C_i(t))}{\partial C_i(t)} * H(t-\tau)d\tau\right)_{i=2} * (p_2 - \gamma_2 + r)}\right]. \tag{17}$$

Notice that I is dependent on the entire concentration trajectory, $c(\tau)$ 0 $\tau \leq t$. This is important because it shows that current emissions contribute to future damages not only due to the finite lifetime of the gases in the atmosphere, but also due to the lagged response of the climate system. We will now systematically explore the effect of assuming specific forms of linear climate response H(t) functions on the index I, given in equation (17).

Case 1: $H(t) = H_0\delta(t)$

In this case the impulse response is assumed to be a dirac delta function of magnitude H_0, implying that there is no lagged response in the climate system. Consequently, climate damages are an instantaneous function of the concentration. This is clearly a physically unrealistic assumption but one that has been used at least twice in the recent literature (Reilly and Richards, 1993; Falk and Mendelsohn, 1993). From equation (17),

$$I = \left[\frac{\alpha_2 * (p_1 - \gamma_1 - r)}{\beta \alpha_1 * (p_2 - \gamma_2 - r)} \right] \tag{18}$$

where α_1 and α_2 represent the instantaneous radiative forcing function for gases 1 and, 2. Notice that the index I for this special case is not time-dependent and can be evaluated parametrically for different values of p_1 and p_2. Tables 1 and 2 provide the value of index I for methane and nitrous oxide. These numbers must be treated only as illustrative, since the delta function assumption is clearly not a valid one; it tends to emphasize the *instantaneous* value of radiative forcing and not the radiative forcing *trajectory* which is ultimately responsible for climate warming and resulting damages. If r, p_1, and p_2 are all equal to zero, and $\beta = 1$, the index I reduces to the GWP with an infinite time horizon.

Table 1. Trace gas index for methane[(a)]

P_1	P_2	Index for CH_4
0	0	29
0	–0.02	0
0	0.02	34
0.02	0	14
0.02	–0.02	12
0.02	0.02	16
–0.02	0	43
–0.02	–0.02	38
–0.02	0.02	51

For CO_2, a lifetime of 120 yr (= 0.00833) and airborne fraction β = 0.6 was assumed. Discount rate of 3% was assumed for all cases.

[(a)]Assuming a lifetime of methane as 10 yr (γ = 0.1) and instantaneous radiative forcing 58 times that of CO_2.

Table 2. Trace gas index for nitrous oxide[a]

P_1	P_2	Index for N_2O
0	0	359
0	–0.02	232
0	0.02	790
0.02	0	172
0.02	–0.02	111
0.02	0.02	378
–0.02	0	546
–-0.02	–0.02	353
–-0.02	0.02	1202

[a]Assuming a lifetime of nitrous oxide as 150 yr (γ = 0.00667) and instantaneous radiative forcing 206 times that of CO_2.

Case 2: H(t) = H_0

The index I for this case is equal to:

$$I = \left[\frac{\left(\int_0^t \frac{\partial R(C_i(t))}{\partial C_i(t)} d\tau \right)_{i=1} * (p_1 - \gamma_1 + r)}{\beta \left(\int_0^t \frac{\partial R(C_i(t))}{\partial C_i(t)} d\tau \right)_{i=2} * (p_2 - \gamma_2 + r)} \right] \tag{19}$$

Notice that, unlike the case in equation (18), the index depends on the *net* radiative forcing at time t, not on the instantaneous radiative forcing alone. Under certain restrictive assumptions the index I in equation (18) can be linked to the GFC index devised by Hammond et al. (1991). Specifically, if t is equal to 1 year, and p_1, p_2, and r are set to zero, then equation (19) effectively reduces to the index devised by Hammond et al. (1991).

Cost-Effectiveness Analysis

A cost-effectiveness analysis is often regarded as a plausible alternative to cost-benefit analysis when benefits are uncertain or unknown or when the value of the benefit stream from a set of actions cannot be explicitly quantified (Tietenberg, 1991). A desirable and obtainable objective is selected, with the implicit assumption that the investments are worthwhile, but a formal cost-benefit criterion is not applied (Morgan and Henrion, 1990).

In the context of global climate change and the problem of determining appropriate abatement strategies, a cost-effectiveness framework has several advantages over a cost-benefit approach. First, the quantification of benefits associated with a particular level of abatement is a daunting task, and one with significant uncertainties. On the other hand, the essential question in a cost-effectiveness approach is one of determining appropriate constraints for the future values of physical variables (for example, global mean temperature). It is likely that this determination may be a less contentious issue in climate negotiations than benefit calculations. ([Indeed, the guidelines for greenhouse abatement in the Framework Convention on Climate Change seek to *". . . achieve stabilization of atmospheric carbon at a level and in a time frame sufficient to: 1) allow ecosystems to adapt naturally to climate change, 2) to ensure that food production is not threatened, and 3) to enable economic development to proceed in a sustainable manner." (UNCED, 1992)])* Second, constraining physical variables (realized change as well as the rate of change) allows us to implicitly value the preservation of natural systems, which most estimates of greenhouse damages fail to do. (For a critical assessment of greenhouse damages, see Ayres and Walters, 1991.) Third, a cost-effectiveness approach could be viewed as an application of the precautionary principle to guard against the possibility of very high future damages. In the following section we set up the C-E formulation for minimizing the net present value of abatement costs, subject to constraints on the temperature trajectory.

Constraints on Realized Warming

In this case we constrain the amount of warming allowed at some time t. Ideally, this is an inequality constraint over the entire trajectory, of the form $T(\tau) \leq T^*(\tau)$, where $T^*(\tau)$ is a desired temperature trajectory. However, inequality constraints are harder to analyze in optimal control problems than equality constraints. For the sake of simplicity, we analyze the equality constrained problem, recognizing that it is possible to find a sufficiently small value of $T^*(t)$ for which the integral path constraint equation (20) is always binding.

$$\int_0^t R(c(\tau)) * H(t-\tau)d\tau - T^*(t) = S(c,t) \quad = 0 \ . \tag{20}$$

The cost-effectiveness analysis is formulated by minimizing the cost of abatement subject to the temperature constraint. One way of solving the constrained optimal control problem is by appending a penalty function to the Hamiltonian and converting it into an unconstrained optimal control problem (Bryson and Ho, 1976).

The Hamiltonian is written as:

$$H = \{A_1(a_1(t)) + A_2(a_2(t))\}e^{-rt} + \lambda_1(t)[-\gamma_1 c_1(t) + \beta(s_1(t) - a_1(t))] + \lambda_2(t)[-\gamma_2 c_2(t) + s_2(t) - a_2(t)] + K\left[\int_0^t R(c(\tau)) * H(t-\tau)d\tau - T^*(t)\right]^2 \ . \tag{21}$$

The coefficient K is a very large number, and if $K \to \infty$, the constraint (equation 20) is satisfied. Equation (22) is very similar to equation (4), and the penalty function is analogous to the damage function $D(c_1,c_2)$ in equation (4). Deviations from the trajectory $T^*(t)$ could lead to large values of the objective function. The first-order conditions are similar to those in the cost-benefit case [equations (5) and (6)], with the last term in equation (21) replacing the damage function $D(c_1,c_2)$. Manipulating the first-order conditions we get the following relationship between the costs of abatement, M_1 and M_2. (The assumption is that the quantity $K[S(c,t)]$ is finite and nonzero for very large values of K.)

$$\frac{\dot{M}_1 - (\gamma_1 + r)M_1}{\beta\left(\frac{\partial T}{\partial c_1}\right)} = \frac{\dot{M}_2 - (\gamma_1 + r)M_2}{\left(\frac{\partial T}{\partial c_2}\right)} \ . \tag{22}$$

Equation (22) is evaluated at $T(t) = T^*(t)$. Assuming linear costs of abatement for the different gases lead to the same functional form for the dynamic trace gas index I, in this case as in equation (17). It can be argued that a cost-effective analysis is equivalent to a cost-benefit analysis where the damages are specified through constraints on climatic variables. Although the form of the index derived for a cost-benefit case is qualitatively similar to that derived from a cost-effectiveness analysis, the actual value of the time-varying index could be significantly different, depending on the form of the damage function used and the chosen cost-effective temperature, $T^*(t)$.

Case 1: $H(t) = H_0\delta(t)$

For the special case where the climatic impulse response is given by $H(t) = H_0\delta(t)$, a constraint on the temperature trajectory is equivalent to a constraint on the radiative forcing trajectory. The Index I, in this case, reduces to a time-invariant value,

$$I = \left[\frac{\alpha_2 * (p_1 - \gamma_1 - r)}{\beta\alpha_1 * (p_2 - \gamma_2 - r)} \right] . \tag{23}$$

Thus, specifying a constraint on the radiative forcing is equivalent to assuming the damage as an instantaneous function of radiative forcing in the cost-benefit case. Refer to Tables 1 and 2 for numerical values of the index for specific greenhouse gases.

POLICY IMPLICATIONS

The desire for comprehensive abatement has been driven by the expectation that more efficient economic solutions may be obtained by considering multiple gases. However, it is necessary to specify several key scientific and economic variables in order to obtain an optimal trade-off between different gases. As demonstrated earlier, the calculation of a trace gas index depends on the dynamic behavior of scientific and economic variables, which are at best uncertain and at worst purely speculative.

Each of the key scientific aspects of the climate problem, i.e., trace gas cycles, radiative forcing, and climate dynamics, plays an important role in index calculations. Most previous indices have focused on trace gas cycles and direct and indirect radiative forcing functions, while largely ignoring the role of climate dynamics—a key determinant of climate damages. Both cost-benefit and cost-effectiveness approaches demonstrated that greenhouse gas indices depend on the entire history of radiative forcing and the lagged climate response. In ongoing work we are seeking to provide a classification of damage functions that will incorporate different physical and economic aspects of the climate change issue.

Climate damages and costs of abatement for individual trace gases are the key economic aspects of index calculations. In addition, specification of business-as-usual scenarios requires predictions of economic and technological growth. The choice of an appropriate discount rate is also a critical issue (Tobey, 1992). Estimating dynamic damage functions is a difficult task; the best that can be done currently is to assume functional forms and carry out the analysis on a parameteric basis (see, for example, Peck and Tiesberg, 1993). While the explicit specification of damages may be avoided using cost-effectiveness analysis, time-varying abatement costs have to be specified for CO_2 as well as non-CO_2 gases. Abatement costs for CO_2 are a topic of much debate due to the disagreement between "bottom up" engineering estimates and "top down" economic estimates (Tobey, 1992). Costs of abatement for other greenhouse gases are sketchy at best (Adams et al., 1992), and little is known about how these costs may change in the future. As shown in Tables 1 and 2, the dynamics of the marginal costs of abatement are key to the calculation of trace gas indices, even in the simplest case.

There has been concern regarding the feasibility of a comprehensive abatement policy due to scientific uncertainties (Grubb et al., 1991) and costs of monitoring and enforcement (Victor, 1991); these are legitimate concerns. However, we contend that since comprehensive abatement requires the evaluation of trade-offs between gases and since the dynamics of abatement costs and damages play an important role in determining an index that represents these trade-offs, an accurate assessment of the feasibility of comprehensive abatement requires, at the very least, a reasonable characterization of the dynamics of greenhouse gas abatement and damages. Even in the illustrative calculations provided in Tables 1 and 2, the indices can have large uncertainties, *even if all scientific uncertainty is assumed away,* i.e., point values are used for uncertain scientific quantities such as trace gas lifetimes and CO_2 airborne fraction. As a result of these uncertainties, we believe that accurate trace gas index calculations are very difficult to perform, and we do not share the optimism of Reilly and Richards (1993).

CONCLUSIONS

In this paper we have provided the framework for the calculation of trace gas indices under two policy framings: cost-benefit and cost-effectiveness analysis. We have showed that, in general, the calculation of indices requires the numerical solutions of a two-point boundary problem. Under simple (albeit unrealistic) assumptions, indices have been calculated. Other proposed indices have been shown to be special cases of this general framework. In future work we plan to numerically estimate time-varying indices for several forms of damage functions and examine the role of key scientific and economic uncertainties in trace gas index calculations.

REFERENCES

Adams, RM, CC Chang, BA McCarl, and J M Callaway. **1992**. The role of agriculture in climate change: A preliminary evaluation of emissions control strategies, pp.273-287. In: *Economic Issues in Global Climate Change.* Westview Press, Boulder, CO.

Agarwal, A and S Narain. **1991**. *Global warming in an Unequal World: A Case of Environmental Colonialism*. Center for Science in the Environment, New Delhi, India.

Ayres, RU and J Walters. **1991**. Greenhouse Effects: Damages, Costs and Abatement, IIASA Technical Report, RR-91-1.

Bryson, AE and Y Ho. **1976**. *Applied Optimal Control.* Hemisphere Publishing Co., New York.

Dickinson, RE. **1981**. Convergence rate and stability of ocean-atmosphere coupling schemes with a zero-dimensional climate model. Atmos. Sci., 38:2112-2120.

Dowlatabadi, H and L Lave. **1993**. Pondering greenhouse policy. Science 259:1381.

Eckaus, R. **1992**. Comparing the effects of greenhouse gas emissions on global warming. The Energy J. 13:25-25.

Enting, IG and GN Newsam. **1990**. Atmospheric constituent inversion problems: Implications for baseline monitoring, J. Atmos.Chem. 11:69-87.

Enting, I and H Rodhe. **1991.** Greenhouse budgets. Nature 349:468.

Epstein, F and R Gupta. **1990**. *Controlling the Greenhouse Effect: Five Global Regimes Compared*. Brookings Institution, Washington, D.C.

Falk, I and R Mendelsohn. **1993**. The economics of controlling stock pollutants: An efficient strategy for greenhouse gases. J. Environ. Econ. Manage. 25:76-88.

Fujii, Y and A Grubler. **1991**. Intergenerational and spatial equity issues of carbon accounts. Energy 16:1397-1416.

Grubb, M J, DG Victor, and WC Hope. **1991**. Pragmatics in the greenhouse. Nature 354:348-350.

Hammond, B, L Moomaw, and E Rodenburg. **1990**. Calculating national accountability for climate change. Environment 33:11-33.

Hammond, B, L Moomaw, and E Rodenbury. **1991**. Greenhouse budgets, Nature. 349:468.

Handel, M. **1991**. Greenhouse Budgets. Nature 349:468.

Harvey, M. **1993**. A Guide to Global Warming Potentials. *Energy Policy*, **January 1993**, pp. 24-33.

Houghton, J. **1990**. Greenhouse gases and aerosols, In: *Climate Change, The IPCC Scientific Assessment*, pp. 7-40. Cambridge University Press, Cambridge, UK.

Kamien, MI and NL Schwartz. **1981**. *Dynamic Optimization: The Calculus of Variations and Optimal Control in Economics and Management*. North Holland Publishing, Amsterdam, The Netherlands.

Lashof, D and DR Ahuja. **1991**. Relative contributions of greenhouse gas emissions to global warming. Nature 344:529-531.

Mitchell, JK. **1992**. Greenhouse Equity. Global Environ. Change 2:82-100.

Morgan, MG and M Henrion. **1990**. *Uncertainty: A Guide to Dealing with Uncertainty in Quantitative Risk and Policy Analysis*. Cambridge University Press, Cambridge, UK.

Nordhaus, WD. **1992**. An optimal transition path for the controlling greenhouse gases emission. Science 258:1315-1319.

Peck, SC and T Tiesberg. **1993**. CO_2 Emmisions control: comparing policy instruments. The Energy J. 21:222-230.

Reilly, J and KR Richards. **1993**. Climate change and trace gas index issue. Environ. Natural Resour. Econ. 3:41-61.

Schneider, SH and SL Thompson. **1981**. Atmospheric CO2 and climate: The importance of the transient response, J. Geophys. Res. 86:3135-47.

Smith, KR. **1993**. Pre-industrial missing carbon and current greenhouse responsibility. In: *Polluted or Pristine?*, Workshop on Pre-Industrial Anthropogenic Impact of the Carbon Cycle, Sept 17-19, 1993.

Smith, KR, JR Swisher, R Kanter, and DT Ahuja. **1993**. Indices for a Greenhouse Gas Control Regime Incorporating Efficiency and Equity, pp. 157-137. In: *Energy Investments and the Environment*, Economic Development Institute, World Bank, Washington, D.C.

Stewart, RB. **1992**. Comprehensive and market based approaches to global climate policy, pp. 24-40. In: *Economic issues in Global Climate Change*, Westview Press, Boulder, CO.

Tietenberg, T. **1991**. *Environmental and Natural Resource Economics*, Harper Collins, New York.

Tobey, JA. **1992**. Economic issues in global climate change. *Global Environmental Change*, September, 1992, pp. 215-228.

USDJ. **1991**. U.S. Task Force on the the Comprehensive Approach to Climate Change, A *Comprehensive Approach to Addressing Potential Climate Change*. U.S. Department of Justice, Washington, D.C.

Victor, DG. **1991**. Limits of market based strtaegies for slowing global warming: The case of tradeable permits. *Policy Sciences*, November 1991.

ASSESSING THE IMPLICATIONS OF CLIMATE CHANGE ON AGRICULTURE

CLIMATE CHANGE AND AGRICULTURAL PRODUCTIVITY

N. J. Rosenberg

Technology Planning Assessment Center, Battelle, Pacific Northwest Laboratories, Washington, D.C.

Key words: *Climate, agricultural productivity*

ABSTRACT

Most of the world's agriculture is practiced out-of-doors, so climate change is of some concern to that industry. Research to date has indicated that impacts (positive or negative) will likely be greatest in regions at the cold and dry margins of current agriculture. At the low end of the range of global warming expected from a 2 × CO_2 radiative forcing, negative impacts may be moderated by the CO_2 fertilization effects and by easily implemented technological adjustments. However, climatic effects on plant growth and productivity are nonlinear. If the worst of the climate scenarios were to occur, CO_2 fertilization and technology might not be sufficient to overcome the losses. Whether or not this occurs depends partially on the nature of a country's agricultural industry. It is generally assumed that developing countries are most vulnerable to negative impacts of climatic change and least able to capitalize on benign changes. However, features of traditional agriculture in developing countries offer opportunities for adaptation to climate change in both developed and developing countries. The above statements seem sensible but are imperfect as planning tools because of lack of knowledge of the kinds of impending climatic changes and their regional distributions.

Interestingly, one may speculate with greater confidence on how actions to mitigate climate change would affect world agriculture. Carbon taxes will increase the cost of energy for agriculture as for all other sectors of the economy; because of its low carbon intensity, natural gas (the raw material for most nitrogen fertilizers) would increase disproportionately in price; carbon-free hydropower would compete more strongly with irrigation agriculture for water supplies; carbon-capturing afforestation would compete with food production for use of the arable lands. Direct and indirect impacts of climate change and possible responses are discussed in greater detail in this paper.

INTRODUCTION

United Nations predictions and other sources indicate that world population could grow to 8.5 billion by 2025 (Keyfitz, 1989) and to 11 billion by the end of the coming century (UNFPA, 1990). As new information becomes available on the effectiveness of population control programs, the rise of virulent diseases, and other factors, these numbers change—becoming sometimes smaller, sometimes larger still. Whatever the numbers actually turn out to be, global agricultural production will have to increase several-fold from present levels if that growing population is to be fed and clothed and if worldwide standards of nutrition are to improve.

The capacity of global agriculture to meet this heavy burden will depend on many factors of which the *direct* effect of climate change is only one, and perhaps a very small one at that. Sustained and increased agricultural productivity and production depend on our ability to manage, conserve, and in some cases, increase the resource base of land, water, and biodiversity on which the future of the industry depends (Crosson and Rosenberg, 1989).

My aim in this paper is to review the state of our understanding of how greenhouse-forced climate change may affect agricultural productivity in the United States and worldwide; how carbon dioxide, the primary contributor to greenhouse warming, may affect crops directly; what tools are available to adapt to whatever degree of climate change may be unavoidable; and how steps taken to reduce greenhouse warming and mitigate or avoid climate change may, by decreasing the availability of land and water to agriculture, negatively affect that industry.

HOW MIGHT CLIMATE CHANGE AFFECT GLOBAL AGRICULTURE?

The Climate Factors that Matter

If climate changes, the time and space distributions of each climatic factor-temperature, precipitation, cloudiness (and hence solar radiation receipts at the surface), humidity and windiness—will be affected. The impacts of these climatic changes will be manifested on agriculture through a group of processes and phenomena that determine how plants grow. These include the timing of phenological events, the rates of photosynthesis, respiration and evapotranspiration, the soil moisture balance, soil fertility conditions, and the incidence of pests: weeds, insects, and diseases.

Table 1 relates climatic factors to the processes they affect or control. The timing of plant development from emergence of the seedling to its readiness for harvest (its phenology) is determined primarily by soil and air temperatures. The rate of photosynthetic conversion of carbon dioxide to simple sugars is determined by solar radiation, temperature, humidity, windiness, soil moisture, and also, of course, by soil fertility. Net primary productivity is determined by the gross photosynthetic rate minus the rate of respiration, which consumes a portion of the photosynthetic product. The latter is responsive primarily to temperature. Soil moisture status is determined by the balance between precipitation (rain, snow) or irrigation and evapotranspiration. The latter is driven by solar radiation, temperature, humidity, and windiness. The CO_2 concentration in the ambient air, which controls stomatal resistance to the diffusion of water vapor and the amount of leaf area covering the soil, is also involved. Soil fertility is, of course, determined by the quantities of chemical fertilizers and organic materials incorporated in the soil and by a range of physical and macro- and microbiological processes that occur in soil. Precipitation and temperature are the primary climatic determinants of these processes. The incidence of pests (weeds, insects, and diseases) is determined by all of the climatic factors and processes described above and by the stocks of weed seeds, insect larvae, fungal spores, etc. in the vicinity of the crops they infest when weather, crop, and soil conditions are right. Understanding how climate change might affect agriculture is, clearly, not a simple matter. But first we need some notion about how climate may change.

How will Climate Change?

Those who attempt to assess the potential impacts of climate change on agriculture (and all other sectors, as well) need some notion of how, when, and where climate may change. In the case of agriculture, information is needed with respect to each of the climatic parameters mentioned above. The primary tool for providing such information is the general circulation model (GCM), versions of which are now being produced by about two dozen modeling groups worldwide. There is general agreement among these modeling groups that the mean temperature of the troposphere will increase from 1.5 to 4.5°C when the radiatively active trace gases (carbon dioxide, CO_2; methane, CH_4; nitrous oxide, N_2O and a few others) reach concentrations equivalent to a doubling of CO_2 alone from its pre-industrial concentration of about 280 vppm

Table 1. Climatic factors controlling important processes and phenomena relating to agricultural productivity.

Process/ Phenomenon	Temperature	Solar Radiation	Precipitation	Humidity	Windspeed	CO_2 Concentration
Phenology	X					
Photosynthesis	X	X	X	X		X
Respiration	X					
Evapotranspiration and Soil Moisture	X	X	X	X	X	X
Soil Fertility	X		X			
Pests	X	X	X	X	X	X

($2 \times CO_2$ hereafter). Such a range of warming should intensify the hydrologic cycle so that both global mean precipitation and evapotranspiration (ET) might increase from 8 to 15%. These conclusions with respect to mean climate change are generally accepted in the scientific community (IPCC, 1990, 1992). However, the GCM differ much more in their projections of changes in cloudiness (and hence solar radiation receipts at the ground), humidity, and windiness. They also differ considerably on the question of how climate changes will be regionally distributed. Figure 1 (from Smith and Tirpak, 1989) illustrates this point by showing the extent to which three GCM differ with respect to the changes that $2 \times CO_2$ would cause in seasonal precipitation in California and the southeastern United States. Clearly, studies of the potential effects of climate warming on agriculture (and other natural systems) produce very different results, depending on the choice of the GCM used to provide the climate change scenario.

The Regression Approach to Climate Impacts

Statistical correlations of expected deviations from current climate conditions were used in early studies to assess impacts of climate change on agriculture. Most often, only the effects of changes in temperature and precipitation were considered, since these are the climatic factors most widely measured and both, of course, are strongly correlated with plant response to climate. The statistical approach to agroecology has a long, honorable history. For example, H.A. Wallace (later Secretary of Agriculture, 1933-1941, and Vice President of the United States, 1941-1945) in the 1920s and 1930s used essentially the same technique to relate annual corn yields in Iowa to mean temperature and precipitation during critical months of the growing season. Later, Thompson (e.g., 1969, 1970) extended this approach to estimate climatic effects to corn and soybean crops in the central United States.

Applications of the statistical approach to climate change questions about grain production were made by Thompson (1975), Bach (1979), Newman (1982), and Blasing and Solomon (1982). They related corn yields in North America to a range of changes in monthly mean temperature, and precipitation, based on interpretation of early GCM experiments of greenhouse-forced climate change. These studies generally concluded that higher temperatures and lower precipitation would move the North American cornbelt northeastward. Blasing and Solomon (1982) (Figure 2) concluded that only small areas of the new

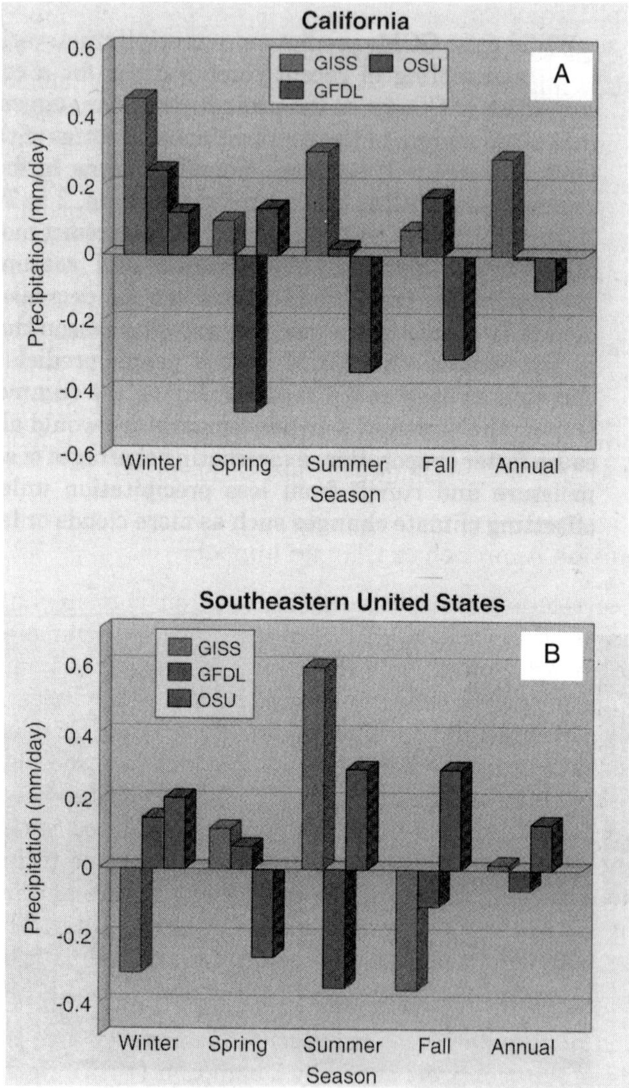

Figure 1. Changes in seasonal precipitation predicted for (A) California and (B) the southeastern United States for $C0_2$ concentrations twice the preindustrial ones. Predictions made by three global circulation models: Goddard Institute of Space Studies (GISS), Geophysical Fluid Dynamics Laboratory (GFDL), and Oregon State University (OSU) (Smith and Tirpak, 1989; reprinted from CAST, 1992).

cornbelt would be productive without irrigation. Waggoner (1983) used a similar statistical approach and concluded that an assumed 1°C warming, coupled with a 10% decrease in precipitation, would significantly reduce yields of the major U.S. crops. Spring wheat yields in North Dakota, for example, would decrease by 12% under these circumstances. Waggoner (1983) went a step further and argued that the so-called "CO_2 fertilization effect" (more on this below) would moderate the impacts of this climate change and adjusted his predicted yields upward by simply reducing all climate change damages by 5%. The regression studies cited above, and many others, have been useful. However, their main weakness—calibration within a range of current climatic variability—does not fit them well for extrapolation outside that range. They point us in the direction of likely change but provide little more guidance than that. (See Rosenberg [1981] for a fuller discussion of the problems associated with the use of regression models for climate change analysis.)

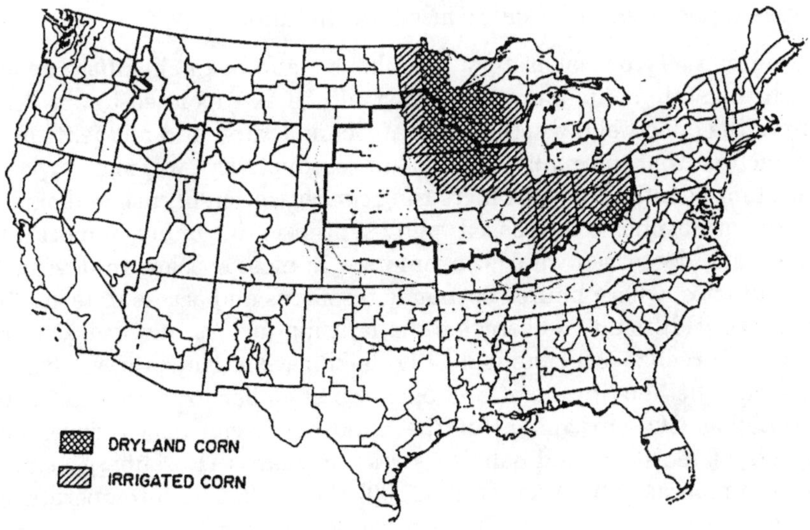

Figure 2. Current and climate-change-altered boundaries of the U.S. cornbelt. Source: Blasing and Solomon (1982).

Direct Effects of Carbon Dioxide Enrichment

By the early 1980s it had been clearly demonstrated under controlled climate conditions that most C3 species of plants (e.g., the small grains, legumes, most tree species) respond to elevated concentrations of atmospheric carbon dioxide with increased rates of photosynthesis and that virtually all C3 and C4 species (e.g., maize, sugar cane, sorghum, millet, some warm-season grasses) demonstrate reduced transpiration rates. The mechanisms involved are described in a number of reviews (e.g., Cure, 1985; Kimball, 1986; Rosenberg et al., 1990; CAST, 1992). The magnitude of this "CO_2 fertilization effect" cannot be known precisely since it varies with species, with weather conditions, with soil fertility, and other factors. However, the literature does show that, on average, a doubling of CO_2 from its recent concentration (340 to 356 vppm), all else being constant, will increase growth and yield about 34% in C3 and about 14% in C4 plants. Reviews of experimental evidence for a wide range of agricultural, woody and weed species have also shown that doubling ambient CO_2 from recent levels reduces transpiration by an average of 34%. The reduction in transpiration is not accompanied by any significant loss in instantaneous photosynthesis and usually results in greater total photosynthesis because of better soil moisture conditions. Water use efficiency (yield per unit of water consumed) increases because of increased photosynthesis and reduced transpiration.

If the findings from laboratory, greenhouse, and open chambers upon which the statements above are based can be extrapolated to agriculture in the open field, where temperature, moisture, and nutrients are the factors that normally limit plant productivity, we may expect important benefits from the increasing concentration of carbon dioxide in the atmosphere: increased photosynthesis in many important species and decreased water consumption in most species. In developing countries, where limited economic resources and access to technology make it difficult to overcome the limiting factors, it will be more difficult to realize the benefits of CO_2-fertilization. However, a laboratory-controlled environment and open-top chamber experiments have shown that CO_2-enrichment of the atmosphere can also reduce the impacts of moisture and salinity stress on plants (Rosenberg et al., 1990). A summary by Allen et al. (1990) shows that high-temperature stress is also alleviated by CO_2-fertilization. The effects on nutrient stress remain less clear at this writing. Thus, there is a good basis of information for building onto Waggoner's simple 5% adjustment that was meant to recognize the existence of a CO_2-fertilization effect. This information base is used with greater or lesser formality in a number of crop growth simulation models now employed in climate change impact studies.

Simulations of Crop Growth Response to Climate Change

The use of process models in climate change studies began, effectively, in the 1980s. The U.S. Environmental Protection Agency's (EPA) report to Congress on potential impacts of climatic change included an analysis of agricultural impacts that relied heavily on process models (Smith and Tirpak, 1989). The study dealt with four U.S. regions: the Great Lakes, the southeastern United States, the Great Plains, and California. Various simulators of crop growth such as CERES-maizen (Jones and Kiniry, 1986), CERES-wheat (Ritchie and Otter, 1985) and Soygro (Jones et al., 1988) were applied using then-current runs of the Geophysical Fluid Dynamics Laboratory (GFDL) and Goddard Institute for Space Studies (GISS) GCM to provide climate change scenarios. The direct effects of CO_2 on photosynthesis and evapotranspiration were considered in these studies, although not, it appears, through the simulations but rather, by appropriate adjustments to the simulation outputs. The two principal findings of this study were that: (1) without the direct effects of CO_2 yields of wheat, soybeans and corn would decline in the Great Lakes, the southeast, and the Great Plains, except in the far north, where the frost-free season would be lengthened; (2) irrigation would moderate losses, as would the CO_2 fertilization effect, but significant increases in irrigation demand would also occur, particularly under the severe GFDL scenarios.

Improving Impact Assessment Methodology

While process models have improved the state of the art somewhat and have provided greater flexibility and depth of analyses as compared to regression methods, it cannot be said that their use alone creates a solid linkage of climate change and agricultural production. Both the regression and process-based simulation assessments were useful in "scoping out" the range of possible effects. However, the utility of all such studies up to the mid-1980s was limited by a number of important methodological defects. Among these were:

- that climate change scenarios lacked the needed detail on temporal and spatial variability (each place within a region was assumed to experience the same climate change, and each day was made warmer, wetter, or drier by exactly the same amount as every other day);

- that climate change effects were calculated for current agriculture rather than for agriculture as it might be in the future, when the climate might actually have changed and when agricultural vulnerability or resistance to climate change will undoubtedly be different than it is today;
- that mitigating effects of the adaptations that might be made with currently available and/or future technologies were not taken into account;
- that interindustry impacts and impacts of changes in agricultural and other sectors on the overall regional economy were not taken into account.

The time was ripe for methodological improvements. A U.S. Department of Energy-sponsored study (the MINK study, Rosenberg, 1993) was undertaken to deal with the most serious of the limitations described above. The MINK study focused on a single, large, agricultural region comprising the states of Missouri, Iowa, Nebraska, and Kansas; it considered the forestry, water resources, and energy sectors, as well as agriculture. Rather than rely on one or more GCM-derived climate scenarios of climate change an "analog" was found in the records of the warm dry decade of the 1930s, the so-called "Dustbowl" era. The Erosion Productivity Calculator (EPIC) of Williams et al. (1984) was used to simulate crop growth and yield for some 50 representative farms under a set of four cases. These were: current agriculture and current (1951-80) climate; current agriculture under the analog climate; anticipated agricultural conditions in 2030 with the current climate and no climate change; and 2030 agriculture with the 1930s analog climate imposed once again. Additionally, the effects of a 100-vppm increase in atmospheric carbon dioxide concentration were tested for each of the four cases. Also, adaptation of two kinds: autonomous (off-theshelf, e.g., changes in crop variety, planting dates, tillage practices, irrigation, etc.) and policy-driven adaptations (new drought-resistant varieties, improved irrigation efficiency, etc.) were applied to the appropriate cases.

While methodology improvement was the primary aim of the MINK study, the following conclusions were reached in analyses for the agricultural sector:

- a return of the 1930s climate would have little effect on wheat yields, but yields of corn, sorghum, and soybeans would decline 15 to 25% if no on-farm adaptations were made;
- a reasonable range of future technological advances modeled through EPIC suggested that crop yields in 2030 will be about 70 to 90% higher than today in the absence of climate change;
- a return of the analog 1930s dustbowl climate in 2030 would reduce yields by about 20% below their potential;
- adaptations and CO_2 fertilization would recover most of this lost yield;
- demand for irrigation water would increase with rising evapotranspiration rates, but streamflows in the major rivers of the region would decline (by 28% in both the Missouri and Upper Mississippi basins, and by 7% in the Arkansas-White-Red River regions). Also, the current depletion of the Ogallala aquifer would be hastened by climate change. The net result would be large-scale abandonment of irrigation in western Nebraska and Kansas and some new (but not offsetting) development of irrigation in the eastern MINK states.
- loss to the regional economy in the worst case (no adaptation, no CO_2 effect, and loss of production ramifying through the entire economy) would be about 10%; in the best case (adaptation, CO_2-effect, and losses in production falling only on exports), the loss would be less than 1%.

Yield changes resulting from the analog climate imposed on current agriculture in the MINK region are shown in Figure 3 (from Easterling et al., 1992). The simulated effects of future technology, carbon dioxide enrichment and adaptations on dry-land corn production are shown in Figure 4 for one of the 50 representative farms of the MINK study.

Conclusions with Respect to U.S. Agriculture

Basically, the EPA and MINK studies reached the same conclusions about climate change and U.S. agriculture. The GCM and analog scenarios of climate change applied in these studies were, with a few

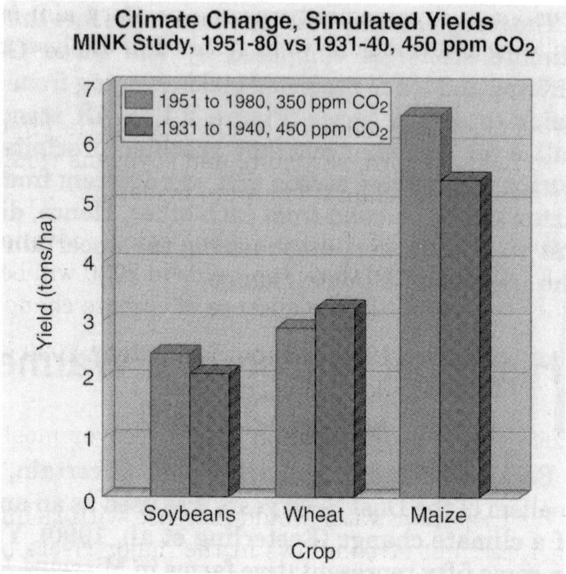

Figure 3. Yields of modern varieties of three crops computed for the weather of 1951 to 1980 and the dust bowl years 1930 to 1939. Yields were computed for 19 locations in Missouri-Iowa-Nebraska-Kansas (MINK) for corn and soybeans and 11 for wheat (Easterling et al., 1993; reprinted from CAST, 1992).

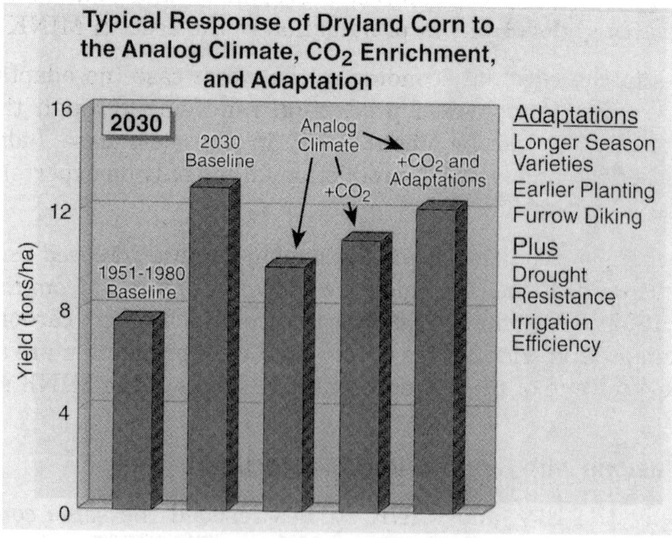

Figure 4. Computed response of dryland corn in the MINK region in 2030. First, 40 years of improved farming and technology raise present yields. Changing to climate of dust bowl years lowers yields; enrichment of CO_2, in air raises yields, and adaptations raise yields further (Rosenberg and Crosson, 1991; reprinted from CAST, 1992).

exceptions, insufficiently severe to greatly disrupt the overall capacity of the United States to feed itself and to remain a significant force in the global trade in agricultural products. But in certain localities, losses in productivity could be considerable and might lead to disruption of social and economic systems. Impacts on regional economies would be severe only in the worst cases, where climate impacts were not offset by CO_2-fertilization effects and where no attempts were made to adapt.

An International View

Thus far, the impacts studies described have dealt with agriculture in the United States. An international view of climate change impacts on worldwide agriculture was undertaken by the International Institute for Application of Systems Analysis (IIASA) and reported by Parry et al. (1988). This study concentrated on the cold and semi-arid margins of present-day agriculture. Various kinds of process models were applied, as were other techniques of agrogeography and agroclimatology. Although a number of climate change scenarios were used only, one, a GISS $2 \times CO_2$ run, was applied to all of the regions studied.

The IIASA study identified most likely first-order effects of climate change on agriculture. These involve changes in the length of the potential growing season and in plant growth rates leading to changes in the required growing season; changes in mean yields; changes in yield variability and in the certainty of expectable yields; changes in yield quality; and changes in the sensitivity of plants to different levels of fertilizer and pesticide applications.

Based on the IIASA study and others, the IPCC reports of 1990 and 1992 reached the following, rather general conclusions:

- It is not yet possible to know whether, on average, global agricultural potential will increase or decrease as a result of climate change.
- Severe negative effects are possible at the regional level and particularly so in some regions of high present-day vulnerability least able to adjust technically to such effects.
- Most vulnerable regions appear to be semi-arid tropical and subtropical, such as western Arabia, the Maghreb, the Horn of Africa, southern Africa, and eastern Brazil.
- Some humid tropical regions, such as southeast Asia and central America, also appear vulnerable.

- Regions that are currently net exporters of grain, such as southern Europe, southern United States, parts of South America and western Australia, may experience water shortages that would cost them productive potential.

The most recent major contribution to the assessment of climate change impacts on global agriculture results from an EPA-sponsored program to model crop responses in 18 countries (Rosenzweig et al., 1993). A uniform methodology was used in all countries to the extent possible. Three 2 × CO_2 GCM scenarios were applied to crop growth models of wheat, rice, maize, and soybeans. The changes in temperature indicated by the GCM employed were near the high end of the 1.5 to 4.5°C IPCC range for 2 × CO_2 climate forcing. The modeling also considered the moderating effects of an increase in ambient CO_2 concentration to 555 ppm, the concentration that might be expected for an effective CO_2 doubling in 2060. Data from CO_2 enrichment experiments provide proportion factors for raising photosynthesis rates and lowering evapotranspiration rates. CO_2 effects were not simulated explicitly in the growth models.

Table 2 from Rosenzweig et al., (1993) summarizes changes in global production of wheat, rice, maize and soybeans under three GCM scenarios without CO_2 fertilization effects and at a CO_2 concentration of 555 ppm. Figure 5 summarizes these results, showing nations in which production of all grains would increase or decrease under the GISS climate change scenario.

Table 2. Simulated changes in wheat, rice, maize, and soybean yields.[a] Source: Rosenzweig et al. (1993)

Crop	GISS[b]	GFDL[b]	UKMO[b]	GISS[c]	GFDL[c]	UKMO[c]
	%	%	%	%	%	%
Wheat	-16	-22	-33	11	4	-13
Rice	-24	-25	-25	-2	-4	-5
Maize	-20	-26	-31	-15	-18	-24
Soy	-19	-25	-57	16	5	-33

[a]Crop yield changes were obtained by weighting site results first by regional production within countries and then by national contribution to total production simulated in the study.
[b]GCM climate change scenario alone.
[c]GCM climate change scenario with direct effects of CO_2.

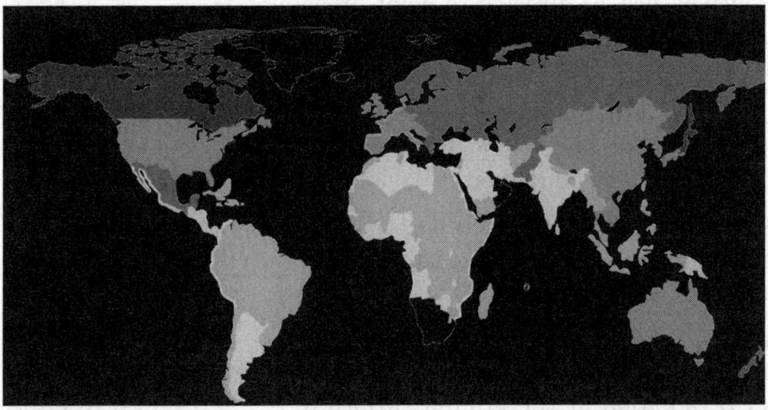

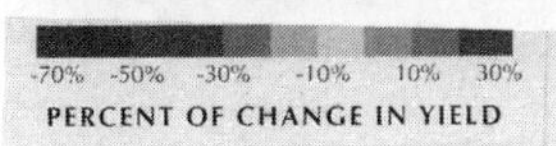

Figure 5. Potential change in grain yield for doubled CO_2 climate change predicted by Goddard Institute for Space Studies global climate model. Physiological CO_2 effects on crop growth taken into account. Source: Rosenzweig et al. (1993).

Findings of the IIASA and the EPA (Smith and Tirpak, 1989) reports figure prominently in the first IPCC (1990) evaluation of potential impacts on agriculture. Those reports and the more recent international study of Rosenzweig et al. (1993) strongly influence the conclusions reached in the 1990 IPCC report and its 1992 special "Summary for Policymakers." The most recent and authoritative (but not necessarily correct) IPCC conclusions with respect to global impacts on agriculture, then, are:

- There is sufficient evidence to indicate that climate change would have important effects on agriculture and livestock, but whether the net global effect would be positive or negative is not yet clear.
- Brazil, Peru, the Sahel, Southeast Asia, the Asian portions of Russia, and China are regions that may suffer severe effects.
- High and midlatitude regions in the northern hemisphere may experience an increase in potential productivity due to a lengthening of the growing season, but this effect is unlikely to open up large new regions to agriculture.
- The greatest damages due to climate change would occur in the semi-arid regions, where drought would be intensified.

- Policy responses directed to plant breeding and changes in agricultural management techniques could lessen the severity of regional impacts.
- Global food production may not be much different from that in the absence of climate change, but the costs of maintaining that level of production may be higher.
- Effects would be most severe at the upper end of the range of mean global temperature increase suggested in the IPCC report (1990): 1.5 to 4.5°C.

And with respect to another aspect of climate change not previously mentioned in this paper the report concludes that increased ultraviolet (UV) B radiation at the ground surface (due to erosion of the ozone layer) would have a negative impact on crops and livestock.

So What can be said with Certainty about Climate Change and Agriculture?

How close are we now to understanding what the regional scale and global impacts of climate change on agriculture may be? The foregoing review suggests that we are probably not too close. The availability of process models has certainly increased the sophistication with which we can analyze climate change effects, but it would be dangerous to place too much confidence in the results of these simulations.

Crop process models used for climate change assessments do not deal explicitly with weeds, diseases, and insect pests but assume that they are controlled. The models do not deal with extreme weather events, such as windstorms and hailstorms. The models are calibrated, often, against experimental crop yields which are generally higher than farmers achieve. Phenology (the timing of development stages) is driven in the models by the accumulation of heat units, so that greenhouse warming causes all crops to mature sooner. This may be a serious oversimplification of what might actually happen in the field. The negative effects of shortening the growing season of summer crops because of higher temperatures and the impacts of decreased water availability due to increased evapotranspiration may be overstated in these models. The benefits to yields of fall-seeded grains through alleviation of cold stress may be overestimated. CO_2 fertilization effects may prove to be more modest under natural conditions where other factors (temperature, soil moisture, nutrients) are less than optimal and where insect and disease problems occur. Finally, the application of models derived largely in the temperate region to tropical agriculture may overstate the importance of small temperature changes.

Despite this apparent lack of assurance in what we know about climate change and agriculture, I am more confident of our ability to adapt once we can predict or actually detect climate change.

WHAT ABOUT ADAPTATION?

The EPA and IIASA studies, and the MINK study in particular, made efforts to simulate the ways in which farmers might respond to climate change and moderate or even overcome yield depressions that would otherwise occur. Adaptations indicated as useful by process model simulations include the use of cultivars requiring either longer or shorter growing seasons, depending on the region and the degree of climatic change. In the high latitudes, for example, the growing season is lengthened by warming, and this allows the use of more productive, long-season cultivars. Such cultivars should be readily available from lower latitudes. Much of the yield loss simulated for summer crops in the mid- and low latitudes is due to a shortening of the effective growing season as crops mature more rapidly in response to higher temperatures. Longer-season cultivars and earlier planting would help to overcome these losses in yield (Easterling et al., 1993; Smith and Tirpak, 1989; Parry et al., 1988; Wilks, 1988; Rosenzweig et al., 1993).

Photoperiod limitations, if any, can be overcome by traditional plant breeding procedures. Where higher summer temperatures, alone or with greater water stress, affect reproductive processes or diminish photosynthesis during the grain-fill period, shorter season or earlier crops may be used. A shift from spring to fall sowing would allow this to be done for the small grains. Fall-sown grains would also benefit from reduced cold-temperature damage.

Development or importation of drought- and heat-resistant cultivars, if they exist, or plant breeding if they do not, would be a logical response to warming and drying. Substitution of, for example, less drought-sensitive soybeans and sorghum for more sensitive corn is another approach.

Where conditions become drier, greater emphasis will likely be placed on moisture-conserving tillage practices. Such practices as stubble mulching, conservation tillage, minimum tillage, furrow diking, and the like are already gaining ground (literally) in the United States because of lower labor requirements and the energy and/or water savings they permit. The rate of adaptation of such techniques would likely increase were climate to become detectably hotter and drier.

The studies of climate change impacts on agriculture cited above, and others, indicate an increase in demand for irrigation water. Certain studies (e.g., MINK, Rosenberg, 1993) indicate a reduction in streamflow and a more rapid depletion of ground water in the future. Therefore, there will be an imperative to improve irrigation efficiency (fraction of the water delivered to the field that reaches and remains in the root zone) and water use efficiency (yield per unit of water consumed in evapotranspiration). Stewart and Nielsen (1990), Rosenberg et al. (1983), and Rosenberg (1992) provide details on a wide variety of management techniques that can improve both kinds of efficiency. In agriculture, as opposed to other sectors, such as forestry, water resource development, power generation, etc., capital investments are relatively small and are made for relatively short periods of time. For example: sowing crops, annually; planting perennial pastures, 5 to 10 yr; planting orchards, 15 to 20 yr; purchasing tractors, 15 yr; development of new varieties, 10 yr. Short time horizons make for relatively rapid adaptations. One exception to this rapid turnaround in agriculture is the development of large-scale irrigation projects, particularly if new dam construction is required. Planning and construction of dams and water distribution systems require much more time (50 yr, in some cases, particularly where there is political opposition to the projects). Given growing environmental concerns, high construction costs, and competing uses for water, it seems likely that, in future, more effort in both developed and developing countries will be devoted to increasing distribution and field irrigation efficiencies than to new construction. (More on this subject below.)

We have yet to experience a noticeable climate change in our part of the world. Thus the “adaptations” described above are anticipatory and based on reasoning about what might be needed if climate changes. However, there is a good body of knowledge about how farmers in both developed and developing countries have responded to situations that are analogous to climate change. Jodha (1989) describes how Indian farmers have responded to shorter- and longer-term climatic and other threats to their production systems. For example, the increased incidence of striga weed in sorghum in the states of Andhra Pradesh and Maharastra led farmers there to rotate sorghum with other crops.

Increased soil salinity and waterlogging following faulty water management in the Chambal irrigation project caused a switch to salinity- and water-tolerant crops and introduction of methods to drain off surplus water. Recurrence of late season drought in several parts of Rajasthan state led to an increased emphasis on crops and varieties with high salvage potential (fodder instead of grains). Recognition of slack potential in the rainy season in high-rainfall, deep-vertisols areas of central India led to introduction of short-duration soybean and mung bean crops in areas traditionally fallow during the rainy season and cropped only afterward. The first three of these "adaptations" are in response to a worsening resource base that Jodha thinks may be a good surrogate for what climate change might do. The last case represents recognition of an opportunity previously foregone and may be a surrogate for new opportunities made possible by climate change.

Expansion of the geographic distribution of a crop is also an extension of its climatic range and is, in a real sense, an adaptation to changed climate if not literally to climate change. One of the best examples of such an expansion is the near-doubling, in the period 1920 to 1980, of the profitable growing range for hard red winter wheat in the North American Great Plains region. This expansion (documented by Rosenberg, 1982), shown in Figure 6, indicates that the crop has moved during this period not only northward to cooler climes but also southward to

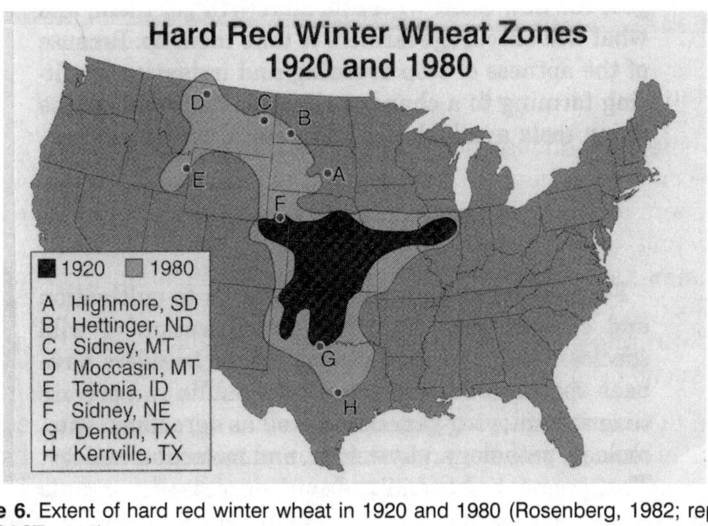

Figure 6. Extent of hard red winter wheat in 1920 and 1980 (Rosenberg, 1982; reprinted from CAST, 1992).

warmer ones. Winter wheat has advantages over the competing spring wheat; it is planted in the normally dry conditions of fall instead of in the wet conditions after spring thaws. It matures before the extreme heat (and often drought) of midsummer. These advantages motivated researchers, extension agents, and farmers to make genetic and soil management improvements that facilitated the dramatic expansion of this crop's geographic extent. (The expansion continues: By 1990, more than 1 million hectares of winter wheat were being grown in the southern portions of the Canadian Prairie Provinces.) In moving from Sidney, Nebraska is to Sidney, Montana the wheat crop was adapted to a 4.2°C decrease in mean annual temperature, an 88-mm (20%) decrease in precipitation, and a 10-day-shorter growing season.

HOW MITIGATION OF CLIMATE CHANGE MIGHT AFFECT AGRICULTURE

Up to this point we have explored how climatic change and CO_2 enrichment of the atmosphere might affect crop production and how agriculture might adapt to their impact. But there is a strong political and environmental sentiment that favors avoiding climate change altogether and, by extension, avoiding the impacts and the need to adapt. Rosenberg and Scott (in press) have explored the question of how the results of attempts to mitigate or avoid climate change might, in themselves, affect agricultural productivity. This section draws heavily on that analysis.

Emissions from the Farm

The whole "greenhouse problem" stems from the increasing atmospheric concentrations of radiatively active trace gases emitted to the atmosphere as the result of human activities. Agriculture is one such activity and, on the global scale, is thought to contribute 15% or so to the human-forced global warming potential through emissions of carbon dioxide, methane, and nitrous oxide. So, one strategy to mitigate global warming and consequent climatic change is to reduce the total of emissions from agriculture.

Carbon dioxide is emitted by agriculture when forests are cleared and prairies broken to provide land for farming and ranching. The decomposing vegetation emits CO_2 and the soils decrease in organic matter content year by year after farming begins until a new and lower equilibrium content is reached. Improved management can restore some carbon to the soil and the conversion of land back to grass or trees should eventually raise the content to a level near that which prevailed in the original ecosystem. (More on this issue below.)

Methane is also emitted as a by-product of animal digestion, with particularly large amounts from ruminant animals. More nutritious forages and chemical feed supplements decrease methanogenesis in the rumen. Use of a bovine growth-stimulating substance, somatotropin, would decrease the quantities of methane emitted per unit of meat or milk produced.

Emissions from rice paddies contribute a large amount of the methane from agriculture. Plant breeding, reductions in the amounts of organic residue returned to the soil and farming improvements, including increased fertilization, could reduce methane emitted per unit of rice produced. A shift to upland rice would also reduce methane emissions.

Nitrous oxide is emitted by soils through both aerobic and anaerobic processes. The application of nitrogenous fertilizers to agricultural soils increases the amounts emitted. Formation of nitrous oxide can be decreased by practices that maximize the amounts of nitrogen taken up by the plant and minimize volatilization to the atmosphere and leaching to depths below the root zone. Improvements in the timing and placement of fertilizers and encapsulation to slow the release of fertilizer help in this regard. Nitrification inhibitors can also help.

If adopted by farmers, the practices described above to diminish methane and nitrous oxide emissions should all contribute to increasing productivity. If it can be shown that these practices are profitable, the rate of adoption will be more rapid than would happen in response to fears of climate change. But certain strategies to diminish emissions from other sectors (energy, transportation, industry) might actually affect agriculture negatively. Afforestation to capture and sequester carbon from the atmosphere could reduce the area of productive land available to agriculture. A sharp reduction of fossil fuel use could, indirectly, lead to decreasing water supplies for irrigation.

Afforestation to Capture CO_2

The aim of afforestation as a climate change mitigation strategy is to capture, through photosynthesis, a large portion of the carbon emitted into the atmosphere by the combustion of fossil fuel, by deforestation, and by biomass burning. The captured carbon can either be sequestered in the roots and boles of trees and in soil organic matter, or the wood can be burned or otherwise converted to provide energy as a substitute for fossil fuels.

Estimates have been made of the amount of land required for afforestation to significantly reduce net carbon emissions to the atmosphere. The estimates vary widely because of varying assumptions about the species to be planted, the biological limits of carbon fixation of the various candidate species, climatic and soil conditions, and time required for a planted stand to reach maturity. Three of these estimates and the assumptions on which they are based are given in Table 3. Two of the estimates take a global view of the problem, and one focuses on the United States.

Table 3. Estimates of land required for afforestation to capture specified portions of the annual net CO_2 emissions to the atmosphere. Source: Rosenberg and Scott (in press).

Source	Total Carbon Removal (GT/yr)	Total Land Area Required (millions of hectares)	Assumptions
Marland, 1988	5.0	500	Production equivalent to best tropical forests
Marland, 1988	5.0	700	Production equivalent to short rotation sycamores in southeastern U.S. (Georgia)
Sedjo & Solomon, 1989	2.9	465	Production equivalent to tree plantations in U.S. southeast and northwest
Moulton & Richards, 1990		28.7[a]	Net U.S. carbon emission = 1.27 GT/yr 10% reduction

[a]Economically marginal and environmentally sensitive croplands, pasture lands, and forest lands on which growth rate of trees could be enhanced. Beyond a 15% reduction in net carbon emissions would require croplands almost exclusively.

Sedjo and Solomon (1989) calculated that, in order to offset the 2.9 GT C that accumulates in the atmosphere each year after natural scavenging, some 465 million hectares must be planted with trees having production equivalent to that of plantation forests in the United States southeast. Marland (1988) calculated that, depending on the productivity of the forests planted, between 500 and 700 million hectares would be needed for the more ambitious goal of removing 5 of the 6.3 to 8.5 GT C that are emitted into the atmosphere yearly. A notion of the scale of land diversion being discussed can be obtained from Figure 7, which shows how large an area is encompassed in Sedjo and Solomon's estimate of 465 million hectares.

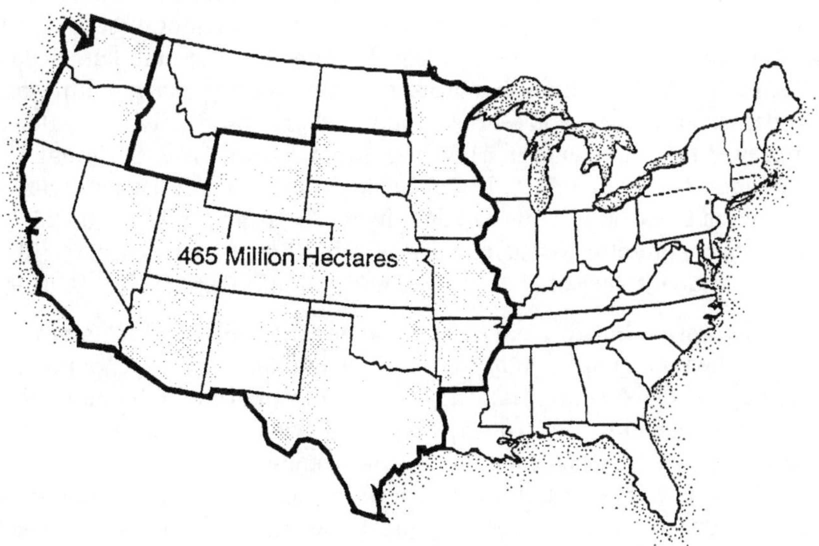

Figure 7. An area of approximately 465 million hectares. Source: Sedjo and Solomon (1989).

Afforestation for sequestering carbon easily captures the imagination. Increasingly, attention is being given to the potential for recycling carbon through production of herbaceous and woody biomass crops (e.g., OTA, 1993). These crops would be grown as feedstocks for chemical manufacture of substances to replace petroleum-derived liquid fuels and natural gas, or for direct combustion as boiler fuel in energy generation. Although possibilities exist for the production of nonwoody biomass on marginal lands, at least in the United States, the better agricultural lands will be needed to produce biomass competitively (Tyson, 1990). A study of the potential for biomass to provide a significant share of the energy needs of The Netherlands (NOVEM, 1992) concludes that if all of that nation's land were devoted to biomass, about 22% of its energy needs could be met. However, at the most, 0.5 million hectares of land could be diverted for biomass, which would provide only about 5% of The Netherlands' energy requirements.

What would be the overall effect of large-scale land diversions away from food production to the sequestering and/or recycling of carbon to mitigate climate change? Moulton and Richards (1990) focused on opportunities for afforestation in the United States to offset portions of our net emission of carbon to the atmosphere. A goal of offsetting 10% of the net U.S. emissions of 1.27 GT C per annum could be accomplished on 28.7 million hectares of economically marginal pasture lands, crop lands, and forest lands where productivity could be improved by management. However, if more than 15% of the net U.S. emissions are to be offset, better agricultural lands must be diverted. To increase carbon sequestration from 15 to about 50% (0.725 GT C/yr), for example, would require 90.7 million hectares (224 million acres), or about 48% of all U.S. cropland (Moulton and Richards, 1990).

This discussion has focused on the temperate zone, for which a number of detailed economic analyses are available. However, plantation forestry to offset accumulations in the atmosphere may be more readily accomplished in the tropics. Sedjo and Solomon (1989) believe that large areas of "degraded lands" in the tropics with few economic uses could be made available for afforestation. Costs of afforestation would be considerably lower in the tropical countries. Recently deforested lands in the tropical rain forest zone are included in the category of "degraded lands." Economics will probably cause many such areas to revert, naturally or through human intervention, to forest.

Information on the uses of U.S. cropland suggests that about two-thirds is devoted to the growing of wheat, corn (maize), and soybeans, with each crop occupying about one-third of that area. Assuming that the land removed from agriculture is proportionately distributed among all crops and has average productivity, Rosenberg and Scott (in press) calculated an approximate maximum amount of lost farm production of these three major crops. (This is a maximum for two reasons: (1) the land most valuable for agriculture would likely remain in agriculture, meaning that average yield would overstate average loss in production on the average hectare diverted into forests; and (2) farmers would likely farm remaining land more intensively, resulting in higher yields.) The loss of 48% of U.S. production would be about 40 million tons of wheat, 70 million tons of corn, and 26 million tons of soybeans (7%, 15%, and 25%, respectively, of world production). Such losses would be a matter of some concern for world food security.

Thus, we see that, in any real world efforts at afforestation, lands of marginal value to agriculture (i.e., highly erodible lands, wetlands, etc.) will be the first committed to this purpose, but if a significant impact is to be made, highly productive agricultural lands will also be required.

Competition for Water

The movement toward reduction of fossil fuel usage will increase demand for sources of energy that do not release CO_2 to the atmosphere or those that, at the least, add only CO_2 that has recently been removed from the atmosphere by photosynthesis. Candidate energy sources that meet this criterion are nuclear fission, wind, solar, geothermal, biomass, and hydroelectricity. This section deals only with the competition between hydroelectric generation and irrigation for water resources.

Surface runoff into streams, rivers, and reservoirs is finite, depending on the vagaries of seasonal and annual precipitation and evaporative demands, that is, on weather and climate. Decades of research into various methods of weather modification have yet to provide evidence of an operational capacity to augment precipitation. Therefore, the supply remains finite, if variable, while demands for water for all purposes increase. Will pressures to reduce use of fossil fuels increase the value of water used for power generation? Will such increased value result in increased allocations of water to hydropower despite growth in competing demands?

Irrigation and hydropower influence one another. Irrigation changes the seasonal pattern of electricity demand, streamflow, and hydropower supply. Irrigation consumes large amounts of electricity for pumping, and irrigation depletes streamflows and thus reduces downstream hydropower (Butcher et al., 1986). The loss of hydropower when water is withdrawn and consumed for irrigation depends on the quantity of water diverted and consumed and the amount of developed head that the water would fall through if left in the stream.

Currently, hydropower is one of the more highly valued uses of water. In many river basins, irrigation is one of the least valued uses, particularly when the water is applied to low-value, extensive crops such as alfalfa and pasture. Electricity generation is a nonconsumptive use of water, and irrigation is one of the most consumptive uses. In most developed river basins, hydropower is a more valuable use of water at the margin than is irrigation. The difference in the relative economic value of water can only grow as the demand for CO_2-clean energy grows and as the costs of fossil fuel rise when anti-climate-change policies take effect.

Most of the trans-Mississippi west, except for the southwest, as well as the Tennessee Valley and Alaska, derive more than a quarter of their power from hydroelectricity and therefore would be particularly vulnerable to desiccation due to natural droughts and/or climate change (Gleick, 1990). The primary users of water for irrigation in the United States are the 17 western states. The west is therefore the region in which the most severe conflicts between hydropower, irrigation, and urbanization are likely to occur.

The current trend of transferring water away from agriculture and to higher-value uses is likely to accelerate for many reasons. Demand for CO_2-clean hydropower would probably accelerate that trend if public policy is aimed at diminishing the likelihood of climate change. If climate change of the kind that the GCM predict for North America, i.e., warmer and drier in summer with less runoff and diminished stream flow, were actually to occur despite efforts at mitigation, these transfers would become more problematic.

To give the latter statements some dimension, I refer again to the MINK study. In the water resources portion of that study, Frederick (1993) compared current (1951 to 1980) streamflow at 15 gaging stations in the Missouri, Upper Mississippi, and Arkansas-White-Red River basins with flows at the same stations during the dry years of the 1930s (see Figure 8). The 15 stations selected were in watersheds in which no man-made changes (e.g., diversions, dams) had been made in the intervening years. The ratios of flows in the 1930s to current flows were applied to the sub-basins represented in order to scale up to the entire river basin. The calculations show that if the climate of the 1930s were to recur, flows today would be reduced from the long-term means by 28%, 28%, and 7% in the Missouri, Upper Mississippi, and Arkansas basins, respectively.

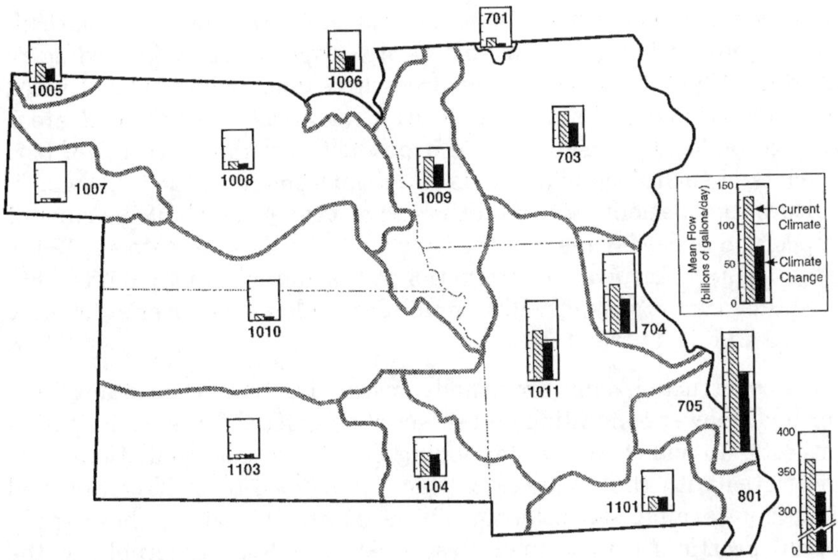

Figure 8. Impact of climate change on mean natural stream flows in selected sub-basins of Missouri, Upper Mississippi, and Arkansas-White-Red River systems (billions of gallons/day). From Rosenberg and Scott (in press) based on data from Frederick (1991).

Frederick (1993) also observed that were the large reservoirs on the upper Missouri to continue operating under current Corps of Engineers assumptions, it would be necessary to curtail hydropower production by about 50%. Of course, the demand for irrigation water would be increasing at the same time: first, because of greater evaporative demand due to higher temperatures and drier air, and second, because the profitability of irrigated agriculture would increase as dryland production falls. The assumption of no change in irrigated acreage and irrigation water applied in amounts determined by climatic conditions leads to an increased demand of 39% in Nebraska and 12% in Kansas, the two MINK states with significant irrigated acreage today.

Would climate change exacerbate or diminish the competition between irrigation and hydropower and in similar or different ways in developed and developing countries? The answers are not obvious but the questions are worth exploration.

Dam construction leads everywhere to inundation of fertile agricultural lands in low-lying floodplains. In tropical Africa (according to Scudder, 1989) the loss of fertile, productive flood-plain land cultivated with traditional techniques is clearly not offset by the small areas gained for formal irrigation as a byproduct of dam construction. It is the loss of such flood-plain lands and displacement of the largely agricultural populations settled on them that prompt much of the current opposition to development of the Narmada river in India (Moore, 1993). This project involves construction of some 30 huge dams, 135 medium-sized dams, and 3000 small dams along the river's 800-mile course and its 41 tributaries.

Once constructed, dams are usually operated to provide constant flows for hydropower generation and seasonally adjusted flows for irrigation projects. In many of the developing countries this regulation often interferes with the practices of traditional flood-plain irrigation and recession-farming downstream of the dams. Of course, these traditional practices can be made less risky and less vulnerable to the vagaries of the weather by controlled releases that permit well-timed flooding. This is unlikely to happen, however, if such releases disrupt hydropower generation. Not all regions would necessarily be drier as the result of climate change. But the picture emerging will not be greatly changed by other assumptions. Even without climate change, that picture is one of much less water available for all purposes, a premium placed on power supplies not derived from the combustion of fossil fuels, and increased (but not necessarily beneficial) demand for water to maintain or even enlarge the irrigation enterprise. On the basis of the foregoing it seems logical to assert that hydropower is likely to out-compete irrigation for increasingly scarce water supplies.

SUMMARY AND CONCLUSIONS

Agriculture will be called on to provide increasingly larger supplies of food, feed, and fiber to meet the needs of a growing world population. Its ability to meet this challenge will depend, among other factors, on the continued availability of land, water, and genetic resources. Greenhouse warming may cause climate to change throughout the world.

Temperatures will rise, precipitation amounts and timing will be altered, as will the patterns of cloudiness and windiness. Crop growth will be affected directly by such changes in climate. Agriculture will be affected indirectly, as well, by changes that occur in the condition of land and water resources.

Available methods for projecting the geographic distributions of climate change and methods for projecting the impacts of climate change on crop growth are imperfect. As a result, it is impossible at this time to know which regions of the world will lose and which will gain in agricultural productivity and competitive advantage from climate change in the coming decades and century.

There are two reasons for this uncertainty: (1) predictions of the regionally distributed effects of climate change differ considerably from one general circulation model to the next; (2) simulation models of plant response to climate change have systematic defects that stem from unrealistic representations of plant phenology (the progression of growth and development) and are unable to deal with episodic events such as windstorms, hailstorms, insect attacks, diseases, etc. Additionally, these simulators do not, with few exceptions, treat the CO_2-fertilization effect (increased photosynthetic rate in C3 plants and decreased transpiration in both C3 and C4 plants) in terms of physiological phenomena and interactions but, rather, make fixed adjustments to calculated yields based on experimental observations of plant response to elevated ambient CO_2 concentrations.

Most early studies of climate change impacts on agriculture in both high-technology modern systems and in the traditional systems still practiced in many of the developing countries failed to consider opportunities for adaptation. More recent analyses have shown that modern and traditional systems provide methods of adaptation that can, at least partially, offset negative climate changes and capitalize on positive ones.

Not only can climate change affect agriculture but even efforts to avoid or mitigate climate change may do so. One mitigation strategy involves reducing emissions of the greenhouse gases, carbon dioxide, methane, and nitrous oxide, from the farm. Activities that reduce carbon dioxide emissions involve improvement of on-farm fossil fuel use and incorporation of organic residues in the soil. Methane emissions per unit of paddy rice production can be reduced through use of new cultivars and fertilization practices. Improved nutrition for bovine animals can reduce their emissions of methane per unit of meat or milk produced. Better fertilization practices and the use of nitrification inhibitors can increase crop production per unit of nitrous oxide emitted from the soil. Most of the strategies proposed to reduce on-farm greenhouse gas emissions should actually increase the productivity and profitability of agriculture.

Two other strategies proposed for mitigating climate change may, on the other hand, impact agriculture negatively. Large-scale afforestation and biomass production have been proposed to draw carbon dioxide from the atmosphere and either sequester it in trees or recycle it as a substitute for some portion of the fossil fuel now being used. Afforestation and biomass plantation on a large enough scale to significantly mitigate climate change will require vast areas of land. Analyses show that marginal lands will not suffice and that good lands will need to be diverted from agriculture for this purpose.

Mitigation strategies also place a premium on non-CO_2- emitting alternatives to fossil fuels. Hydropower generation emits no CO_2 and is a high-value use of water. Irrigation is a low-value use. Competition between hydropower and irrigation is already intensifying for reasons having nothing to do with climate change. Hydropower will compete even more strongly for water resources as its price rises in response to cutbacks in fossil fuel use. Irrigation will be the loser.

REFERENCES

Allen, SG, SB Idso, BA Kimball, JT Baker, LH Allen, Jr., JR Mauney, JW Radin, and MG Anderson. **1990**. *Effects of Air Temperature on Atmospheric CO_2 - Plant Growth Relationships*, Report DOE/ER-0450T (TRO48). U.S. Department of Energy, Office of Energy Research, Washington, DC.

Bach, W. **1979**. The Impact of Increasing Atmospheric CO_2 Concentrations on the Global Climate: Potential Consequences and Corrective Measures. Environ. Int. 2:215-228.

Blasing, TJ and AM Solomon. **1982**. *Response of the North American Corn Belt to Climatic Warming*, Publication 2134. Oak Ridge Laboratory, Environmental Sciences Division, Oak. Ridge, TN.

Butcher, WR, PR Wandschneider, and NK Whittlesey. **1986**. Competition between Irrigation and Hydropower in the Pacific Northwest,

Chapter 2, pp. 25-66. *In: Scarce Water and Institutional Change*. KD Frederick, (ed.). Resources for the Future, Washington, DC.

CAST. **1992**. *Preparing U.S. Agriculture for Global Climate Change*, Task Force Report No. 119. Council for Agricultural Science and Technology, Ames, IA.

Crosson, PR and NJ Rosenberg. **1989**. Strategies for Agriculture. (Special Issue: Managing Planet Earth). Sci. Am. 260:128-135.

Cure, JD. **1985**. Carbon Dioxide Doubling Responses: A Crop Survey, pp. 99-116. In: *Direct Effects of Increasing Carbon Dioxide on Vegetation*. BR Strain and JD Cure (eds.), DOE/ER-0238. U.S. Department of Energy, Carbon Dioxide Research Division, Washington, DC.

Easterling, WE, III, PR Crosson, NJ Rosenberg, MS McKenney, LA Katz, and KM Lemon. **1993**. Agricultural impacts of and responses to climate change in the Missouri-Iowa-Nebraska-Kansas region. Clim. Change 24:23-61.

Frederick, KD. **1993**. Climate change impacts on water resources and possible responses in the MINK region. Clim. Change 24:83-115.

Gleick, PH. **1990**. Vulnerability of Water Systems, Chapter 10, pp. 223-240. *In: Climate Change and U.S. Water Resources,* PE Waggoner (ed.). J. Wiley & Sons, New York.

IPCC. **1990**. *The IPCC Scientific Assessment,* Working Group I. JT Houghton, GJ Jenkins, and JJ Ephraums (eds.). Intergovernmental Panel on Climate Change, Cambridge University Press, Cambridge, UK.

IPCC. **1992**. *Climate Change 1992: The Supplementary Report to the IPCC Scientific Assessment*. JT Houghton, BA Callander, and SK Varney (eds.). Intergovernmental Panel on Climate Change Cambridge University Press, Cambridge, UK.

Jodha, NS. **1989**. Potential Strategies for Adapting to Greenhouse Warming: Perspectives from the Developing World, Chapter 11, pp. 147-158. *In: Greenhouse Warming: Abatement and Adaptation,* NJ Rosenberg, WE Easterling III, PR Crosson, and J Darmstadter (eds.). Resources for the Future, Washington, DC.

Jones, CA and JR Kiniry (eds.). **1986**. CERES-Maize: A Simulation Model of Maize Growth and Development. Texas A & M University Press, College Station, TX.

Jones, JW, KW Boote, SS Jagtap, G Hoogenboom, and GG Wilkerson. **1988**. *Soybean Crop Growth Simulation Model. User's Guide*, SOYGRO V5:41, p. 53. Florida Agricultural Experimental Station, J. No. 8304, IFAS, University of Florida, Gainesville, FL.

Keyfitz, N. **1989**. The Growing Human Population. Sci. Am. 260:118-126. (Special Issue: Managing Planet Earth).

Kimball, BA. **1986**. Influence of Elevated CO, on Crop Yield. Vol. II (Physiology, Yield, and Economics). pp. 105-115. *In: Carbon Dioxide Enrichment of Greenhouse Crops,* HZ Enoch and BA Kimball (eds.). CRC Press, Boca Raton, FL.

Marland, G. **1988**. *The Prospect of Solving the CO_2 Problem Through Global Reforestation*, pp. 66. DOE/NBB-0082 (TRO39). U.S. Department of Energy, Washington, DC.

Moore, M. **1993**. India's Lifeline or Man-Made Disaster? *Washington Post,* p. A12.

Moulton, RJ and KR Richards. **1990**. *Costs of Sequestering Carbon Through Tree Planting and Forest Management in the United States*. p. 46, GTR WO-58. Forest Service Report, U.S. Department of Agriculture, Washington, DC.

Newman, JE. **1982**. Impacts of Rising Atmospheric Carbon Dioxide Levels on Agricultural Growing Seasons and Crop Water Use Efficiencies, Vol. II, Part 8. In: *Environmental and Social Consequences of a Possible CO_2-Induced Climate Change*. U.S. Department of Energy, Carbon Dioxide Research Division, Washington, DC.

NOVEM. **1992**. *The Feasibility of Biomass Production for the Netherlands Energy Economy.* Report 9210, pp. 100 & Annex. Netherlands Agency for Energy and the Environment, Apeldoorn, The Netherlands.

OTA. **1993**. Potential Environmental Impacts of Bioenergy Crop Production, OTA-BP-E-118, pp. 71, Office of Technology Assessment. U.S. Government Printing Office, Washington, DC.

Parry, ML, TR Carter, and NT Konijn. **1988**. *The Impact of Climatic Variations on Agriculture.* Kluwer Academic Publishers, Dordrecht, The Netherlands.

Ritchie, JT and S Otter. **1985**. Description and Performance of CERES-Wheat: A User-Oriented Wheat Yield Model. In: *ARS Wheat*

Yield Project, ARS-38:159-175, WO Willis (ed.). U.S. Department of Agriculture, Washington, DC.

Rosenberg, NJ. **1982**. The increasing CO, concentration in the atmosphere and its implication on agricultural productivity; II. Effects through CO_2- induced climate change. Clim. Change 4:239-254.

Rosenberg, NJ. **1992**. Adaptation of agriculture to climate change. Clim. Change 31:385-405.

Rosenberg, NJ, ed. **1993**. Towards an integrated assessment of climate change: The MINK Study. Clim. Change 24:1-173.

Rosenberg, NJ and PR Crosson. **1991**. The MINK Project: A new methodology for identifying regional influences of, and responses to, increasing atmospheric CO_2 and climate change. Environ. Conserv. 18:313-322.

Rosenberg, NJ and MJ Scott. Implications of Policies to Prevent Climate Change for Future Food Security. Global Environ. Change (in press).

Rosenberg, NJ, BL Blad, and SB Verma. **1983**. *Microclimate: The Biological Environment,* p. 495, John Wiley, New York.

Rosenberg, NJ, BA Kimball, P Martin, and CF Cooper. **1990**. From Climate and CO_2 Enrichment to Evapotranspiration, Chapter 7, pp. 151-175. In: Climate Change and U.S. Water Resources, PE Waggoner (ed.). John Wiley, New York.

Rosenzweig, C, ML Parry, G Fischer, and K Frohberg. **1993**. *Climate Change and World Food Supply,* Research Report No. 3, p. 28. University of Oxford, Environmental Change Unit, Oxford, UK.

Scudder, T. **1989**. Conservation vs. development: River basin projects in Africa. Environment 31:4-9, 31:27-32.

Sedjo, RA and AM Solomon. **1989**. Climate and Forests, Chapter 7, pp. 105-119. *In: Greenhouse Warming: Abatement and Adoption,* NJ Rosenberg, WE Easterling III, PR Crosson, and J Darmstadter (eds.). Resources for the Future, Washington, DC.

Smith, JB and D Tirpak (eds.). **1989**. *The Potential Effects of Global Climate Change on the United States.* U.S. Environmental Protection Agency, Office of Policy, Planning, and Evaluation, Washington, DC.

Stewart, BA and DR Nielsen (eds.). **1990**. *Irrigation of Agricultural Crops*. American Society of Agronomy, Madison, WI.

Thompson, LM. **1969**. Weather and technology in the prediction of corn in the U.S. cornbelt. Agron. J. 61:453-456.

Thompson, LM. **1970**. Weather and technology in the production of soybeans in the central United States. Agron. J. 62:232-236.

Thompson, LM. **1975**. Weather variability, climate change, and grain production. Science 188:535-541.

Tyson, S. **1990**. *Biomass Resource Potential of the United States*. Solar Energy Research Institute, Golden, CO.

UNFPA. **1990**. *The State of World Population, 1990. Choices for the New Century,* p. 34, United Nations Population Fund, United Nations, New York.

Waggoner, PE. **1983**. Agriculture and a Climate Changed by More Carbon Dioxide. In: *Changing Climate*, Report of the Carbon Dioxide Assessment Committee, Board of Atmospheric Sciences and Climate. National Academy Press, Washington, DC.

Wilks, DS. **1988**. Estimating the consequences of CO_2 - induced climatic change on North American grain agriculture using general circulation model information. Clim. Change 13:19-42.

Williams, JR, CA Jones, and PT Dyke. **1984**. A modeling approach to determining the relationship between erosion and soil productivity. Trans. Am. Soc. Agric. Eng. 27:129-144.

CHANGE IN PRECIPITATION DUE TO CO_2 INCREASE AND ITS IMPACT ON AGRICULTURAL PRODUCTION IN CHINA

H. -X. Cao

Chinese Academy of Meteorological Sciences,
Beijing, Peoples' Republic of China

Key words: *Greenhouse effect, climate change precipitation simulation, climate impact, monsoon*

ABSTRACT

Simulations of monthly precipitation in China were compared with actual observations by using the following general circulation models: UKMO, NCAR, GISS, GFDL and CCC. Conclusions, mainly from the simulation using the UKMO model, are as follows:

1. The simulated seasonal variation in precipitation for the current climate is in broad agreement with observations, particularly with respect to the large-scale distribution pattern. However, for specific localities, the discrepancy between simulated and observed precipitations is significant.
2. The model was then used to simulate the climate of China for a world with a doubled concentration of atmospheric CO_2. In the simulation, some parts of China became wetter and some drier. Parts of China that are currently subject to drought (e.g., northwest China) became extraordinarily dry in the simulations. Based on simulated changes in precipitation and temperature, the effects of a $2 \times CO_2$ climate on the production of rice, wheat, and maize are examined. These effects show a pronounced regional variation.

INTRODUCTION

Most parts of China have a "monsoon" climate, comprising great seasonal variation in both the geographical distribution and intensity of precipitation. The zonally distributed precipitation moves northward during spring and summer, becoming heavier. In fall and winter, precipitation recedes southward, weakening as it moves. The seasonal variation in precipitation affects agriculture, forestry, and even socioeconomic factors.

Climatic effects due to the accumulation of carbon dioxide (CO_2) have been investigated in many studies. Most studies focus on the global greenhouse effect, and several papers on regional climate changes have been published (Rind, 1989; Cao et al., 1992; Cao, 1993; Zhao and Ding, 1990). In this paper, our main goal is to determine the spatial and temporal changes in precipitation as a result of CO_2-doubling. This will provide a basis on which to study the impact of CO_2-doubling on climate, allowing policymakers to plan a response strategy.

In comparison with temperature and pressure, precipitation has much greater variability, therefore, simulating precipitation is quite difficult. Studies of simulations generated by the United Kingdom Meteorological Office (UKMO), the U.S. National Center for Atmosphere Research, (NCAR), the Goddard Institute for Space Studies, USA (GISS), the Geophysical Fluid Dynamics Laboratory, USA (GFDL), and Canadian Climate Centre (CCC) models are compared, concentrating on simulated monthly precipitation over China and adjoining areas. Because precipitation simulated with the UKMO model is in good agreement with actual observations, our investigation into the seasonal variation in precipitation and changes due to CO_2-doubling is based on the results simulated with the UKMO model.

COMPARISON OF SIMULATED AND OBSERVED PRECIPITATION

Precipitation simulated with the five models listed above is compared with observations by Shea (1986). The most accurate was that generated with the UKMO model. Figure 1 shows simulations and observations for January. The NCAR and GFDL simulations for January demonstrate an overestimated precipitation rate (mm/day). The GISS and CCC simulations for January show heavier precipitation in West China and less in East China, just the opposite of the observations. We cannot expect a global climate model to make a detailed simulation of precipitation in a region as large as China, because the resolution of the model is still quite coarse. Besides, precipitation in China is scarce in January, adding to the difficulty of simulation.

Precipitation simulated with the UKMO model is also accurate in July. The simulated rain zone is approximately zonal, stretching 5 mm/day over the Yangtze River northward. The observed precipitation reached even farther north. Simulation in Northwest China, North China, and Northeast China are in good agreement with observations. However, the NCAR model overestimates the precipitation in July. In the GISS simulation, the rate of precipitation (5 mm/day) extends much more to Northeast China. In simulations with the GFDL and the CCC models, the rain zone stretches too far north and west in comparison with

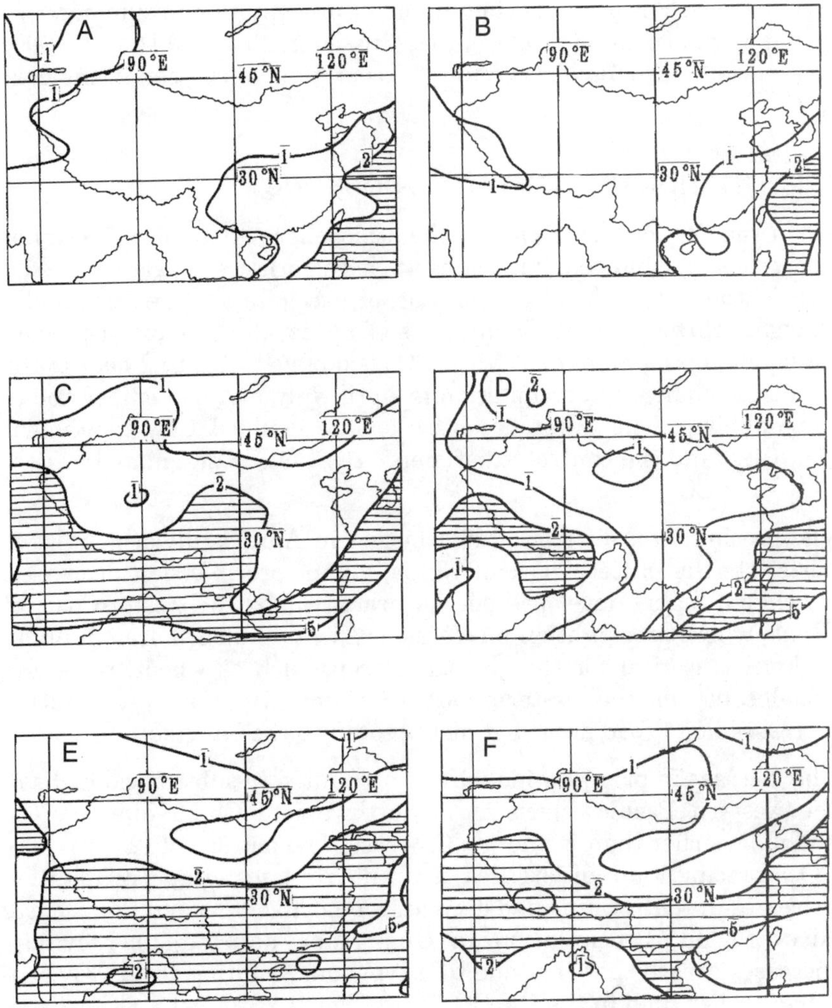

Figure 1. Observation and simulations of precipitation mm/(day) in January: (A) the United Kingdom Meteorological Office; (B) Chart from Shea (1986); (C) the Goddard Institute for Space Studies, USA; (D) Canadian Climate Centre; (E) the U.S. National Center for Atmospheric Research; (F) the Geophysical Fluid Dynamics Laboratory, USA.

observations. The accurate simulation for precipitation in summer generated by the UKMO model is also discussed in Zhao and Ding's (1990) paper (see Figure 5). We therefore focus on that simulation in the next section.

PRECIPITATION CHANGE ON CO_2-DOUBLING

When the CO_2 concentration in the atmosphere is doubled, precipitation zones are changed, but the basic characteristics of seasonal variation in the geographical distribution of precipitation are not greatly changed. However, in different parts of China, the amount of precipitation changes a great deal when CO_2 is doubled. Figure 2 depicts the simulated change in precipitation in April, July, and September due to CO_2-doubling. The difference between the doubled CO_2 experiment ($2 \times CO_2$) and the control experiment ($1 \times CO_2$) is designated as ($2 \times CO_2 - 1 \times CO_2$).

The precipitation difference distribution in April is divided into two parts: (1) the eastern part of China, where precipitation decreases, which is a highly developed agricultural area; (2) the western part of China, where precipitation increases, is a less developed area. Rainfall in April is critical for the growth of crops such as wheat, rape, and broadbeans, in the eastern part of China. In May, precipitation increases in a broad area as it moves south to the Yangtze River.

The increase in precipitation may be attributed to subtropical highs in the western Pacific which move northwest to China due to CO_2-doubling earlier than in the $1 \times CO_2$ case. The pattern of ($2 \times 0_2 - 1 \times CO_2$) precipitation differences in June reverses, i.e., precipitation increases in south China and decreases in north China and the Yangtze River region. Precipitation in Guangdong and Fujian provinces increases greatly, possibly due to a typhoon, resulting from CO_2-doubling, earlier than in the $1 \times CO_2$, case. The ($2 \times CO_2 - 1 \times CO_2$) distribution of precipitation is similar in July and August, i.e., precipitation increases over most of China and decreases from northwest China to the upper Yangtze River valley. CO_2-doubling causes serious drought, because these areas are usually short of precipitation. The lack of rain extends southward in September after CO_2-doubling. This means that the special rainy season vanishes in September over and around Sichuan Province, the southern part of Shanxi Province, and the eastern part of Gansu Province. [The season is called the Huaxi Qiuyu in Chinese (the rainy season over west China in Autumn)].

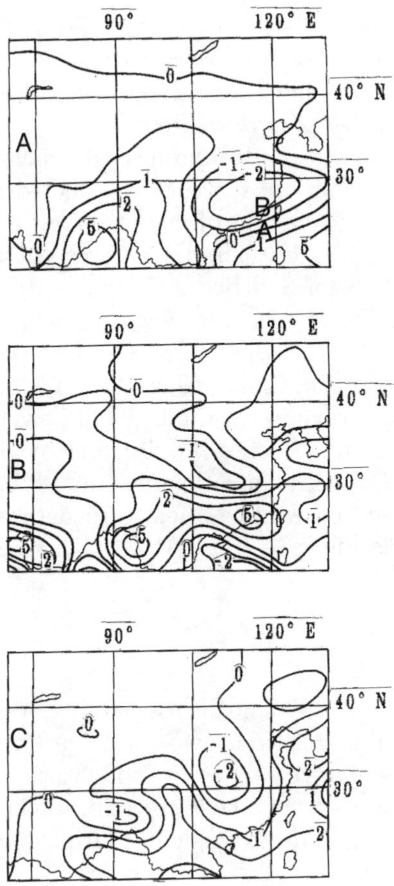

Figure 2. Changes in simulated precipitation (A, April; B, July; C, September) as a result of CO_2-doubling.

IMPACT ON AGRICULTURE

Scenarios of simulated climates show that temperature in China will increase, accompanied by a small increase in precipitation. Therefore, in a warmer winter, growing season becomes longer, enhancing growth of crops. However, growth is also determined by other factors such as an effective water supply and the growing processes and development features of crops.

Other studies of the impact of climate change from CO_2-doubling have been made by Gao et al. (1991, 1993). The effects of climate change on crops such as wheat, rice, and corn are quite different in various regions of China; for some regions, the effect is positive; for other regions, it is negative. Applying regressive analysis between meteorological elements and yields, meaningful results have been obtained. For example, rice production will increase: the meteorological production increment per mu (Chinese unit of measure for a land area: 15 mu = 1 hectare) will be 6%, 2%, 1.6%, and 1.8%, respectively, in comparison with that in the 1980s for Southwest China, South China, East China, and Central China. For corn, meteorological production increments per mu will be 7%, 1%, 2%, and 3.7%, respectively, in comparison with that in the 1980s for Northeast China, North China, Northwest China, and Southwest China. For corn in the Central and Lower Yangtze River valley and South China, increments will decrease by 3% and 20%, respectively. When the CO_2 concentration in the atmosphere is doubled, meteorological yields of winter wheat will decrease 97.20 kg/ha in North China, and 19.80 kg/ha in Northwest China compared with that in the 1980s.

CONCLUSIONS

Some parts of China will become drier and others will become wetter as a result of CO_2-doubling. Some regions that are now dry, for example Northwest China, will have persistent drought, threatening people and the environment. Conversely, where precipitation becomes abundant as a result of CO_2-doubling, for example in south China, there will be terrible floods.

It would seem that a warmer climate due to elevated CO_2 would benefit agricultural production. However, the effects of changes in precipitation are complicated, particularly because precipitation in most parts of China would decrease as a result of doubling CO_2. Therefore, rice planting would decrease and wheat could not acquire enough water.

ACKNOWLEDGMENTS

I am grateful to John F.B. Mitchell, Paul Witefield, and Joe R. Lavery for their great help when I worked in the Intergovernmental Panel on Climate Change, Working Group I, England, where the simulation charts were processed. I thank my student, Gou-Lin Feng, for typing the manuscript; and my colleagues, Shu-Feng Wang, Zong-Xiu Liu, and Qi- Kuang Wu, for preparation of the figures.

REFERENCES

Cao, H-X, JFB Mitchell, and JR Lavery. **1992**. Simulated diurnal range and variability of surface temperature in a global climate model for present and doubled CO_2 climates. *J. Climate 5*:920-943.

Cao, H-X. **1993**. Simulated seasonal movement of precipitation zone in rainy season over China and its change on CO_2-doubling. *J.Environ.-Sci. (China) 5*:150-157.

Gao, S-H, J-P Guo, and Y-R Pan. **1993**. Impacts of climate change on wheat development and production in the Northern China. *J.Environ. Sci. (China) 5*: 176-185.

Gao, S-H et al. **1991**. Possible impacts of greenhouse effect due to CO_2 increase in the atmosphere on rice and corn productions (in Chinese). In: *The Experiment and Study of Impacts of Climate Change in China on Agriculture*, pp. 96-99. China Meteorological Press, Beijing.

Rind, D, R Golgberg, and R Tuedy. **1989**. Change in climate variability in the 21st century. Clim. Change 14:5-37.

Shea, DJ. **1986**. *Climatological Atlas*. 1950-1979, NCAR/TN-269+STR.

Zhao, Z-c. and Y-H Ding. **1990**. Sensitivity experiments and assessment of climate changes in China induced by greenhouse effect. J. Environ. Sci. (China) 2: 73-84.

ASSESSING ECOSYSTEM RESPONSES TO CLIMATE CHANGE

ECOSYSTEM RESPONSES TO CLIMATE CHANGE

G. H. Orians

Department of Zoology and Institute for Environmental Studies, University of Washington, Seattle, Washington

Key words: *Physiology, migration, range shifts, evolution, productivity*

ABSTRACT

Climate changes pose potential threats to biological diversity, the functional integrity of ecosystems, and the ability of those ecosystems to provide.the goods and services upon which humanity depends. Predictions of climate change are still imprecise, particularly at local levels, but predictions of ecosystem responses to those changes are even more uncertain. Nevertheless, because the ecological effects of all environmental changes are functionally altered through individual organisms, an approach that combines models of responses of individual organisms with models of population dynamics and interactions among populations can provide a solid foundation upon which to build powerful predictive theories of ecosystem responses. I will compare models of physiological and behavioral responses of different types of organisms to the physical environment and explore how those models can inform our understanding of species distributions, population crashes, and interspecific interactions. These population responses, in turn, affect such ecosystem processes as primary production, energy transfers among trophic levels, and efficiency of nutrient cycling, the processes that generate ecosystem services. I will also discuss the factors that limit the rates at which species can respond evolutionarily to changing climates and suggest which species are likely to do so and at what rates. Finally, I will discuss the rate at which ranges of organisms will have to change H they are to track the movement of climatic zones across the face of the Earth. I will show that, for many organisms, the required rate is much faster than the probable maximum rate.

INTRODUCTION

The likelihood, during the next century, of substantial climate changes, driven by atmospheric increases of "greenhouse gases," has stimulated a wide variety of interests and concerns. Good estimates of the consequences of climate change are necessary if we are to devise appropriate mitigation strategies and to prepare to adapt to those changes that will occur despite mitigation efforts (NAS, 1991). Unfortunately, projections of climate change are still spatially coarse, and different general circulation models (GCM) give different results. Nonetheless, all models agree that the global average temperature will rise and that warming will be greater at high latitudes than at low latitudes. Most models also predict that the interiors of continents will be drier, whereas coastal regions may receive increased precipitation (Schneider, 1993).

Considerable uncertainty also surrounds the likely effects of climate change on living organisms and the ecosystems of which they are a part. There are millions of different species, each of which will respond differently to climate change. In addition, because carbon dioxide is a nutrient for photosynthetic plants, plant functioning changes as a result of increases in atmospheric concentrations of carbon dioxide even in the absence of climate change. For other organisms, increases increases in concentrations of carbon dioxide exert their effects indirectly via their influences on plants and climate change. Because the consequences of increased carbon dioxide concentrations and accompanying climate change are expressed indirectly through complex intra- and interspecific interactions, results from short-term studies of single species under controlled conditions cannot safely be extrapolated to predict long-term effects on complex ecological communities. The existence of these uncertainties makes possible a wide range of estimates of the ecological effects of greenhouse-gas-driven climate change (Idso, 1992; Fajer and Bazzaz, 1992).

In this paper, I summarize what is known about the direct responses of plants to changes in concentrations of atmospheric carbon dioxide. I review efforts to assess the consequences of climate change for organisms living under natural conditions and explore the ecosystem-level consequences of those changes.

RESPONSES OF PLANTS TO CO_2 FERTILIZATION

Green plants and algae use light energy to transform CO_2 and water into simple sugars. Because all other organic molecules are derived from these sugars, photosynthesis is the ultimate energy source for the lives of most other organisms. Provided that temperature, moisture, and nutrients are not limiting, higher concentrations of CO_2 should increase rates of plant photosynthesis, particularly species using the C_3 pathways (Woodward, 1990). Many short-term experiments have demonstrated that individuals of both agricultural and wild plants exposed to concentrations of 650 to 700 ppm of CO_2—the levels expected a century from now under the Intergovernmental Panel on Climate Change's (IPCC's) Business as Usual Scenario (Houghton et al., 1990)—grow faster than those grown under current atmospheric concentrations (Oechel and Strain, 1985). Wild plants have increased their vegetative growth by an average of 13% and their reproductive output by an average of 31%; grain yields have increased, on average, by 34% (Bazzaz, 1990; Mooney et al., 1991; Strain and Cure, 1985). Also, most plants grown in higher CO_2 concentrations consume less water per unit leaf area because more CO_2 diffuses into a leaf for each molecule of water lost. Therefore, the length of time the stomata—the openings in the leaves through which water exits and CO_2 entersneed to be open is inversely correlated with the atmospheric CO_2 concentration. Water loss rates have been reduced by as much as 30% for soybeans and 50% for maize under laboratory experiments using CO_2 concentrations double the current ones (Carlson and Bazzaz, 1989). Increased atmospheric concentrations of CO_2 may also protect plants from damage from some atmospheric pollutants because fewer of those molecules enter the interiors of leaves (Carlson and Bazzaz, 1989).

Most plant species so far tested increase their growth rates in CO_2-rich atmospheres, but species do not respond equally. Specifically, plants that photosynthesize using the C_3 pathways (wheat, rice, most cool-weather herbs, all trees) usually increase their growth rates much more in CO_2-rich environments than those species using the C_4 pathways (corn, sugarcane, some hot-desert shrubs, many hot-climate grasses), because photosynthetic rates of C_4 plants are CO_2-saturated at current atmospheric concentrations (Woodward, 1990). Also, within each group there is considerable interspecific variation in responses to enrichment, the reasons for which are unclear. These variations are increased when plants are grown together in mixtures at high densities.

Although a CO_2-rich atmosphere clearly increases short-term growth rates of plants under conditions of abundant nutrients and favorable temperatures, the long-term effects of CO_2 enrichment cannot be deduced by simple extrapolation from such experiments. Few experiments have been performed in the field on large scales for periods of several years or more because of the technical difficulties and expense of conducting experiments at those scales. The CO_2 enrichment of an intertidal marsh grass ecosystem and enrichment of sour orange trees on experimental plots increased growth rates over several growing seasons (Ziska et al, 1990; Curtis et al., 1989; Idso and Kimball, 1992), but growth rates of Alaskan tundra sedges declined after an increase during the first 2 yr of enrichment (Tissue and Oechel, 1987). Similarly, growth rates of eastern North American deciduous trees tested under greenhouse conditions declined after an initial increase (Norby et al., 1992; Bazzaz and Miao, 1992). Although the database is still small, the increased growth rates theoretically possible under CO_2 enrichment have been sustained only in nutrient-rich ecosystems. In nutrient-poor ecosystems, other factors have quickly become limiting and acted to return growth rates to their pre-enrichment levels even if high CO_2 levels were maintained.

PLANT RESPONSES TO CHANGING CLIMATES

Any changes in plant performance due to the fertilization effect of increased CO_2 will be modified by changes in climate that accompany enrichment. Temperature has numerous effects on plant growth due to its pervasive role in the regulation of biochemical reaction rates, morphogenetic processes, and matter and energy exchange with the environment (Long and Woodward, 1988). However, the effects on growth of increasing temperature and increasing CO_2 may operate through different pathways and have contrasting results. Coleman and Bazzaz (1992) demonstrated that CO_2 and temperature interact strongly to influence resource acquisition and growth in two annual plant species. Plant responses to increased CO_2 are also influenced by light, nutrients, SO_2, exposure, and competition (Zangerl and Bazzaz, 1984; Bazzaz and Garbutt, 1988; Bazzaz et al., 1989). These complexities may help to explain some of the conflicting results obtained in different experiments. They also emphasize that the long-term effects of changing climates will vary among species and will depend on changes in many environmental factors.

During short-term studies of plants' physiological responses to changing CO_2 concentrations and temperature, the plants do not have time to evolve. However, if the plants of a region are subjected over long time periods to CO_2-enrichment and changed climates, they are certain to evolve adaptations to those new conditions. Stomatal densities in several species of high-latitude plants changed during past glacial cycles concurrently with changes in atmospheric CO_2 concentrations (Beerling et al., 1991). Similar evolutionary changes are to be expected in the future, but the rates of such changes will be so slow, especially among long-lived perennial plants, that they can be neglected during the time frame of the next century. Plant responses to climate change during and after the last ice age were dominated by shifts in range rather than by evolution (Bradshaw and McNeilly, 1991; Huntley, 1991). Similar responses are to be expected during the next century, modified by the fact that climate changes are likely to be unusually rapid, and many barriers to migration exist today. Annual crop plants, however, could be selected to respond to CO_2 enrichment at rates that might increase the ability of agriculture to adapt to global climate change.

ANIMAL RESPONSES TO CHANGING CLIMATES

Available evidence suggests that the maximum concentrations of atmospheric CO_2 predicted for the coming century will not directly influence the physiology of animals. However, climate change will make the world less hospitable to some species and more hospitable to others. A promising approach to predicting the consequences of climate change for animal populations and interactions among species are mechanical models that link environmental variation, organismal physiology, and population dynamics (Dunham, 1993). These models are based on the premise that environmental factors influence birth, death, and migration rates by affecting the physiological ecology of individual organisms and, as a result, their growth and reproductive rates (Dunham et al., 1989; Huston et al., 1988). A mechanistic approach begins by developing biophysical models of how variations in environmental factors affect the physiology and activity of individuals. In the next step, these patterns are integrated across individuals to derive population-level patterns. Finally, feedback dynamics, such as density dependence and interspecific interactions are incorporated into the models to predict the responses of populations to environmental changes. The models also explain why those responses happened. Heritable variations in responses to environmental changes that result in greater survival and reproductive success will be favored by natural selection. Therefore, these models offer insights into probable evolutionary adaptations to changes in future environments (Dunham, 1993).

General energy budget models for small ectotherms combine climatic input variables with the biophysical and physiological properties of the organism to determine the distribution of available thermal microclimates (Grant and Porter, 1992). This distribution constitutes the operative biophysical environment. All activities (thermoregulation, foraging, digestion, mate acquisition, predation) and inactivity are constrained to occur within this distribution (Grant and Dunham, 1988; Grant, 1990). With suitable input parameters, a model of this type has been run for *Sceloporus undulatus,* a small lizard that is widespread in eastern and southwestern United States (Grant and Porter, 1992). Using data from 11 sites across the range of the species, the predicted maximum annual energy available for allocation to eggs, according to the simulations, is generally close to the observed values. A similar model developed for *Sceloporus merriami* in Big Bend, Texas, projected that increases in air temperature of +2°C or +5°C would reduce the time the lizards could be active during the day in summer, thereby lowering reproductive rates enough to threaten survival of the population in some areas (Dunham, 1993).

These models show that individual responses to interacting environmental gradients involve significant age, sex, site, and seasonal interactions among life history characters. By modeling performance of a species under different environmental conditions, such models can throw light on how individuals may respond to changing climates and how their constrained responses may express themselves in altered growth, reproduction, and survival rates and, ultimately, evolutionary changes (Adolph and Porter, 1993).

RESPONSES OF ENDOTHERMS TO CLIMATE CHANGES

Endotherms maintain relatively constant body temperatures by altering the properties of their insulating layers (changing the amounts of fat, adjusting angles of fur or feathers, seasonal changes in quantities of insulation, by increasing heat production by shivering or by nonshivering thermogenesis at low temperatures and by evaporative cooling at high temperatures). The most important components of a homeotherm's response to changes in ambient temperature are: (1) the lower lethal temperature below which metabolic heating is insufficient to replace heat loss, (2) the lower critical temperature below which metabolic heat production must rise to offset heat loss if body temperature is to be stabilized, (3) the upper critical temperature above which metabolic rate rises as the animal uses metabolic power for cooling, and (4) the upper lethal temperature at which the animal dies. Between the lower and upper critical temperatures is a thermoneutral zone within which the animal does not need to expend energy above its basal metabolic rate to maintain its temperature. The consequences of elevated ambient temperatures for endothermic animals are expressed primarily in reduced activity times during the day as the animals seek out and remain in cooler sites above or below ground, and in increased metabolic expenditures and water loss for cooling.

RATES OF EVOLUTION OF TOLERANCES TO HIGH TEMPERATURES

Many small animals have high reproductive rates and short generation times, therefore, they have the potential for rapid evolutionary change. Ability to evolve rapidly could be extremely important for survival during climate change because species with poor dispersal abilities are unlikely to be able to shift ranges rapidly enough to keep up with predicted rates of climate change. Genetic variation in temperature sensitivity, a prerequisite for evolutionary change in response to climate change, has been documented for many animals (Hoffmann and Blows, 1993). Lynch and Lande (1993) modeled the rate of evolutionary change in a population subjected to a gradual, directional environmental change. They show that the lag in adaptation by a large, sexually reproducing population is directly proportional to the rate of environmental change and inversely proportional to the genetic vairance in the polygenic trait that governs the adaptation to the environment. The maximum rate of change that a population can tolerate decreases with decreasing population size, decreasing genetic variance, and increasing environmental variability. Laboratory experiments with *Drosophila* have demonstrated that thermal sensitivity can evolve rapidly under both natural and artificial selection (Huey and Kingsolver, 1993).

Lynch and Lande's model has been used to predict the maximum rates of environmental change that can be tolerated by animals having different thermal performance curves. If genetic variation is independent of the breadth of the performance curve, populations with intermediate values of performance breadth can tolerate the greatest rates of environmental change. Both theoretical and experimental studies suggest that small, genetically relatively uniform populations are more likely to experience difficulty in evolving rapidly under climate change than are larger, more variable populations. This knowledge can be used to identify populations that are most likely to need assistance in either moving or adapting to climate change.

Rates at which Species Ranges Can be Altered

When climate changes, the ranges of most species must also change, because regions that were formerly suitable become unsuitable and formerly inhospitable areas become favorable. The magnitude of range shifts under projected climate changes and the speeds at which species' ranges can shift are therefore matters of substantial concern. Rapid spreads in ranges by introduced species, such as house sparrows (*Passer domesticus*) and starlings (*Sternus vulgaris*) in North America, indicate that under suitable conditions, the ranges of some species can change much more rapidly than will be required under plausible climate change scenarios. However, because rapid changes in the ranges of native species are rare, results from introduced species may not typify responses of most species.

One way to assess abilities of species to shift ranges is to examine responses to past climate changes, recognizing that those changes were much slower than the ones predicted for the coming century. Median global surface air temperature has increased by approximately 4°C during the past 18,000 yr, primarily in response to changes in the distribution of insolation reaching the earth, the melting of ice sheets, and the warming of sea-surface temperatures. This magnitude of change approximates that projected for the next century, although the rate of change was realized over a much longer time period. The record of the past provides useful information about the natural variability of the climate system and how vegetation responded to large climate changes. Several investigators have incorporated such information into projections from GCM to simulate future changes in vegetation patterns (Solomon, 1986; Solomon and West, 1985; Botkin et al., 1991).

Under conditions projected by all three major GCM scenarios [Goddard Institute of Space Studies (GISS), Geophysical Fluid Dynamics Laboratory (GFDL), and Oregon State University (OSU)], the ranges and centers of abundances of dominant species are expected to change dramatically. For example, the ranges of sedges, spruce, birch, pines, oaks, and prairie forbs (nongrasses), species representative of the major vegetation regions of eastern North America, are predicted to be displaced northward by at least 100 km (Overpeck et al., 1991). The GFDL-based results are the most dramatic because that model predicts the greatest summer warming in eastern North America (about 8.5°C). The GISS and OSU results (both projecting about 3.5°C maximum warming)also result in large northward shifts. All three models predict that populations of spruce and northern pines will decrease in abundance. Areas dominated by mixed conifer-hardwood forests would become increasingly deciduous in nature. Ranges of oaks and southern pines may shift northward as much as 500 km. The general result is that regional drying will favor the spread of more xeric vegetation at the expense of more mesic vegetation. The differences between the vegetation distribution for the simulated future climate and the present one are as large as between distributions 7,000 to 11,000 yr ago and those of today. Thus, range changes during the next several hundred years may be larger than the total amount of change that took place during the past 10,000 yr.

Will most species be able to shift their ranges at the rates that appear to be necessary under projected climate changes? Animals with good dispersal abilities, such as most birds and insects, are likely to be able to shift ranges at high rates. By contrast, most mammals, reptiles, and amphibians typically move only short distances during their lives and they have limited abilities to cross barriers. Today's landscapes are replete with highways, urban developments, and cropland that are effective barriers to dispersal of many of these species. Pioneer tree species with small, wind-dispersed seeds, may be able to achieve the necessary rates of movement, but many of the dominant trees of eastern forests have large seeds that fall directly under the tree (oaks, hickories, beeches) and are dispersed only short distances by mammals and birds. Few seeds of these species move farther than 1 km from the adult that produced them. Several decades of growth are necessary before a tree reaches reproductive age. In combination, these facts suggest that the ranges of these species are unlikely to shift faster than a few kilometers per century, an order of magnitude slower than the rates required to keep up with projected climate changes. If populations of these species are not to experience precipitous declines, human intervention to move seeds may be necessary (Graham and Grimm, 1990).

POPULATION INTERACTIONS

As living organisms respond to CO_2 enrichment and climate change, they are altered in ways that affect their interactions with their mutualists, competitors, predators, and pathogens. As indicated previously, species differences among plants in responses to CO_2 enrichment result in shifts in abundances in favor of plants better able to respond to enrichment (Williams et al., 1986, 1988). The implications of these changes for community composition and dynamics are as yet relatively unexplored. In communities dominated by long-lived perennial plants, changes may be slow to express themselves and, therefore, not obvious in the short-term.

The effects of CO_2 enrichment on herbivores are likely to be mediated through changes in the chemical compositions of plant tissues (Ayres, 1993). Plants grown in CO_2-rich atmospheres have lower concentrations of proteins in their leaves than those grown under today's CO_2 concentrations (Lincoln et al., 1986; Fajer et al., 1989). This lowers the nutritional quality of those plant tissues so that insect herbivores eat more of those leaves, presumably to obtain enough protein for growth and development (St. Omer and Horvath, 1983; Garbutt and Bazzaz, 1984). However, insects eating protein-poor leaves grow more slowly, take longer to mature, and have reduced fecundity (Lincoln et al., 1986; Fajer et al., 1989; Osbrink et al., 1987; Johnson and Lincoln, 1990). The consequences of these changes are difficult to predict. They may, but do not necessarily, result in reduced herbivore populations and lowered predator populations. Some of these changes (e.g., in agricultural systems) might be regarded as beneficial from a human perspective, but others will not be.

Little is known about the effects of temperature changes on plant allocation patterns or resistance to herbivores (Ayres, 1993). Rates of photosynthesis, respiration, and growth all vary with temperature. Of particular importance for herbivores, the length of time during which high-quality immature foliage is available, should be inversely correlated with temperature (Pollack, 1990). The consequences of elevated temperature are likely to be negative for endothermic herbivores whose growth rates do not increase with temperature, but insect herbivores, which do grow faster at higher temperatures, may not be adversely affected. If the temperature sensitivity of an herbivore's growth rate differs from that of the plants on which it feeds, significant effects on population dynamics are possible. For example, in a greenhouse experiment, larva of a geometrid moth, *Epirrita autumnata*, responded more rapidly to elevated temperature than the birches upon which they fed. As a result, the rate of growth of the moth population increased up to 2.9 times even though temperatures were elevated by only 1°C (Ayers, 1993).

A likely consequence of changing climate is the loss of some species from the ecosystems in which they currently live. If one or more species is lost from a system, a variety of consequences are possible for the remaining ones. The removal of a keystone species—one that exerts major influences on community dynamics and whose functional role in the community cannot be replaced by some other species—could result in major population changes, perhaps even extinctions, in many other members of its community (Orians and Kunin, 1990). One approach to predicting the effects of species loss is to construct large numbers of plausible food webs, then "surgically remove" species and analyze the consequences for the remaining ones (Pimm, 1979, 1980, 1992). A major conclusion from both general theoretical analyses and empirical studies is that predator-controlled systems are likely to be especially sensitive to removal of their main predators. Systems with these properties typify places where animals compete with plants for space, as do hard-bottom substrates in coastal marine systems. Similarly, in terrestrial environments, large mammals may exert powerful control over vegetation succession. Mammals probably exerted powerful control over vegetation in much of the world until recent times; but today, except for the widespread influence of domestic mammals, communities with these dynamics are confined primarily to Africa (Orians and Kunin, 1990).

Both common sense and experience suggest that some species are much more prone to extinction than others. Species that are particularly prone to extinction include top carnivores, the largest members of ecological guilds (species that use the same resources in similar ways), species with poor dispersal and colonization ability (intolerant of habitat fragmentation), local endemics, species with colonial nesting habits, migratory species, rare species, species dependent on unreliable resources, and species with little recent evolutionary experience with disturbances (Terborgh, 1974; Terborgh and Winter, 1980). Unfortunately, human alterations of landscapes are already endangering many species having these characteristics. Climate change will add yet another stress to the list.

ECOSYSTEM RESPONSES

Prediction of ecosystem responses to changes in the abundances and properties of their constituent species is difficult because of the large number of interspecific interactions, the existence of nonlinear responses of organisms to environmental variables, and the long lag times before responses to changes stabilize. Predictions of future productivity that take into account all of these variables have not yet been attempted. However, models that simulate the relationships of photosynthetic plants to temperature, moisture, and mineral nutrients are being explored. For example, investigators at the Ecosystems Center, Marine Biological Laboratory at Woods Hole, Massachusetts, are simulating the probable effect of climate change, in combination with increases in atmospheric concentrations of CO_2, on ecosystem processes. Of particular importance are changes in the net primary production (NPP), the difference between the carbon fixed during photosynthesis and that released to the atmosphere during respiration.

Under conditions of doubled CO_2 but no change in climate, the Terrestrial Ecosystem Model (TEM) predicts that total terrestrial NPP will increase by 16%. Northern ecosystems, such as tundra and boreal forests, show little response; tropical forests show a considerable increase; and temperate forests show an intermediate increase. The model predicts that the availability of inorganic nitrogen in the soil will continue to limit NPP in most northern regions. However, if both cliamate and CO_2 levels change, the predicted increase in total terrestrial NPP ranges between 20 and 26%, depending on the climate model, and is similar for vegetation at all latitudes. The increases in NPP at high latitudes under these conditions are due to the increased availability of inorganic nitrogen in the soils stimulated by higher temperatures. In tropical regions where vegetation is generally less limited by nitrogen availability, the increases are dominated by the effect of elevated CO_2 (McGuire et al., in press; Melillo et al., in press).

Predictions from the various GCM models are sensitive to expected change in other factors such as cloud cover. For example, TEM run under projections of three of the GCM predict an increase in NPP in southeast Asia; under the OSU model, TEM predicts decreases in NPP in much of the region. This is because the OSU model predicts an increase in cloudiness of 12.5% for tropical forests in the region, whereas the other models do not predict a change in cloud cover.

The marked differences in the TEM predictions for NPP in southeast Asia from the various GCM highlight the sensitivity of predictions to what is and is not included in the models. In general, predictions of changes in cloud cover are very unreliable or are missing. In addition, most models do not account for the effects of a deteriorating climate on performance by long-lived plants in the places they live. Nor do they include lags in population responses to climate change or nonlinear changes in soil nitrogen availability. Incorporation of those factors into models causes large departures from predicted responses during transitions to an equilibrium under an atmosphere with a CO_2 concentration twice that of today (Pastor and Post, 1993). Changing climates may also alter frequencies of disturbances such as tropical storms and fires (Clark, 1988). We can only guess how these disturbances and interactions will modify predictions from models that fail to include them.

MARINE PRODUCTIVITY AND CLIMATE CHANGE

The productivity of marine ecosystems is sensitive to both climate change and atmospheric concentrations of CO_2. Studies of changes in carbon isotopes, species composition of planktonic foraminiferan assemblages, and direct measurement of the organic carbon content of marine sediments, have revealed that major changes in marine productivity accompanied the temperature and CO_2 changes during Quaternary ice-age cycles (Williamson and Holligan, 1990). These studies show that the downward flux of biogenic carbon from the upper ocean was about twice as high during glacial periods as it is now. This caused a reduction in atmospheric CO_2 that was enhanced by circulation changes that increased the alkalinity of polar surface waters.

Global warming is expected to result in warmer, more stable surface layers, especially in temperate oceans that currently experience winter mixing and summer stratification. The surface primary production of such regions is two or three times higher than that of permanently stratified, nutrient-poor subtropical and tropical waters. Therefore, climate warming is predicted to lower primary production as a result of reduced upward mixing of nutrients, lower downward fluxes of organic matter, and increase the efficiency of nutrient cycling in surface waters. All three of these effects would tend to decrease oceanic uptake of carbon in temperate oceans, with possible feedback to rates of climate change. In addition, human activity is causing extreme reductions in populations of marine predators, some of which function as keystone species (Paine, 1993).

CONCLUSIONS

Changes in climate and atmospheric CO_2 concentrations of the magnitude projected for the coming century have happened during recent millennia, although current high rates of change appear to be unprecedented. The world's biota have persisted through past changes, but current rates of environmental change are more rapid than they have ever been before, and the human activities that are driving these changes are also altering the earth in ways that make it more difficult for organisms to adapt to those changes. In the past, as climates changed, range shifts by organisms were accomplished in relatively continuous environments. Today, however, suitable habitats are highly fragmented, and many barriers to dispersal extend across landscapes. Humans today coopt about one-third of global terrestrial primary production and the proportion is steadily rising (Vitousek et al., 1986). The world in which organisms must adapt to climate change is dramatically different from any that has existed in the past, and the differences, for the most part, hinder adaptations. The massive extinctions that are being caused by these changes are likely to continue even in the absence of climate change, but they are certain to increase as this stress is added. Therefore, the probability that continued increases in atmospheric concentrations of CO_2 will result in major changes in the distribution, composition, and functioning of ecosystems is high. The consequences of these changes are sufficiently great that a risk-aversive public policy is warranted (NAS, 1991).

REFERENCES

Adolph, SC and WP Porter: **1993** Temperature, activity, and lizard life histories, Am. Natur. 142,273-295.

Ayres, MP. **1993**. Plant defense, herbivory, and climate change, pp. 75-94. In: *Biotic Interactions and Global Change*, Kareiva, PM, JG Kingsolver, and RB Huey (eds.). Sinauer Associates, Sunderland, MA.

Bazzaz, FA. **1990**. The response of natural ecosystems to the rising global CO_2 levels. Ann. Rev. Ecol. Systematics 21:167-196.

Bazzaz, FA and K Garbutt. **1988**. The response of annuals in competitive neighborhoods: Effects of elevated CO_2. Ecology 69:937-946.

Bazzaz, FA and SL Miao. **1992**. Enhancement of growth in high CO_2 environments declines in 2nd-year tree seedlings. Nature.

Bazzaz, FA, K Garbutt, EG Reekie, and WE Williams. **1989**. Oecologia 79:223-235.

Beerling, DS, WG Chaloner, B Huntley, A Pearson, and MJ Tooley. **1991**. Tracking stomatal densities through a glacial cycle: Their significance for predicting the response of plants to changing atmospheric concentrations of carbon dioxide. Global Ecol. Biogeogr. Lett. 1:136-142.

Botkin, DB, DA Woodbv, and RA Nisbet. **1991**. Kirtiand's Warbler habitats: A possible early indicator of climatic warming. Biol. Conserv. 56:63-78.

Bradshaw, AD and T McNielly. **1991**. Evolutionary response to global climate change. Ann. Botany (Suppl.) 67:5-14.

Carlson, RW and FA Bazzaz. **1989**. The effects of elevated CO_2 concentrations on growth, photosynthesis, transpiration, and water use efficiency of plants, pp. 609-623. In: *Environmental and Climatic Impact of Coal Utilization*, JJ Singh and A Deepak (eds.). Academic Press, New York.

Clark, WS. **1988**. Effects of climate change on fire regimes in northeastern Minnesota. Nature 334:233-235.

Coleman, JS and FA Bazzaz. **1992**. Effects of elevated CO_2 and temperature on growth and resource-use of co-occurring C_3 and C_4 annuals. Ecology 74:1244-1259.

Curtis, PD, BG Drake, PW Leadley, W Arp, and D Whigham. **1989**. Growth and senescence of plant communities exposed to elevated CO_2 concentrations on an estuarine marsh. Oecologia 78:20-26.

Dunham, AE. **1993**. Population responses to environmental change: operative environments, physiologically structured models, and population dymanics, pp. 95-119. In: *Biotic Interactions and Global Change*. Kareiva, PM, JG Kingsolver, and RB Huey (eds.). Sinauer Associates, Sunderlnad, MA.

Dunham, AE, BW Grant, and KL Overall. **1989**. Interfaces between biophysical and physiological ecology and the population ecology of terrestrial vertebrate ectotherms. Physiol. Zool. 62:335-355.

Fajer, ED and FA Bazzaz. **1992**. Is carbon dioxide a 'good' greenhouse gas? Effects of increasing carbon dioxide on ecological systems. Global Environ. Change 2:301-310.

Fajer, ED, MD Bowers, and FA Bazzaz. **1989**. The effects of enriched carbon dioxide atmospheres on plant-insect herbivore interactions. Science 243:1198-1200.

Garbutt, K and FA Bazzaz. **1984**. The effects of elevated CO_2 on plants. III. Flower, fruit and seed production and abortion. New Phytol. 98:433-446.

Graham, RW and EC Grimm. **1990**. Effects of global climate change on the patterns of terrestrial biological communities. Trends Ecol. Evolution 5:289-292.

Grant, BW. **1990**. Tradeoffs in activity time and physiological performance for a thermoregulating desert lizard, *Sceloporus merriami*. Ecology 71: 2323-2333.

Grant, BW and AE Dunham. **1988**. Thermally imposed time constraints on the activity of the desert lizard *Sceloporus merriami*. Ecology 69:167-176.

Grant, BW and WP Porter. **1992**. Modeling global macroclimatic constraints on ectotherms energy budgets. Am. Zool. 32:154-178.

Hoffman, AA and MW Blows. **1993**. Evolutionary genetics and climate change: Will animals adapt to global warming? pp. 165-178 In: *Biotic Interactions and Global Change*. Sinauer Associates, Sunderland, MA.

Houghton, JT, GJ Jenkins, and JJ Ephraums. **1990**. *Climate Change: The IPCC Scientific Assessment*. Cambridge University Press, Cambridge, UK.

Huey, RB. **1982**. Temperature, physiology, and the ecology of retiles, Vol. 12, pp. 25-91. In: *Biology of the Reptilia*, C Gans and FH Pouch (eds.). Academic Press, New York.

Huey, RB and JG Kingsolver. **1993**. Evolution of resistance to high temperature in ectotherms. Am. Natur. 142 (Suppl.):S21-S46.

Huntley, B. **1991**. How plants respond to climate change: Migration rates, individualism and the consequences for plant communities. Ann. Botany (Suppl.) 67:15-22.

Huston, M, D DeAngelis, and W Post. **1988**. New computer models unify ecological theory. BioScience 38:682-691.

Idso, SB. **1991**. Carbon dioxide and the fate of Earth. Global Environ. Change 1:178-182.

Idso, SB and BA Kimball. **1992**. Effects of atmospheric CO_2 enrichment on photosynthesis, respiration and growth of sour orange trees. Plant Physiol. 99:341-343.

Johnson, RH and DE Lincoln. **1990**. Sagebrush and grasshopper responses to atmospheric carbon dioxide concentration. Oecologia 84:103-110.

Lincoln, DE, D Couvet, and N Sionit. **1986**. Response of an insect herbivore to host plants grown in carbon dioxide enriched atmospheres. Oecologia 69:556-560.

Long, SP and FL Woodward, (eds.). **1988**. *Plants and Temperature*. Symposia of the Society for Experimental Biology 42.

Lynch, M and R Lande. **1993**. Evolution and extinction in response to environmental change, pp. 234-250. In: *Biotic Interactions and Global Change*, Kareiva, PM, JG Kingsolver, and RB Huey (eds.). Sinauer Associates, Sunderland, MA.

McGuire, AD, LA Joyce, DW Kicklighter, JM Melillo, G Esser, and CJ Vörösmarty. Productivity response of climax temperate forests to elevated temperature and carbon dioxide: A North American comparison between two global models. Climate Change, in press.

Melillo, JM, AD McGuire, DW Kicklighter, B Moore III, CJ Vörösmarty, and AL Schloss. Global climate change and terrestrial net primary production. Nature, in press.

Mooney, HA, BG Drake, RJ Luxmoore, WC Oechel, and LF Pitelka. **1991**. Predicting ecosystem responses to elevated CO_2 concentrations. BioScience 41:96-104.

NAS. **1991**. *Policy Implications of Greenhouse Warming. Mitigation, Adaptation, and the Science Base*. National Academy of Sciences Press, Washington, D.C.

Norby, RJ, CA Gunderson, SD Wullschleger, EG O'Neill, and MK McCracken. **1992**. Productivity and compensatory responses of yellow-poplar trees in elevated CO_2. Nature 357:322-324.

Oechel, WC and BR Strain. **1985**. Native species responses to increased carbon dioxide concentration, pp. 117-154. In: *Direct Effects Increasing Carbon Dioxide on Vegetation*, BR Strain and JD Cure (eds.). NTIS, Springfield, VA.

Orians, GH and WE Kunin. **1990**. Ecological uniqueness and loss of species, pp. 146-184. In: *The Preservation and Valuation of Biological Resources*, GH Orians, GM Brown, Jr., WE Kunin, and JE Swierzbinski (eds.). University of Washington Press, Seattle, WA.

Osbrink, WLA, JT Trumble, and RE Wagner. **1987**. Host suitability of *Phaseolus lunata for Trichoplusia ni* (Lepidoptera: Noctuidae) in controlled atmospheres. Environ. Entomol. 16:6349-644.

Overpeck, JT, PJ Bartlein, and T Webb 111. **1991**. Potential magnitude of future vegetation change in eastern North America: Comparisons with the past. Science 254:692-695.

Paine, RT. **1993**. A salty and salutary perspective on global change, pp. 347-355. *Biotic Interactions and Global Change*, Kareiva, PM, JG Kingsolver, and RB Huey (eds.). Sinauer Associates, Sunderland, MA.

Pastor, J and WM Post. **1993**. Linear regressions do not predict the transient responses of eastern North American forests to CO_2-induced climate change. Climate Change 23:111-120.

Peters, RH. **1983**. *The Ecological Implications of Body Size*. Cambridge University Press, New York.

Pimm, SL. **1979**. The structure of food webs. Theoret. Pop. Biol. 16:144-158.

Pimm, SL. **1980**. Food web design and the effect of species deletion. Oikos 35:139-149.

Pimm, SL. **1992**. *The Balance of Nature*? University of Chicago Press, Chicago, IL.

Schneider, SH. **1993**. Scenarios of global warming, pp. 9-23. In: *Biotic Interactions and Global Change*, Kareiva, PM, JG Kingsolver, and RB Huey (eds.). Sinauer Associates, Sunderland, MA.

Solomon, AM. **1986**. Transient response of forests to CO_2 induced climate change: Simulation modeling experiment in eastern North America. Oecologia 68:567-579.

Solomon, AM and DC West. **1985**. *Rep. DOE/ER-0236*. U.S. Department of Energy, Washington, D.C.

St. Omer, L and SM Horvath. **1983**. Elevated carbon dioxide concentrations and whole plant senescence. Ecology 64:1311-1314.

Strain, BR and JD Cure. **1985**. *Direct Effects of Increasing Carbon Dioxide on Vegetation*. U.S. Department of Energy, Washington, D.C.

Terborgh, J. **1974**. Preservation of natural diversity: The problem of extinction prone species. BioScience 24:715-722.

Terborgh, J and B Winter. **1980**. Some causes of extinction, pp. 119-133. In: *Conservation Biology: An Evolutionary-Ecological Perspective*, ME Souléand BA Wilcox (eds.). Sinauer Associates, Sunderland, MA.

Tissue, DT and WC Oechel. **1987**. Response of *Eriophorum vaginatum* to elevated CO_2 and temperature in the Alaskan tussock tundra. Ecology 68:401-410.

Vitousek, PM, P Ehrlich, A Ehrlich, and P Matson. **1986**. Human appropriation of the products of photosynthesis. BioScience 326:368-373.

Williams, WE, K Garbutt, and FA Bazzaz. **1988**. The responses of plants elevated CO_2: V. Performance of an assemblage of serpentine grassland herbs. Environ. Exp. Botany 28:123-130.

Williamson, P and PM Holligan. **1990**. Ocean productivity and climate change. Trends in Ecol. Evol. 5:299-303.

Woodward, FI. **1990**. Global change: Translating plant ecophysiological responses to ecosystems. Trends Ecol. Evol. 5:308-311.

Woodward, FI, GB Thompson, and IF McKee. **1991**. The effects of elevated concentrations of carbon dioxide on individual plants, populations, communities and ecosystems. Ann. Botany 67 (Supplement) 23: 23-38.

Zangerl, AR and FA Bazzaz. **1984**. The response of plants to elevated CO_2 II. Competitive interactions among annual plants under varying light and nutrients. Oecologia 62:412-417.

Ziska, LH, BG Drake, and S Chamberlain. **1990**. Long-term photosynthetic response in single leaves of a C_3 and C_4 salt marsh species grown at elevated atmospheric CO_2 in situ. Oecologia 83:469-472.

IMPACTS OF VEGETATION CHANGE ON REGIONAL CLIMATE AND DOWNSCALING OF GCM OUTPUT TO THE REGIONAL SCALE

J. H. Copeland,[1] T. N. Chase,[1] J. Baron,[2,3] T.G.F. Kittel,[3,4] R. A. Pielke[1]

[1]National Park Service Water Resource Division

[1]National Resources Ecology Laboratory, Colorado State University, Fort Collins, Colorado

[4]Climate System Modeling Program, University Corporation for Atmospheric Research Boulder, Colorado

Key words: *Regional climate, climate change, RAMS, vegetation change, downscaling, Colorado, central grasslands*

ABSTRACT

We have begun generating high spatial resolution climate change scenarios of the Central Grasslands Biogeographic Region in order to better understand the sensitivity of the Grasslands' climate to global forcing. We are using the Colorado State University regional Atmospheric Modeling System (RAMS) to perform the downscaling simulations from global to meso-scale. The model configuration covers the continental United States from the Pacific Coast to the Appalachians and from Mexico to Canada with a grid separation of 30 km. The output of RAMS simulations will be used as input for regional vegetation modeling.

Before attempting climate change scenarios, we need to verify that the model is acceptable of recreating the current climate. Extensive testing of RAMS in an operational mode with high resolution data has proven very successful. The climate change scenarios will use GCM output as boundary conditions, which has an average grid separation of 400 to 600 km. It is essential to determine *a priori* that the coarse resolution GCM output has enough information content to generate the appropriate weather systems. We show results of RAMs climate verification simulations for July. These simulations will use NMC analyses as boundary conditions degraded to the resolution of a GCM. To verify the model-generated climatology, we will compare it with analyses of actual station observations for the same period.

This work is being carried out as part of the National parks Service's Global Climate Change Research Program. It's aim is to examine potential climate change scenarios and assess the vulnerability of vegetation in and adjacent to our National Park areas.

INTRODUCTION

The problem of attempting to assess climate change on any scale is complicated by the many feedbacks between the various components of the biosphere. Neglecting one or more of these feedbacks can lead to simulations of potential climate change which may have magnitude and/or sign errors. This is assuming that the scenarios can be verified. In order to assess the possible regional effects of global climate change we have taken the twofold approach of assessing regional sensitivity and downscaling global scenarios.

We will first discuss the importance of local surface characteristics in defining the climate of a particular region. Several examples of regional sensitivity assessments will be presented next. In the third section, various downscaling methods will be reviewed, followed by a section on our work in this area to date. Finally, we close with a summary and suggestions for further research.

INFLUENCE OF SURFACE CHARACTERISTICS

Regional sensitivity assessments involve simulations of typical atmospheric circulations which define the local climate. These circulations are either generated or modified by local gradients of surface characteristics (topography, vegetation, soil moisture, etc.). The topographical effects have long been known and studied. It is only recently that the effects of vegetation on local circulations have received a similar level of attention. The influence of surface characteristics are most pronounced in situations of weak synoptic forcing. This has been the basis for the sensitivity studies presented, although in the future more synaptically active scenarios will also be examined.

Colorado Treeline Study

In order to evaluate possible regional effects of a changing global climate we have performed a series of model simulations, using the Colorado State University Regional Atmospheric Modeling System (RAMS) for each of two initial conditions representing a change in treeline in the Colorado Rocky Mountains (a hypothesized effect of global warming). The model uses three nested grids. The largest grid has a 125-km spacing and covers most of the continental United States in an effort to provide RAMS with as much large-scale input as possible. The points on the second grid are spaced at 25 km. The third grid is centered on Rocky Mountain National Park and includes adjacent areas of the Continental Divide and the Great Plains. Points in the third grid are spaced at 6.25 km. A cumulus parameterization (Weissbluth, 1991) that is applicable at a wide range of horizontal scales and is effective in dealing with partially resolved convection has been incorporated into the model.

This set of experiments modeled a typical late-July day where a large high-pressure region over the western United States minimized dynamically forced convection. Flow from the Gulf of Mexico supplied moisture to the Colorado region. Vegetation was modeled using values of albedo and roughness length appropriate to three vegetation types: shortgrass steppe, montane forests, and alpine tundra. Vegetation distribution was defined as a function of topographic height. The control case (Low Treeline) represents treeline as it is observed today, with a second case (High Treeline) representing an altered local vegetation distribution due to a higher treeline. For example, the height where grassland changed to forest was raised from 1982 m to 2745 m, and that where forest changed to tundra from 3050 m to 3660 m (Table 1). This change in vegetation distribution caused a 0.05 increase in the average albedo in the finest grid and an average decrease in surface roughness of 0.12. Both cases were run for 12 hr.

Table 1. Albedo and roughness length values used to simulate the High Treeline and Low Treeline cases.

Region	Albedo	Roughness	Low (m)	High (m)
Alpine tundra	0.2	0.01	>3050	>3660
Forest	0.1	1.0	>1982 <3050	>2745 <3660
Grasslands	0.3	0.05	<1982	<2745

Figure 1A shows the total number of points experiencing precipitation in the finest grid for each case and is therefore a comparison of the spatial extent of precipitation for each hour. In Figure 1B, total precipitation is averaged over the number of points where precipitation existed at the time indicating precipitation magnitude. The two cases have approximately the same spatial extent of precipitation until about 8 hr into the simulation, when the two cases diverge; the Low Treeline case displays a much larger number of points (up to 200 more points at hour 11) than the High Treeline case. The accumulated precipitation averaged over these points is comparable until the 5th hour, when the High Treeline case shows a larger average precipitation over approximately equal numbers of points. After 8 hr the number of points in the Low Treeline case with precipitation begins to increase over the High Treeline case, while the average precipitation at those points also increases sharply, giving the Low Treeline case higher total precipitation when averaged over the entire domain.

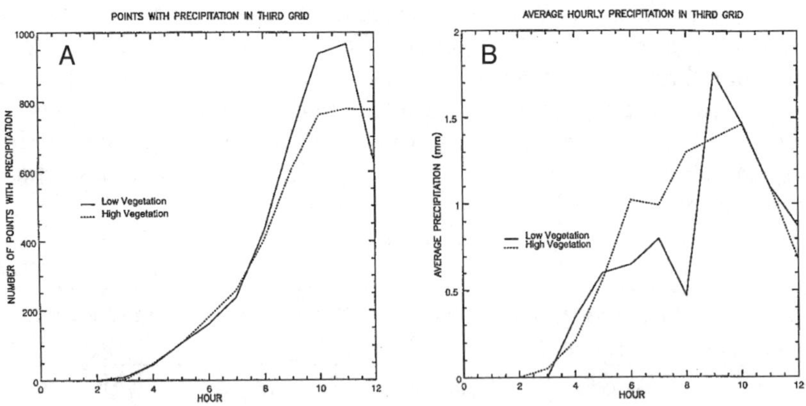

Figure 1. A, Total number of points experiencing precipitation in finest grid: Low Treeline (solid line), High Treeline (dashed line). B, Average hourly precipitation in finest grid for precipitating grid cells: Low Treeline (solid line), High Treeline (dashed line).

The accumulated precipitation over the entire simulation for the Low Treeline case and the High Treeline case (Figure 2) shows that, in general, precipitation patterns are similar for the two simulations. The Low Treeline case shows larger magnitudes in most areas, especially in the southwest quadrant of the domain. This is likely due to the Low Treeline case having large areas of differing vegetation; in the High Treeline case the vegetation is more homogeneous, with grasslands dominant. Discontinuities in albedo can cause mesoscale circulations to develop that can enhance precipitation and enlarge zones of convergence (Knowles, 1993). This can be seen in the near-surface wind fields (Figure 3) for The 9th hour of the simulation. This is when the two cases diverge the most. The Low Treeline case displays larger and stronger zones of convergence at this time. This gives rise to large amounts of precipitation, which are not generated in the High Treeline case. These comparisons indicate that mesoscale changes can have a strong effect on regional weather and, possibly, on local climate.

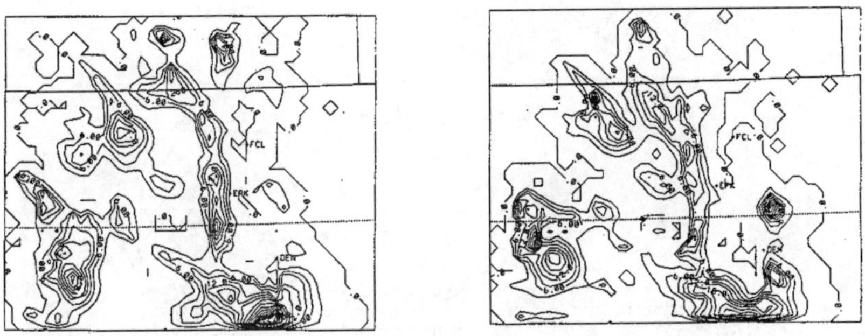

Figure 2. Total accumulated precipitation in finest grid: Low Treeline (left), High Treeline (right).

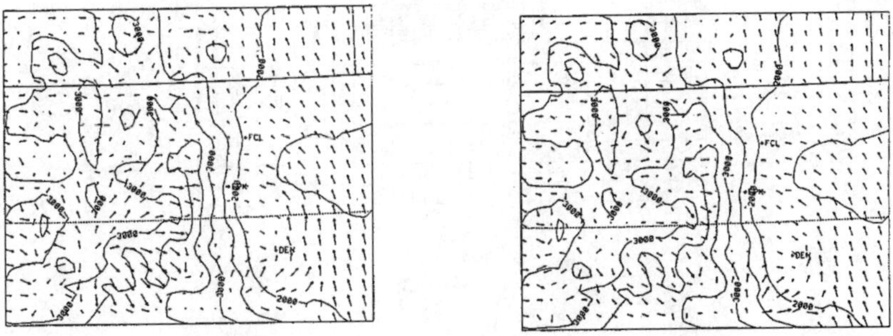

Figure 3. Near-surface, 109.8-m, wind field vectors for finest grid at hour 9: Low Treeline (left), High Treeline (right).

Savannah River Study

The next example of vegetation change influencing local circulations is not as a feedback of a global climate change but rather as a potential input due to human practices. These results are from a study of how vegetation distribution impacts pollutant dispersion around the Savannah River Laboratories (Finley, personal communication, 1993). The RAMS was used with three nested grids (32-km, 8-km, and 2-km horizontal grid separation, Figure 4), centered over the laboratory site. The initial scenario conditions were for May 24, 1990—a typical late-spring high-pressure system located over the southeastern United States, with

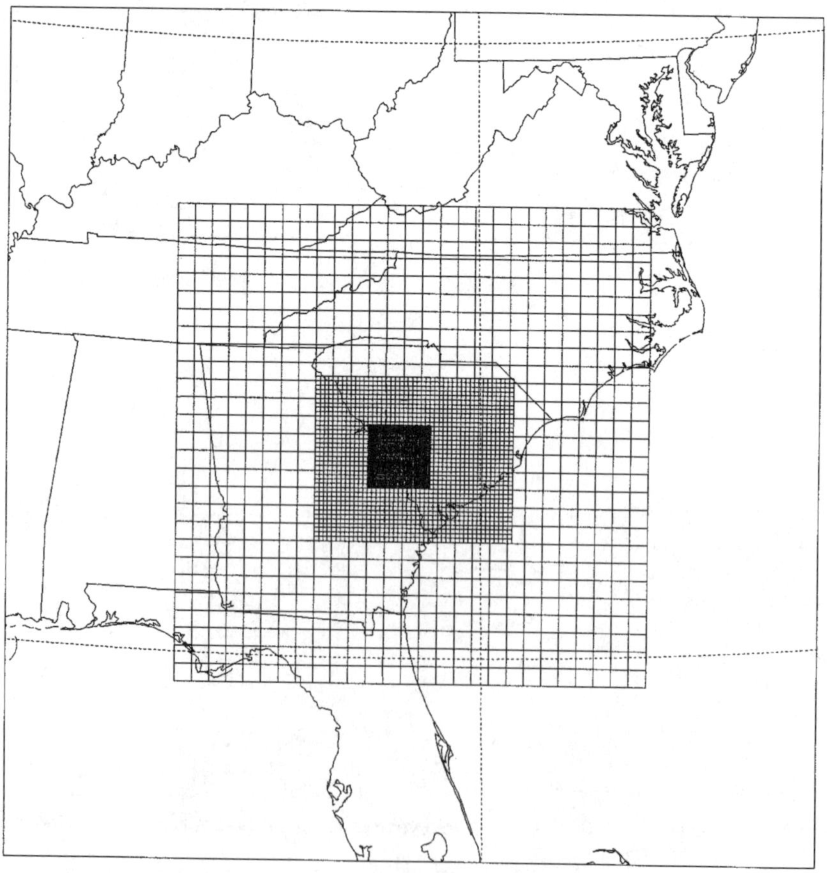

Figure 4. Numerical grids used in Savannah River simulations; grid separations: 32 km, 8 km and 2 km.

weak synoptic forcing. Two cases were run: the first with a bare soil surface, and the second with vegetation derived from the U.S. Geological Survey (USGS) 1-km land-cover database (Loveland et al., 1991). Figure 5 shows the 320-m winds overlaid on topographic contours for both runs. The differences between the runs is most noticeable along the river valley and the highlands to the north.

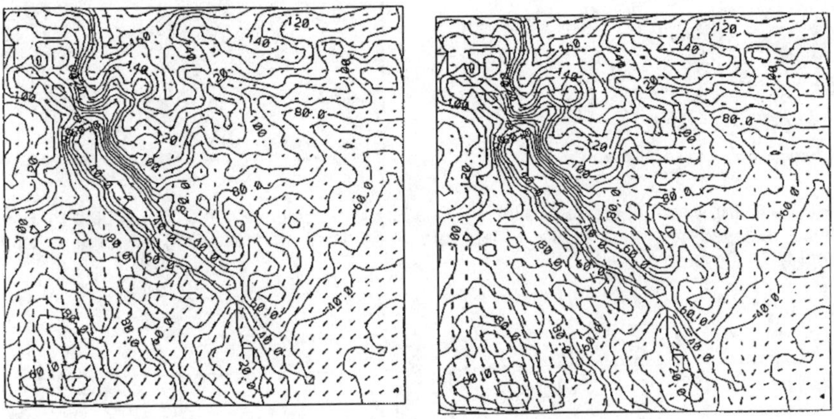

Figure 5. Horizontal wind field on finest grid at 320 m above surface. Local time, 3:00 p.m. Topography contour, 10 m. Bare soil case (left), vegetation case (right).

The output from RAMS was used to drive a Lagrangian particle dispersion model (LPDM). The particles were released near the center of the laboratory site. Figure 6 shows the particle clouds 4 hr after release. The different dispersion patterns are very evident, with the bare soil run maintaining a tight cloud that has headed to the southwest. The vegetation case has a much more diffuse particle cloud that has spread out along the river valley to the south of the laboratory.

DOWNSCALING

Numerous general circulation models (GCM) have been used for years to examine questions related to climate change and sensitivity. One of the shortcomings of these models is their inability to resolve the fine-scale structures of regional climate and weather systems. Figure 7 shows the grid separation, or equivalent grid separation for spectral

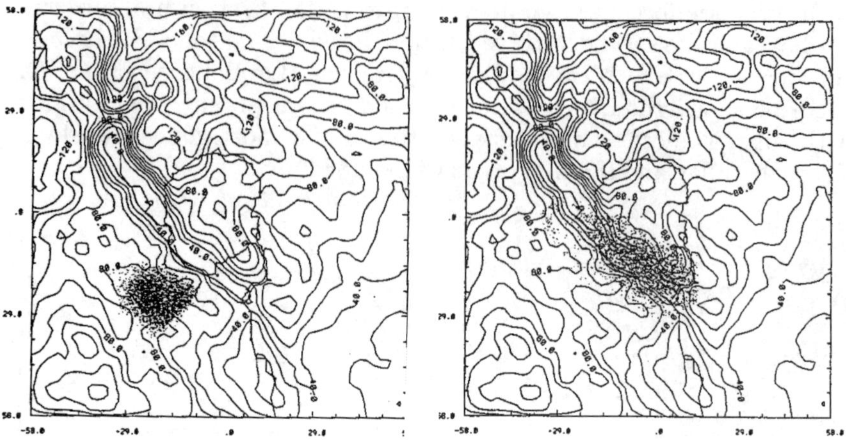

Figure 6. Top view of particle positions 4 hr after start of release; local time, 6:00 p.m. Particles released from near center of laboratory site. Bare soil case (left), vegetation case (right).

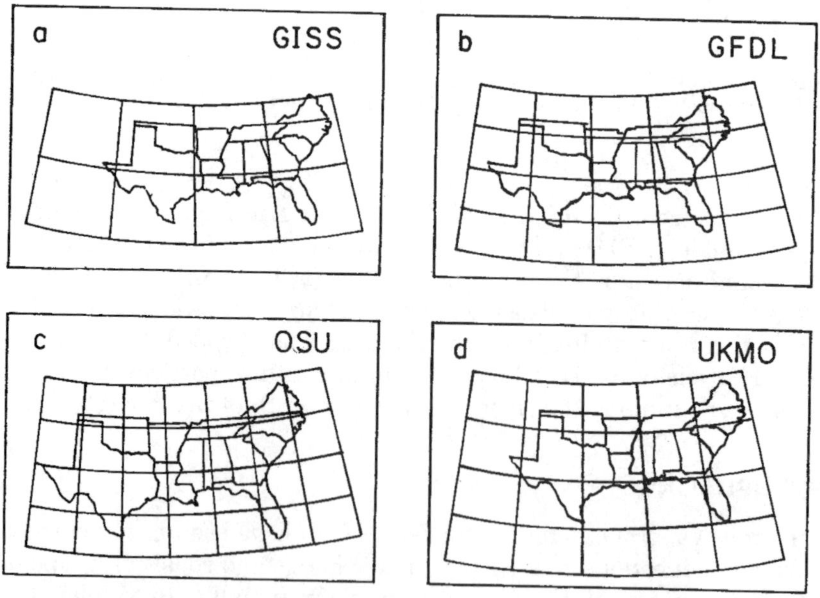

Figure 7. Horizontal resolution of four GCM across southern United States (from Cooter et al., in press).

models, of several GCM over the southeastern United States (Cooter et al., in press). The minimum resolvable scale is four times the grid separation; it is even argued that the minimum skillful resolution is at least eight grid intervals (Grotch and MacCracken, 1991). The atmospheric circulations responsible for the distribution of regional scale climate are the result of systems and gradients that are at the most several hundred kilometers, well below the resolution of current GCM. Surface topographical features and heterogeneities in vegetation and soil moisture which have significant regional impacts in generating and/or modifying atmospheric circulations are also unresolved by the GCM. In order to interpret GCM results on the regional scale one must invoke an interpolation method. Three methods in use are direct, statistical, and dynamic interpolation.

Direct interpolation is the extraction of GCM data that geographically correspond to a climate station of interest. The problem with this method is that the GCM grid is generally unrepresentative of any climate station within its area. Figure 8 (Dumenil, 1993) shows that only by increasing resolution do the trends and magnitudes of precipitation show slight improvement. Statistical interpolation is the use of correlation models to relate regional structures to their parent, large-scale features. Von Storch et al. (1993) used a canonical correlation technique to retrieve mean wintertime Iberian Peninsula precipitation from North Atlantic sea-level pressure patterns. This allowed them to obtain mean precipitation structures that were not resolved by the Max Planck Institute GCM. Dynamic interpolation is accomplished with one-way nesting. Output from GCM model runs can be used as initial and boundary conditions for a limited-area model. The limited-area model, with its finer spatial scale (10- to 100-km grid separation) is better able to resolve the important regional circulations. The western United States is ideal for this approach since the complex topographical forcing which is missing in GCM due to their insufficient spatial resolution (Giorgi, 1990) plays a major role in the regional climate of this area. We have chosen the dynamic interpolation approach. A two-way nested global regional modeling system would allow for the upscale transfer of information in addition to the downscaling we are looking for, but it suffers from being computationally expensive.

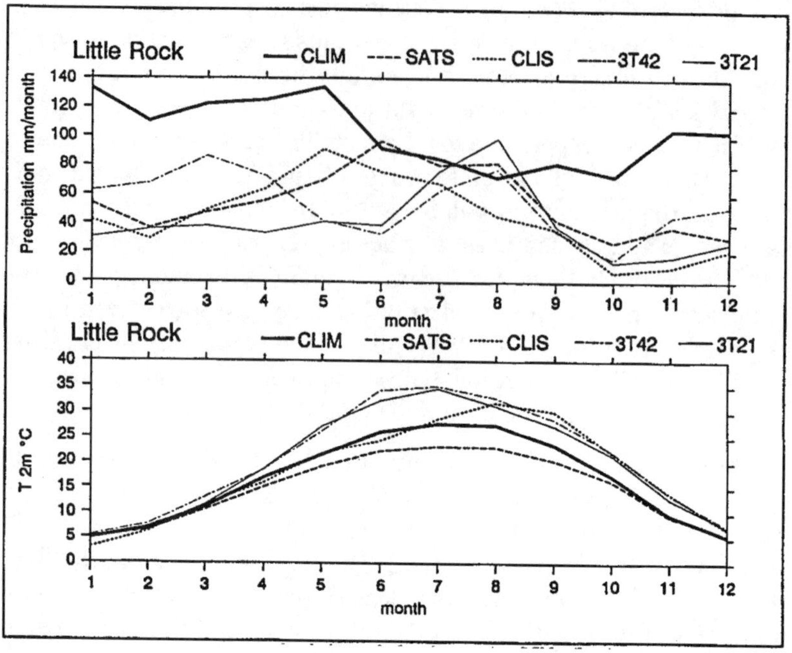

Figure 8. Climatology of Little Rock, Arkansas, compared with the results of the Max Planck Institute model. Little Rock (bold solid), T21 (solid), T42 (dash-dot), CLIS, ideal soil moisture (dot) and SATS, saturated soil (dash). (From Dumenil, 1993.)

RAMS Modifications

We are modifying RAMS (Pielke et al., 1992) to perform downscaling simulations from global to mesoscale in order to better understand the sensitivity of the Central Grasslands climate to global forcing. The model configuration covers the conterminous United States from the Pacific Coast to the Appalachians and from Mexico to Canada, with a grid separation of 30 km. Prior to attempting climate change scenarios we need to verify that the model is capable of recreating the current climate. Extensive testing of RAMS in an operational mode with high-resolution data has proven very successful (Thompson, 1993). The climate change scenarios will use GCM output, which has an average grid separation of 400 to 600 km, as boundary conditions. It is essential to determine *a priori* that the coarse-resolution GCM output has enough information content to generate the appropriate weather systems. The

RAMS climate verification simulations use NMC analyses as boundary conditions, degraded to the resolution of a GCM.

In order to maintain a feasible time frame for the regional climate scenarios we needed to constrain both the grid configuration and the complexity of physical parameterizations. The grid configuration has a constant horizontal grid separation of 30 km on a stereographic grid centered roughly 100 km northwest of Denver, Colorado. There are 20 vertical levels from 119 m above ground level (AGL) up to 22895 m AGL. The main modifications to RAMS have been to the station height quantities and radiation and precipitation processes.

The lowest atmospheric level resides within the surface layer. In order to make comparisons between model and observed quantities we have used a similarity theory formulation (Louis, 1979) to derive atmospheric quantities of wind speed (at 10 and 2 m AGL) and temperature and humidity (both at 1.8 m AGL), valid at standard observational heights. Precipitation processes fall into two broad categories: grid-scale resolvable and subgrid-scale parameterized (convection). The convective parameterization we use is a modified Kuo scheme (Tremback, 1990). From the climate perspective, one is generally interested in the accumulation and surface distribution of precipitating species. With this in mind, we have chosen to use a much-simplified, grid-resolvable precipitation scheme. This scheme is based on a condensational dump-bucket with a cloud-top temperature rainout efficiency (Rhea, 1978; Thompson, 1993). The current scheme accounts only for liquid precipitation; work is in progress toward a snow-scheme and snow-cover module.

The RAMS radiation parameterizations allow for a choice of a clear-sky shortwave/longwave package (Mahrer and Pielke, 1977) or one which explicitly accounts for atmospheric condensate fields (Chen and Cotton, 1983). For climate modeling purposes the Chen-Cotton scheme is excessively computationally expensive, and the clear-sky scheme, though fast, neglects important cloud-shading effects. To alleviate these problems we are using a modified clear-sky scheme, in which the downward longwave flux is set to the blackbody flux, determined by the cloud-bottom temperature. The downward shortwave flux is reduced by an albedo based on cloud thickness (Neiburger, 1949). Both the radiation and precipitation schemes work well in an operational form of RAMS (Thompson, 1993).

Important aspects of our simulations are the effects of fine-scale structures in topography and vegetation and soil characteristics. To calculate the atmospheric fluxes of heat, moisture, and momentum, RAMS has a multilevel soil/vegetation module (McCumber and Pielke, 1981; Tremback and Kessler, 1985), which in our case uses 11 vertical levels. Data for this module come from the USGS vegetation database (Loveland et al., 1991), aggregated to Biosphere-Atmosphere Transfer Scheme classifications.

SUMMARY

Several important points bear reiteration. Local spatial variations in topography and vegetation have an important role in determining the regional climate by generating local circulations or modifying existing ones. Changes in vegetation characteristics can be an important feedback in climate change. Downscaling of GCM climate change results is necessary for interpretation on regional and smaller scales. This can be accomplished through statistical or dynamic interpolation methods; direct interpolation is unsuited for the regional scale. Further work is needed in examining the sensitivity of climate to vegetation, particularly in situations with stronger synoptic flow. Finally, there is a need for improved coupling between the various components of the biosphere for climate scenario simulation.

ACKNOWLEDGMENT

Much of this research was carried out under National Park Service contracts CA1268-2-90CSU-06 and CA1268-2-9004, which support climate change research in and around our national parks.

REFERENCES

Chen, C and WR Cotton. **1983**. A one-dimensional simulation of the stratocumulus-capped mixed layer. Boundary-Layer Meteorol. 25:289-321.

Cooter, EJ, BK Eder, SK LeDuc, and LTruppi. General circulation model output for forest climate change research and applications. Southeastern Experiment Station Technical Bulletin, in press.

Dumenil, L. **1993**. Near surface systematic errors in GCMs - dynamics vs. land surface processes. GEWEX News, winter 92/93:2-3.

Giorgi, F. **1990**. Simulation of regional climate using a limited area model nested in a general circulation model. J. Climate 3:941-963.

Grotch, SL and MC MacCracken. **1991**. The use of general circulation models to predict regional climate change. J. Climate 4:286-303.

Knowles, JB. **1993**. The influence of forest fire induced albedo differences on the generation of mesoscale circulations. M.S. Thesis, Department of Atmospheric Science, Colorado State University, Ft. Collins, CO, 85 pp.

Louis, JF. **1979**. A parametric model of vertical eddy fluxes in the atmosphere. Boundary-Layer Meteorol. 17:187-202.

Loveland, TR, JW Merchant, DO Ohlen, and JF Brown. **1991**. Development of a land-cover characteristics database for the conterminous U.S. Photogram. Eng. Remote Sensing 57:1453-1463.

Mahrer, Y and RA Pielke. **1977**. A numerical study of the airflow over irregular terrain. Beitr. Physik Atmosph. 50:98-113.

McCumber, MC and RA Pielke. **1981**. Simulation of the effects of surface fluxes of heat and moisture in a mesoscale model - Part 1: Soil layer. J. Geophys. Res. 86:9929-9938.

Neiburger, M. **1949**. Reflection, absorption and transmission of insolation by stratus cloud. J. Appl. Meteorol. 6:98-104.

Pielke, RA, WR Cotton, RL Walko, CJ Tremback, ME Nicholls, MD Moran, DA Wesley, TJ Lee, and JH Copeland. **1992**. A comprehensive meteorological modeling system - RAMS. Meteorol. Atmosph. Physiol. 49:69-91.

Rhea, JO. **1978**. Orographic precipitation model for hydrometeorological use, PhD dissertation, Atmos. Sci. Paper No. 287, Colorado State University, Fort Collins, CO, 199 pp.

Thompson, G. **1993**. Prototype real-time mesoscale prediction during 1991-1992 winter season and statistical verification of model data, MS thesis, Atmos. Sci. Paper No. 521, Colorado State University, Fort Collins, CO, 105 pp.

Tremback, CJ. **1990**. Numerical simulation of a mesoscale convective complex: Model development and numerical results, PhD dissertation, Atmos. Sci. Paper No. 465, Colorado State University, Fort Collins CO, 247 pp.

Tremback, CJ and R Kessler. **1985**. A surface temperature and moisture parameterization for use in mesoscale models, pp. 433-434. Preprints, 7th AMS Conference on Numerical Weather Prediction, June 17-20, Montreal, Quebec, Canada, American Meteorological Society, Boston, MA.

Von Storch H, E Zorita, and U Cubasch. **1993**. Downscaling of global climate change estimates to regional scales: An application to Iberian rainfall in winter. Journal of Climatology, 6:1161-1171.

Weissbluth, MJ. **1991**. Convective parameterization in mesoscale models, Ph.D. Dissertation. Department of Atmospheric Science, Colorado State University, Fort Collins, CO, 211 pp.

ASSESSING THE EFFECTS OF CLIMATE CHANGE ON WATER RESOURCES

APPLICATION OF TOPOGRAPHICALLY DISTRIBUTED MODELS OF ENERGY, WATER AND CARBON BALANCE OVER THE COLUMBIA RIVER BASIN: A FRAMEWORK FOR SIMULATING POTENTIAL CLIMATE CHANGE EFFECTS AT THE REGIONAL SCALE

D. P. Turner[1] and D. Marks[2]

[1]ManTech Environmental Technology, Inc., Corvallis, Oregon

[2]U.S. Geological Survey, US EPA Environmental Research Laboratory, Corvallis, Oregon

Key words: *Water balance, carbon balance, Columbia River basin, climate change, regional scale*

ABSTRACT

The investigation of many issues relating to global change is dependent on spatially distributed models of land surface processes. These models may originate as point models, and the methodology for "scaling up" to the regional level remains a significant research issue. Here we describe an approach to running a carbon and water balance model (Forest-BGC) over a gridded digital elevation surface for the Columbia River Basin. Estimates were made at each 1-km^2 grid point at a daily time step over a 2-week period (May 1–15, 1990) for climate variables required by the model, including solar radiation, minimum and maximum air temperature, and precipitation. The daily times series for the climatic drivers was developed from interpolations of regional meteorological data. To initialize the run, surfaces for vegetation properties, including biome type and leaf area index (LAI) as well as soil factors, including water-holding capacity and carbon content, were also prepared. The spatial distribution of vegetation type was from land cover maps derived in part from satellite remote sensing. The LAI surface was developed using a water balance index based on spatially distributed precipitation and potential evapotranspiration. The model simulated the daily water balance and tracked the soil moisture status at each grid point. Photosynthesis and plant respiration were calculated to estimate vegetation carbon balance. This modeling framework will provide an opportunity to examine scaling issues as related to regional analyses and will be used in evaluating potential effects of interannual variability and long-term climate change on the regional carbon and water balance.

INTRODUCTION

A critical issue in the application of spatially distributed models to global change problems is that of "scaling up" existing point models to relevant spatial and temporal domains. Although various approaches to regional applications have been considered (King, 1991), that of running the relevant model over a grid of points across the landscape is desirable in many cases. These model-based simulations require development of methodologies and models to distribute climatic drivers, such as solar radiation, precipitation, temperature, humidity, and wind, as well as vegetation characteristics such as vegetation type and leaf area index.

One potential source of distributed climate data is the General Circulation Model (GCM). However, given the limitations of the current GCM in replicating the current climate at the regional scale (Cushman and Spring, 1989), an approach to developing climatic inputs based on time series data from a network of meteorological stations is needed. The availability of high-resolution digital elevation models has given impetus to this approach (Dolph et al., 1992; Marks et al., 1993). For land cover characteristics, the availability of satellite imagery has opened up the possibility of creating digital land cover maps at reasonably high spatial resolution.

In this paper we describe the development of databases to initialize and drive a model of carbon and water balance (Forest-BGC; Running and Coughlan, 1988) over the Columbia River Basin at a 1-km^2 spatial resolution. Forest-BGC has been run in a spatially distributed mode previously (Running et al., 1989) and has been used in climate change analysis (Running and Nemani, 1991). In this application, we use alternative approaches to creating many of the climatic and land surface input variables.

The spatial resolution of 1 km^2 was chosen, in part, because that is the approximate resolution of the satellite-mounted Advanced Very High Resolution Radiometer (AVHRR) sensor which has been used for mapping land cover (Loveland et al., 1991). The time step of 1 day is a compromise between the objectives of retaining the fundamental biophysical and physiological processes to as great a degree as is possible and still running the model at the regional scale (Meentenmeyer, 1989).

METHODS

For the purposes of model development and testing, databases to initialize and drive the model for a 2-week period (May 1–14) in water year 1990 (October 1, 1989 to September 30, 1990) over the Columbia River Basin were produced. At a spatial resolution of 1 km^2, the Basin contains 671,579 grid cells. With an elevation range from sea level to over 3000 m, and annual precipitation ranging over two orders of magnitude, the Basin covers a great diversity of land forms and vegetation types.

The Model

Forest-BGC was designed for regional analysis of the carbon, nitrogen, and water cycles. A few simple compartments—including foliage, stems and roots—are used rather than representations of individual trees. The focus is on incorporating basic plant physiology and biophysics in a very general fashion in order to simulate effects of climatic change on ecosystem mass flux. Processes including photosynthesis, maintenance respiration, and evapotranspiration, are simulated at a daily time step, whereas allocation of assimilated carbon among the plant parts and decomposition are simulated annually. Species are not identified; however, the model may be run in three modes based on life form (conifer, broadleaf, or grass), with physiological variables such as maximum stomatal conductance varying by life form (Running and Hunt, 1993). Comparison of model outputs, for above-ground net primary productivity and site water balance, to observed patterns of carbon and water flux have indicated reasonable agreement for a variety of locations (Running and Coughlan, 1988; Running and Nemani, 1988; Hunt et al., 1991; Running and Hunt, 1993.

DATA INPUTS

Climatic Variables

Daily times series data from the National Climate Data Center and Snotel meteorological station networks (described in USDA Soil Conservation Service, 1988; Earth Info, 1988) provided the basis for distributed temperature and precipitation surfaces. There were approximately 1000 stations in the Columbia Basin (Figure 1). The digital elevation model (DEM) for this analysis was based on a 15 arc-second grid developed by the U.S. Geological Survey from 3 arc-second

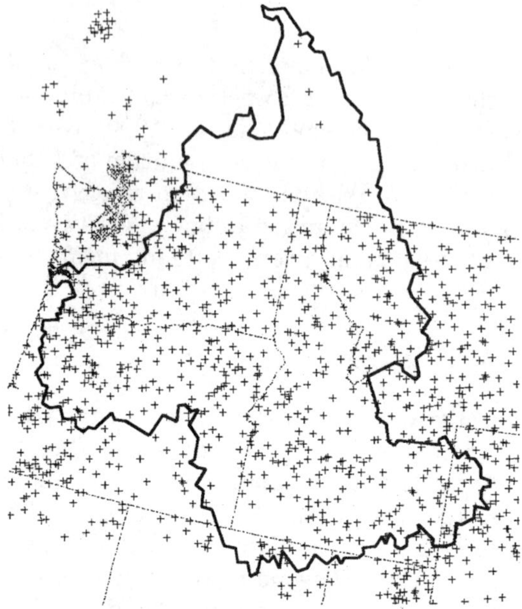

Figure 1. Geographic distribution of network of meteorological stations.

digital topographic elevation data (Defense Mapping Agency, 1981). This was reprojected into an Albers equal area projection at 1 km using a bilinear interpolation algorithm (Figure 2).

The daily climatic variables needed to drive the Forest-BGC model are minimum and maximum temperature, dewpoint temperature, precipitation, and solar radiation. Daily meteorological station data for maximum and minimum temperatures were corrected for elevational effects and mapped over the DEM, using the assumption of locally adiabatic conditions (Marks et al., 1993). The dewpoint temperature was assumed to be the minimum daily temperature following Hungerford et al. (1989).

The approach for preparing daily precipitation surfaces is based on the logic in the PRISM model (Daly and Neilson, 1992). PRISM is a moving window regression model which operates on monthly precipitation data. For each grid cell, the model determines a geomorphic facet (such as a portion of the west side of the Cascade Mountains) for which there

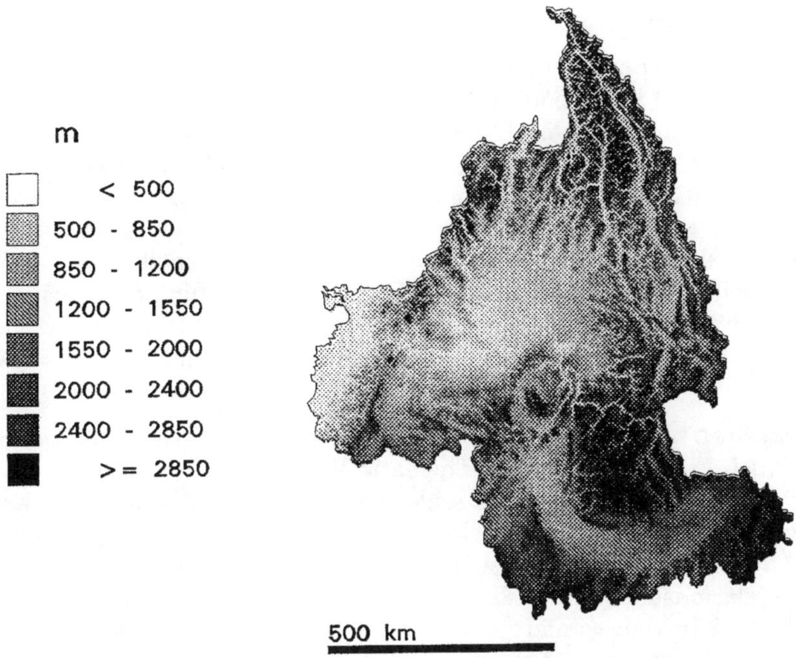

Figure 2. Digital elevation map for Columbia River Basin. Spatial resolution, 1 km.

are sufficient meteorological stations to construct a regression of precipitation against elevation. The daily data are too variable to use the regression approach. Thus, for the daily precipitation surfaces, the proportion of the monthly precipitation observed on a given day for the network of meteorological stations was used to create proportionality surfaces, and these were applied to the monthly precipitation values.

Clear-sky short-wave solar irradiance (0.3 to 1.1 nm) was calculated using the topographically corrected spectral radiation model described by Dubayah et al. (1990), and modified for regional application by Marks et al. (1991). The principal model inputs are elevation and surface albedo (Channel 1, AVHRR). In this study, adjustment for cloud-cover effects was based on differences between daily minimum and maximum temperature (Bristow and Campbell, 1984).

Site Variables

Vegetation Type. The multiple vegetation classes in the land cover map (1-km spatial resolution) of Loveland et al. (1991), which is based on the temporal Normalized Difference Vegetation Index (NDVI) signal from the AVHRR sensor, were aggregated to three biome-level classes, designated as conifer, shrubland, and grassland. Comparison of the forest cover estimates from the satellite-based land cover map with estimates from a ground-based forest inventory (Waddell et al., 1989) revealed close correspondence in Oregon and Washington (Turner et al., 1993a). For the Canadian portion of the Columbia Basin, the map of Rowe (1972) was used as the basis for approximating vegetation type.

Leaf Area Index (LAI). Output from Forest-BGC is strongly influenced by LAI. Field studies at the Thematic Mapper scale (~30 m; Peterson et al., 1987) and the AVHRR scale (~1 km; Running et al., 1989) have indicated promise for developing remote-sensing-based algorithms which could be used for initializing LAI over forested regions. However, the relationship of indices such as NDVI to LAI tends to be asymptotic, and LAI in cool coniferous forests of the Pacific Northwest may be toward the high end of the potential LAI range; thus the sensitivity of LAI to NDVI may be low in parts of this region.

An alternative approach relies on an index of site water balance to estimate potential LAI. Much current theory on the relationship of climate to vegetation type and LAI holds that site LAI will rise to a level which just uses the available water (Woodward, 1987; Nemani and Running, 1989; Neilson et al., 1992). Therefore, an index of site water balance should correlate with LAI and could provide the basis for predicting LAI. Grier and Running (1977) found evidence for such a relationship in the coniferous forests of the Pacific Northwest.

In a preliminary examination for this paper, a water balance index (WBI) was created over the Pacific Northwest and used to examine the WBI-LAI relationship. A WBI surface (10-km spatial resolution) was constructed for the U.S. portion of the Pacific Northwest by subtracting the growing season (May to September) sum of potential evapotranspiration (PET) from the growing season sum of precipitation. Long-term average monthly precipitation surfaces were derived from the PRISM model, and monthly PET surfaces were from Dolph et al. (1992). Measured LAI (projected) using an allometric approach, with approximate locations, were from Peterson et al. (1987) and ranged from 0.7 in juniper forests of eastern Oregon to 13.4 on the Oregon coast. This range approximates that for coniferous forests throughout North America. With the exception of a high elevation site, the relationship between field measurements and the corresponding WBI was nearly linear (7 sites, $r^2 = 0.67$. $P = 0.02$). The LAI for conifer areas above 1300 m was based on a relationship between a heat sum (the sum of the maximum daily temperatures during the growing season) and the observed LAI at two high-elevation sites. An LAI limit of 8 was imposed on the final LAI map because of uncertainties about the high LAI reported for the Pacific Northwest coniferous forests using the allometric approach (Marshall and Waring, 1986). Lacking a WBI surface for the Canadian portion of the Basin, a fixed LAI of 6 was assigned to all conifer grid cells. Shrub and grassland grid cells were assigned LAI of 1.0 and 1.5, respectively, throughout the Basin.

Foliar and Stem Carbon and Nitrogen. Foliar carbon was a function of LAI and the specific leaf area conversion factors in Biome-BGC (Running and Hunt, 1993). Foliar nitrogen was set at 1.5% of foliar carbon; however, literature studies (e.g., Zinke and Stangenberger, 1979) may allow improved estimates for this variable. Foliar carbon influences the maintenance respiration and total net primary productivity, and leaf N concentration potentially influences the photosynthetic rate. Since one of our initial interests is in the distribution of potential net primary productivity, the assumption was made that all stands were 50 yr of age, and a stem biomass of 5 kg carbon m^{-2} was assigned to all conifer grid cells. Ultimately, approaches based on satellite remote-sensing will be used to identify forest type (Z. Zhu, Southern Forest Experimental Station, Starkville, MS, personal communication) and stand age class (e.g., Cohen and Spies, 1992) for which representative stand-level carbon budgets have been prepared (Turner et al., 1993b). Stem nitrogen content was set at a fixed proportion (0.0035) of stem carbon.

Soil Water-Holding Capacity. Soil water-holding capacity is a function of soil texture, organic carbon content, rock fragment content, bulk density, and soil depth. Quantitative algorithms relating these variables to water-holding capacity have been developed (Rawls et al., 1982) and have been evaluated using data from the U.S. National Soil Survey Laboratory (Kern, in press a). The spatial patterns of these variables are available in digitized form from the United Nations (UN) Soil Map of the World [1:5M scale] (UN, 1972-1978; Kern, 1991) and this information was used to develop a soil water-holding capacity map.

For this preliminary analysis, the soil water content was set at its maximum at the beginning of the test period.

Carbon and Nitrogen in the Soil and Litter. Soil organic carbon content can be estimated by reference to the carbon content of specific taxonomic units, as derived from laboratory data for representative pedons and the spatial distribution of soil taxonomic units from the U.N. Soil Map of the World. The soil organic matter content at a 1:5M scale has been developed for North America by Kern (1994) using this approach and was used in this analysis. Soil nitrogen was set at a fixed proportion of soil carbon (0.05), however, the Zinke et al. (1986) database offers opportunities for improved estimates. Representative values for litter carbon and nitrogen for the three biomes were assigned based on previous measurement programs (Vogt et al., 1986). These soil and litter variables are relevant only to the annual time step component of Forest-BGC where decomposition is estimated.

RESULTS

The spatially distributed average maximum daily temperature over the 14-day simulation (Figure 3) shows a response to elevation and distance from the Pacific Coast. The highest values are found in the interior of the Columbia Basin; the lowest values at the high elevations of the northern Rocky Mountains. The map of summed precipitation (Figure 4) displays the orographic effects of the Cascade and northern Rocky Mountain Ranges and the rain shadow in the central and southern parts of the Basin. The scarcity of meteorological station data in the Canadian portion of the Basin created significant artifacts, notably an east-west band of artificially low precipitation, which can be corrected only by acquiring data from more stations.

The average clear-sky solar radiation for May 1 (Figure 5) was 21,000 kJ m^{-2}, and the average final solar radiation was 17% less. Observations from the Solar and Meteorological Surface Observation Network (National Climate Data Center) offer the possibility of validating these estimates and creating cloudiness coefficients from direct measurements. Soil water-holding capacity (Figure 6) ranged from 6 to 23 cm. The vegetation map (Figure 7) shows the extensive coverage of the coniferous forest biome and the extent of the grassland and shrubland biomes in the drier areas. Potential Leaf Area Index (Figure 8) reflects both the vegetation type and climate.

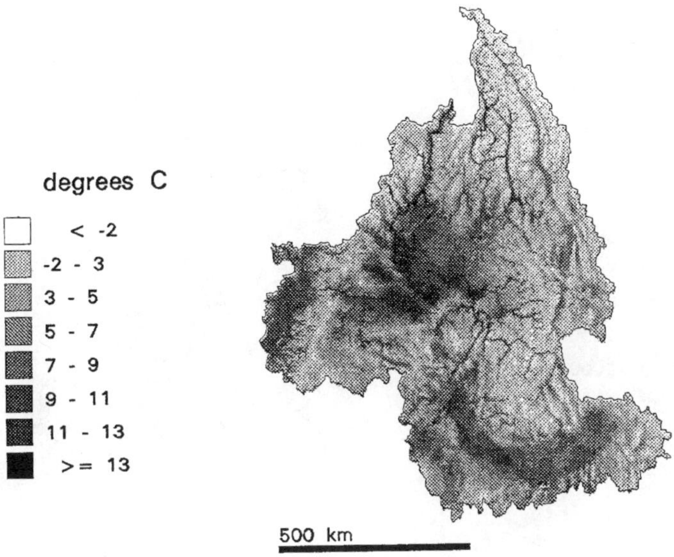

Figure 3. Average daily maximum temperature (May 1 to 14, 1990).

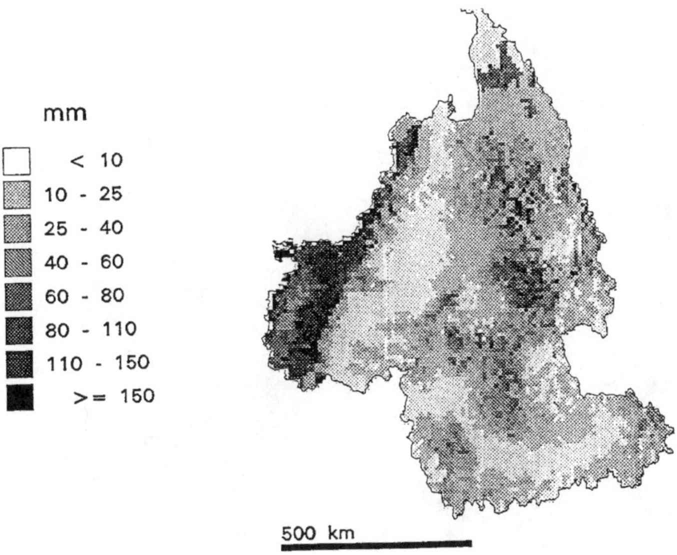

Figure 4. Summed precipitation (May 1 to 14, 1990). Spatial resolution, 10 km.

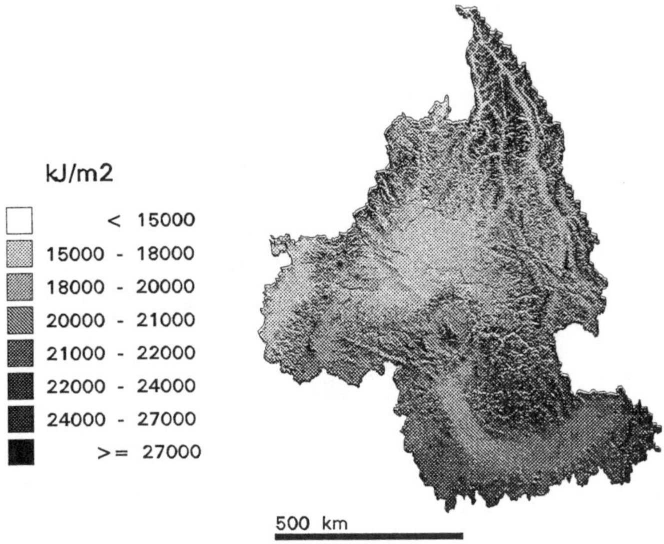

Figure 5. Clear-sky solar irradiance (May 1, 1990).

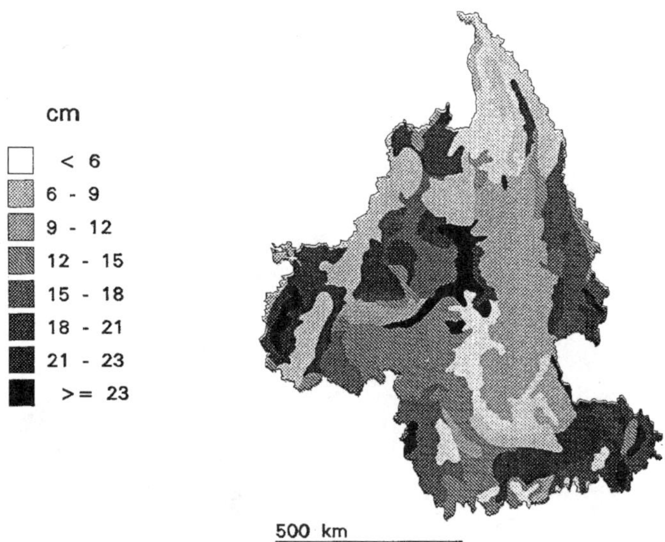

Figure 6. Soil water-holding capacity.

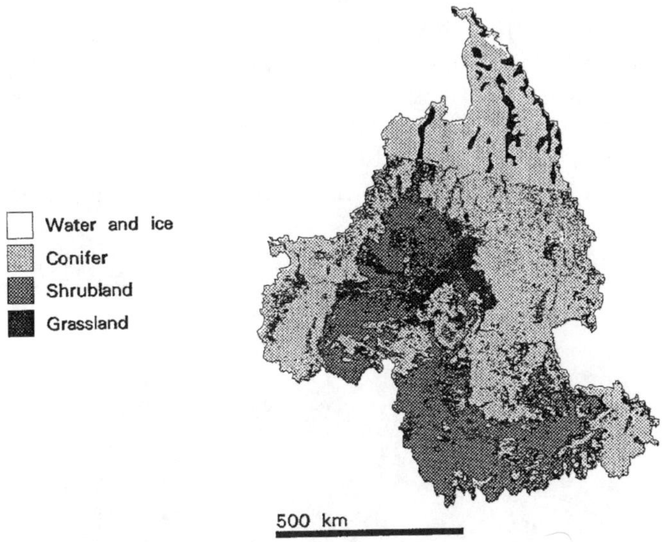

Figure 7. Vegetation type, derived from Loveland et al. (1991). Spatial resolution, 1 km.

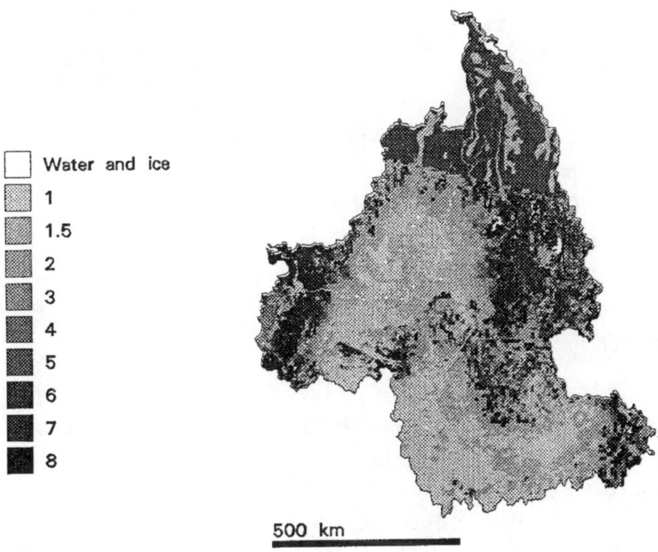

Figure 8. Potential leaf area index (projected).

Simulated net photosynthesis (Figure 9) and actual evapotranspiration (Figure 10) for the first 2 weeks in May integrate the effects of climate and vegetation. Both appear to be strongly influenced by LAI and temperature. Areas of relatively high temperature may have low ET over this period because of low LAI. The gradients of net photosynthesis and evapotranspiration in Canada, despite the fixed LAI of the conifer grid cells, thus gives an indication of their sensitivity to temperature.

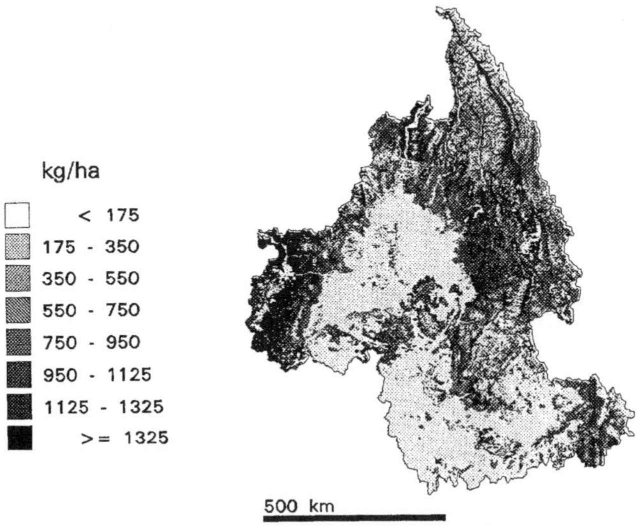

Figure 9. Summed net photosynthesis (May 1 to 14, 1990).

DISCUSSION

Spatially distributed ecosystem models which have been applied at the continental to global scale have been run primarily at coarse spatial resolutions (e.g., ½ × ½ degree of longitude and latitude, ~50 km in the temperate zone) with a time step of 1 month (McGuire et al., 1993). At a monthly time step, important components of hydrologic cycle, such as interception and percolation through the soil, must be extensively parameterized. Productivity must likewise be tightly constrained, based on observed annual means rather than being more strictly

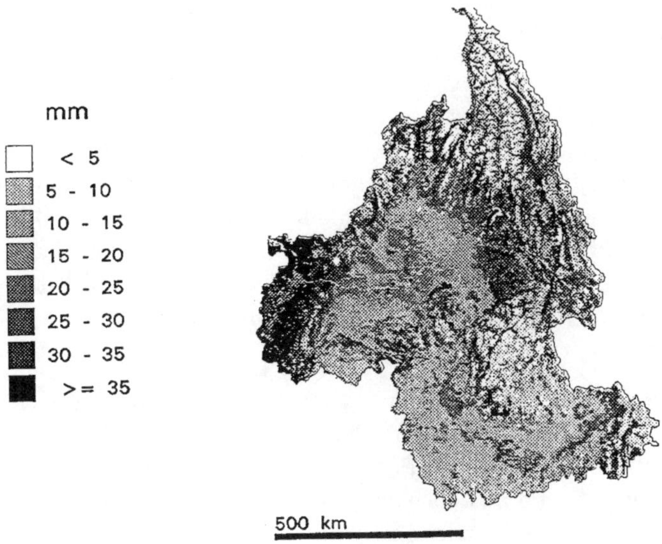

Figure 10. Summed evapotranspiration (May 1 to 14, 1990).

process-driven. At coarse spatial resolutions, virtually all effects of slope and aspect which are evident in complex terrain are also lost, and much of the spatial heterogeneity in vegetation is ignored. The optimal spatial and temporal resolution for a given analysis depends, of course, on the particular application, but for modeling climate change impacts at the regional scale, particularly in complex terrain, a spatial resolution of 1 km and a temporal resolution of 1 day may often be appropriate. At finer spatial and temporal resolutions, the requirements for data input and storage become an increasing constraint. The issues related to combining input data layers developed at different spatial resolutions are also significant and can at least begin to be explored within the modeling framework we have developed.

The ecosystem models applied at the global scale have also often depended on long-term average historical data for their climate drivers. In contrast, we have focused on time series data. The benefit of employing time series data is that comparisons may be made using climate data from relatively warm or cool years. The combination of GCM outputs and the analysis of interannual climate variability has recently been proposed as an optimal approach to creating regional climate change scenarios (Robock et al., 1993).

Regional simulations of carbon and water balance using time series data also creates the opportunity for validating the water balance components of the model with measured runoff and the phenological characteristics, such as snow cover and leaf area, with remote-sensing imagery. Further opportunities for model calibration and validation are offered by field measurement programs such FIFE (Sellers et al., 1988) where critical variables such as soil water status and leaf water potential are monitored, along with the full complement of meteorological data needed to drive process-based models.

Simulations of ecosystem response to future environments will require attention to a number of specific issues which have not been addressed here. Vegetation type and LAI are likely to change along with climate, and global vegetation models such as BIOME (Prentice et al., 1992) and MAPSS (Neilson, 1993) will be needed for projecting the distribution of vegetation types. Incorporating the biochemical aspects of plant response to CO_2 (Farquhar and von Caemmerer, 1982) in the photosynthesis algorithms of the physiological models may also be necessary in order to treat the projected levels of atmospheric CO_2. The importance of soil nitrogen mineralization will have to be examined closely since, in some models, a projected increase in nitrogen mineralization determines the sign of the productivity response to climate change (McGuire et al., 1993).

A variety of specific applications are relevant to the type of modeling framework described in this paper. These include evaluating potential changes in forest productivity on national forest carbon budgets (Turner et al., 1993b) and risk assessments with regard to climate change impacts on tree species with known geographic distributions. Any type of application will require increased attention to calibration and validation of the model or models. Field-based programs which combine on-site measurements with remote-sensing imagery will contribute greatly toward that end.

ACKNOWLEDGMENTS

We thank S. Running for sharing Forest-BGC, C. Daly for assistance with PRISM, and R. Dodson, D. Jacobsen, J. Kern, and J. Kimball for help in assembling the required databases and executing the simulation.

The research described in this manuscript has been funded in part by the U.S. Environmental Protection Agency. This manuscript has been prepared at the EPA Environmental Research Laboratory in Corvallis, Oregon, through contract number 68-C6-0006 to ManTech Environmental Technology, Inc. It has been subjected to the Agency's peer and administrative review, and approved for publication. Mention of trade names or commercial products does not constitute endorsement or recommendation for use.

REFERENCES

Bristow, KL and GS Campbell. **1984**. On the relationship between incoming solar radiation and daily maximum and minimum temperature. Agric. For. Metelorol. 31:159-166

Cohen, WB and TA Spies. **1992**. Estimating structural attributes of Douglas-fir/Western Hemlock forest stands from Landsat and SPOT imagery. Remote Sens. Environ. 41:1-17.

Cushman, RM and PN Spring. **1989**. Differences among model simulations of climate change on the scale of resource regions. Environ. Manage. 13:789-795

Daly, C and RP Neilson. **1992**. A digital topographic approach to modeling the distribution of precipitation in mountainous terrain., pp. 437-454. In: Interdisciplinary Approaches to Hydrology and Hydrogeology. American Institute of Hydrology, Minneapolis, MN.

Defense Mapping Agency. 1981. Product Specifications for Digital Terrain Elevation data, 2nd edition. D.M.A. Aerospace Center, St. Louis Air Force Station, St. Louis, MO, p. 23

Dolph, J, D Marks, and GA King. **1992**. Sensitivity of the regional water balance in the Columbia River Basin to climate variability: Application of a spatially distributed water balance model, pp. 233-266. In: *Watershed Management: Balancing Long Term Sustainability and Environmental Change*, R.J. Naiman and J.R. Sedell (eds.). Springer-Verlag, NY.

Dubayah, R, J Dozier, and F Davis. **1990**. Topographic distribution of clear-sky radiation over the Konza Prairie, Kansas. Water Resour. Res. 26:679-691.

Earth Info. 1988. *Climate Data Users Manual*. Denver, CO.

Farquhar, GD and S von Caemmerer. **1982**. Modeling of photosynthetic response to environmental conditions, In: *Physiological Plant Ecology II: Water Relations and Carbon Assimilation*, O.L. Lange, P.S. Nobel, C.B. Osmond, and H. Ziegler (eds.). Encyclopedia of Plant Physiology, New Series, Vol. 12B. Springer-Verlag, Berlin, Germany.

Grier, CC and SW Running. **1977**. Leaf area of mature northwestern coniferous forests; Relation to site water balance. Ecology 58:893-899.

Hungerford, RD, RR Nemani, SW Running, and JC Coughlan. **1989**. MTCLIM: A mountain microclimate simulation model, Research Paper INT-414. USDA Forest Service, Intermountain Research Station, Odgen, UT. 52 pp.

Hunt, ER, Jr., FC Martin, and SW Running. **1991**. Simulating the effects of climatic variation on stem carbon accumulation of a ponderosa pine stand: Comparison with annual growth increment data. Tree Physiol. 9:161-171.

Kern, JS. Estimation of soil water retention modeled from texture, organic matter, and bulk density. J. Soil Sci. Soc. Am. (in press a).

Kern, JS. Global geography of soil water retention modeled from texture, organic matter, and bulk density. J. Soil Sci. Soc. Am. (in press b)

Kern, JS. **1994**. Spatial patterns of soil organic carbon in North America. J. Soil Sci. Soc. Am. (in press b). 58:439-455.

King, AW. **1991**. Translating models across scales in the landscape, pp. 479-517. In: *Quantitative methods in landscape ecology: The Analysis and Interpretation of Landscape Heterogeneity*, M.G. Turner and R.H. Gardner (eds.). Springer-Verlag, New York.

Loveland, TR, JW Merchant, DO Ohlen, and JF Brown. **1991**. Development of a land-cover characteristics database for the conterminous U.S. Photogram. Eng. Remote Sens. 57:1453-1463.

Marks, DG, R Dubayah, and K Longley. **1991**. Modeling the topographic and spectral variability of clear-sky solar radiation at regional to continental scales, Vol 3, p. 1711. In: *Proceedings of 1991 IGARSS*

Symposium, Remote Sensing: Global Monitoring for Earth Management. Epsoo, Finland.

Marks, DG, GA King, and J Dolph. **1993**. The implications of climate change on the water balance of the Columbia River Basin. Climate Res. 2: 203-213.

Marshall, JD and RH Waring. **1986**. Comparison of methods of estimating leaf-area index in old-growth Douglas-fir. Ecology 67:975-979.

McGuire, AD, LA Joyce, DW Kicklighter, JM Melillo, G Esser, and CJ Vorosmarty. **1993**. Productivity response of climax temperate forests to elevated temperature and carbon dioxide: A North American comparison between two global models. Clim. Change 24:287-310.

Meentemeyer, V. **1989**. Geographical perspectives of space, time, and scale. Land Ecol. 3:163-173.

Neilson, RP. **1993**. Vegetation redistribution: A possible biosphere source of CO_2 during climatic change. Water, Air, Soil Pollut. 70:659-673.

Neilson, RP, GA King, and GJ Koerper. **1992**. Toward a rule-based biome model. Land Ecol. 7:27-43.

Nemani, R and SW Running. **1989**. Testing a theoretical climate-soil-leaf area hydrologic equilibrium of forests using satellite data and ecosystem simulation. Agric. For. Meteorol. 44:245-260.

Peterson, DL, MA Spanner, SW Running, and KB Teuber. **1987**. Relationship of thematic mapper simulator data to leaf area index of temperate coniferous forests. Remote Sens. Environ. 22:323-341.

Prentice, IC, W Cramer, SP Harrison, R Leemans, RA Monserud, and AM Solomon. **1992**. A global biome model based on plant physiology and dominance, soil properties and climate. Biogeography 19:117-134.

Rawls, WL, DL Brankensiek, and KE Saxton. **1982**. Estimation of soil water properties. Trans. ASAE 25:1316-1320.

Robock, A, RP Turcon, MA Harwell, TP Ackerman, R Andressen, H Chang, and MVK Sivakumar. **1993**. Use of general circulation model output in the creation of climate change scenarios for impact analysis. Clim. Change 23:293-335.

Rowe, JS. **1972**. *Forest Regions of Canada*. Department of the Environment, Canadian Forestry Service, Ottawa, Canada. 172 pp.

Running, SW and JG Coughlan. **1988**. A general model of forest ecosystem processes for regional application. Ecol. Model. 42:125-154.

Running, SW and ER Hunt, Jr. **1993**. Generalization of a forest ecosystem process model for other biomes, Biome-BGC, and an application for global-scale models, pp. 151-158. In: *Scaling Processes between Leaf and Landscape Levels*, J.R. Ehleringer and C. Field (eds.). Academic Press, San Diego, CA.

Running, SW and RR Nemani. **1988**. Relating seasonal patterns of AVHRR vegetation index to simulated photosynthesis and transpiration of forests in different climates. Remote Sens. Environ. 24:347-351.

Running, SW, and RR Nemani. **1991**. Regional hydrologic and carbon balance responses of forests resulting from potential climate change. Clim. Change 19:349-368.

Running, SW, RR Nemani, DL Peterson, LE Band, DF Potts, LL Pierce, and MA Spanner. **1989**. Mapping regional forests evapotranspiration and photosynthesis by coupling satellite data with ecosystem simulation. Ecology 70:1090-1101.

Sellers, PJ, FG Hall, G Asrar, DE Strebel, and RE Murphy. **1988**. The first ISLSCP field experiment. Bull. Am. Meteorol. Soc. 69:22-27.

Turner, DP and GJ Koerper, H Gucinski, CE Peterson, and RK Dixon. **1993a**. Monitoring global change: comparison of forest cover estimates using remote sensing and inventory approaches. Environ. Monit. Assess. 26:295-305.

Turner, DP, JJ Lee, GJ Koerper, and JR Barker. **1993b**. The forest sector carbon budget of the United States: Carbon pools and flux under alternative policy options, EPA/600/3-93/093. U.S. EPA Environmental Research Laboratory, Corvallis, OR. 202 pp.

UN. 1972-1978. Soil Map of the World, Vols. I-X. Food and Agriculture Organization and Educational, Scientific and Cultural Organization, United Nations, Rome, Italy.

USDA. 1988. Snow Surveys and Water Supply Forecasting, Agriculture Information Bulletin 536. Soil Conservation Service, U.S. Government Printing Office, Washington, D.C.

Vogt, KA, CC Grier, and DJ Vogt. **1986**. Production, turnover, and nutrient dynamics of above-and belowground detritus of world forests. Adv. Ecol. Res. 15:303-377.

Waddell, KL, DD Oswald, and DS Powell. **1989**. Forest Statistics of the United States, PNW-RB-168, Pacific Northwest Research Station. U.S.D.A. Forest Service Resource Bulletin, Portland, OR.

Woodward, FI. **1987**. *Climate and Plant Distribution*. Cambridge University Press, London, UK.

Zinke, PJ, and AG Stangenberger. **1979**. Ponderosa pine and Douglas-fir foliage analyses arrayed in probability distributions. In: *Proceedings of the Forest Fertilization Conference*, September 25-27, 1979, Union, WA. British Columbia Ministry of Forests, Vancouver, B.C.

Zinke, PJ, AG Stangenberger, WM Post, WR Emanuel, JS Olson. **1986**. Worldwide organic soil carbon and nitrogen data, ORNL/CDIC-18/NDP-018. Carbon Dioxide Research Division, U.S. Department of Energy, Oak Ridge National Laboratory, Oak Ridge, TN, 141 pp.

ASSESSMENT OF THE WATER RESOURCES EFFECTS OF CLIMATE CHANGE IN THE COLUMBIA RIVER BASIN

J. C. Sias,[1] D. P. Lettenmaier,[1] and L. W. Vail[2]

[1]University of Washington, Seattle, Washington

[2]Pacific Northwest Laboratory, Richland, Washington

ABSTRACT

The potential effect of regional climate change on the water resources of the Columbia River Basin is evaluated using a combination of a deterministic hydrological model, a stochastic stream flow model, and a reservoir system simulation model. The potential effect of prescribed changes in temperature and precipitation (and, therefore, evapotranspiration) on stream flow is modeled by first calibrating the hydrologic model with historical climate data for the five index catchments. The index catchments, which range in size from 210 to 2850 mi^2, are the North Fork American River, Washington; the Middle Fork Flathead River, Montana; Reynolds Creek, Idaho; Camas Creek, Oregon; and the Salmo River, British Columbia. The hydrologic model is a combination of the National Weather Service model (NWSRFS) for snow accumulation and ablation, and TOPMODEL for soil moisture accounting. The stochastic stream flow model is SPIGOT, developed by J.R. Stedinger, and others, at Cornell University.

The hydrological models are calibrated using data mostly from the 1970 to 1989 period. The stochastic disaggregation model, which relates index catchment flows to upstream nodes of the reservoir system simulation model, is calibrated using stream flow data from the period 1948 to 1988. Stochastic model calibration is performed using historical stream flows at each node and the index catchments. However, the base (current climate) reservoir system simulation is performed using simulated (from the hydrological model) stream flows for the five index sites for a baseline period. This removes the effects of biases in the hydrological and stochastic simulation models. Both the baseline and altered climate reservoir system inflows produced in this way were used as input to a simplified, 14-node water resources management model developed at Pacific Northwest Laboratory. The reservoir system simulations provide some indication of how the performance of the existing reservoir system would be affected in an altered climate, and of how operating rules for the system may need to be modified. Generally, the most significant effect of a warmer climate would be the loss of natural storage in mountain snowpacks, and reduction in the ability of the reservoir system to maintain stream flows in the late summer and fall, due to a shifting of the tributary hydrographs from spring to winter.

SIMULATION OF DISTRIBUTED LAND SURFACE FLUXES AND RUNOFF FROM A MOUNTAINOUS DRAINAGE BASIN*

M.S. Wigmosta,[1] L.W. Vail,[1] and D.P. Lettenmaier[2]

[1]Pacific Northwest Laboratory, Richland, Washington

[2]University of Washington, Seattle, Washington

ABSTRACT

A distributed, physically based hydrologic model that includes canopy interception, evaporation, transpiration, and snowmelt, as well as runoff generation at the 90-m scale, was applied to the Middle Fork Flathead River watershed in northwestern Montana.

Digital elevation data are used to model topographic controls on incoming solar radiation, air temperature, precipitation, and downslope water movement. Transpiration is simulated, based on a Penman-Monteith formulation that incorporates local net solar radiation, surface meteorology, soil type and current moisture status, and species-dependent leaf area index and stomatal resistance. Snow accumulation and ablation is modeled using an energy balance approach that includes the effects of local topography and vegetation cover. The 3,000-km^2, snowmelt-dominated study basin ranges in elevation from 900 to 3,000 m. The model was calibrated using 5 yr of recorded precipitation and stream flow. The model was verified against two additional years of runoff and against Advanced Very High Resolution Radiometer-based spatial snow-cover data at the 1-km2 scale available for 1990 and 1991 from the National Operational Hydrologic Remote Sensing Center, National Weather Service, NOAA. The snowcover data, available at roughly 2-week intervals, were used to access model simulation of distributed snowpack dynamics.

*Pacific Northwest Laboratory is operated by Battelle Memorial Institute for the U.S. Department of Energy under Contract DE-AC06-76RLO 1830.

TECHNOLOGY, ENERGY, AND ECONOMICS

CLIMATE CHANGE AND THE ENERGY SYSTEM: REGIONAL INFORMATION

W. S. Breed and R. H. Ball

Office of Environmental Analysis,
U.S. Department of Energy, Washington, D.C.

Key words: *Climate change, impacts assessment, energy sector, energy supply, energy demand*

ABSTRACT

Recent efforts to assess the effects of global climate change on the energy sector demonstrate the difficulties in reconciling the gap between global-scale predictions and regional data needs. However, the details of climate that are critical to each effect have yet to be determined.

Analysis of energy supply and demand adds sectorial twists, as energy use is pervasive in society. Considerations which are found to be either important or unpredictable from models include: differences in building shells, appliance characteristics and growth rates, peak demands on the system in response to climate change, and potential change in the sector characteristics due to mitigation activities or policies. Application of regional impact information for policy and planning requires detailed treatment of long planning lead-times and the long lifetimes of the technologies.

This paper summarizes interesting examples from some of the assessments that have been done to date. For energy demand, air-conditioning penetration and use, humidity, and increased irrigation are the important drivers. On the energy supply side, the output of thermal plants is largely unaffected by climate changes; however, the impacts of mitigation policies on this output and the effects of climate change on hydropower are uncertain. The vulnerability of the renewable energy supply to changes in climate is unknown; for example, wind, biomass, and geothermal energy may be impacted. In less-developed countries, deforestation and drought may be crucial for traditional biomass energy; increasing demand from industry and appliances are also important.

This analysis will point out the need for better regional data and integrated assessments, and identify issues for future design and analysis needs. Key research needs are discussed.

INTRODUCTION

Global climate change (GCC) could affect energy supply and demand in a number of ways. Energy supply systems comprise various activities, such as exploration, mining, agriculture, conversion, water management, and transportation. Energy is used intensively in many human activities, although only a few of these, especially space conditioning, appear to be highly climate-sensitive. Impacts on energy demand also affect the price and availability of supply. Hence, the types of potential GCC impacts on energy activities are quite varied. The general types of potential impacts were discussed in the IPCC Work Group II 1990 Impacts report and the 1992 Supplement (IPCC, 1990, 1992).

These and other studies have indicated some of the potentially important impacts that might occur and have estimated the magnitude of effects for some of them. The IPCC 1992 Supplement summarized several regional and national studies of key energy impacts. Salient conclusions (Ball and Breed, 1992) were:

1. Climate-sensitive energy uses include space conditioning, crop drying, and irrigation pumping. Detailed, region-specific studies are needed to assess aggregate demand impacts in temperate regions.
2. Hydroelectric power generation is one of the most sensitive energy sources. Impacts on power generation potential require careful analysis of detailed climate changes and hydrology in each basin.
3. Other centralized renewable energy systems, such as solar and biomass energy, are still in developmental stages and their vulnerabilities have not been investigated in detail. The impact on wind and solar systems depends, in part, on characteristics of the overall energy supply and demand system.
4. Traditional biomass energy (wood and crop wastes) forms a large fraction of the energy supply in many developing nations, especially in Africa. Hence, this energy supply is sensitive to the extent that forests and crops are vulnerable to GCC.
5. Thermal power plants can be affected by the availability of cooling water, the vulnerability of water intake structures to streamflow changes, the sensitivity of evaporative cooling systems to wet-bulb temperature, or to impacts that can simultaneously affect a large number of facilities in an interconnected electric grid.
6. In most developed countries, GCC impacts on energy supply and demand activities are generally regarded as likely to be relatively modest, not the most important, and amenable to adaptation. These impacts may be more important in developing countries, which are

more dependent on renewable energy and, perhaps, have less latitude for adaptation.

In discussing scales of analysis, we need to distinguish between the smallest units of analysis or "pixels" at which processes may need to be described and the size of overall domains that may be used to delimit a detailed case study or model domain. Some effects have natural study domains, such as major water basins for analyzing impacts on hydropower. Even in that case, we recognize that a larger study domain or coupling of domains may be relevant, because utility grids and interconnections may span major water basins, while competing water users have economic ties far beyond any individual basin. Of course, the pixel scale for hydrological effects will be sub-basin or even smaller scales, depending on the pertinent processes. It appears inevitable that water supply interactions should be analyzed on at least a water basin level, but the economic adjustments to impacts have to be viewed on even larger scales. Some of the biggest analytical challenges for GCC analysis concern the micrometeorological-scale pixel effects for which weather and climate changes cannot be readily modeled or even described by historical weather data.

The regional character or spatial and temporal scales of impacts may be determined by several factors, including: (1) the scales of weather and climate changes; (2) inherent scales of variation in the affected natural and man-made systems; (3) scales of organization of energy supply systems, such as utility grids; (4) linkages among systems, such as economic relationships. Due to those factors, the linkages between supply and demand of energy and the interactions among different economic activities affected by GCC, no single process scale of effect or spatial domain is likely to fit perfectly for analyzing all GCC impacts.

CURRENT SITUATION AND POLICY NEEDS

A key question is whether the study of climate change and impact assessment can offer utility planners any information on adaptation that would improve their system planning. Given current uncertainties and the nature of probabilistic forecasts for climate changes and impacts, could they improve system robustness with a reasonably expected return on the investment? Would it help to know that the probability distribution for weather events has become more broad, even if the direction of changes in mean value is unknown? For example, if one tells a hydropower developer that there is a greater probability for more extreme drought sequences than before, and also for more extreme wet periods, how would this change his reservoir planning and operation? Some recent experiences draw attention to extremes in climate variability, vulnerabilities and robustness in the system, and current policymakers' and planners' concerns.

Climate extremes experienced in the United States during the summers of 1988 and 1993 provide some evidence for the vulnerability of energy systems (particularly electric generating plants) to extreme events. Drought in the summer of 1988 resulted in decreases in available hydropower at a number of sites in North America. For example, by May of 1988 the Tennessee Valley Authority (TVA) had reservoir levels only 76% of normal and losses of about 46% of normal hydropower generation. However, some systems in North America experienced higher than normal hydropower availability due to higher than normal precipitation. Net hydropower losses, estimated at about 17,000 MW, resulted in economic losses. However, the North American electric system was able to cope with the combined losses in hydroelectric capacity, together with a comparable loss in thermal power and higher than normal peak loads (Energy Economist, 1988).

In the summer of 1993, different regions of the United States experienced drought, widespread floods, record high and low temperatures, and severe storms. Electricity demand in the mid-Atlantic states increased 20.8% from the previous year. Floods in the midwest caused problems for coal transport. Although coal prices rose and several coal plants were forced to shut down, alternative routes for coal transport, stockpiles, and fuel switching enabled electric systems to meet demands. One nuclear power plant in a flood area was forced to shut down due to safety concerns about the capability to evacuate communities. This situation also occurred at a Florida nuclear plant in 1992 in association with hurricane Andrew (Cutter Information Corp., 1993).

A recent survey of electric utilities found that only 33% of utility executives say that global warming has an impact on their company decisions, and it ranked only 17th in importance among environmental issues (Utility Environment Report, 1993). This suggests that the possibility of future changes in climate would not be an important consideration in system planning, although historical weather variations are likely to be factored in. However, some recent analyses point out the need, in long-range planning, to consider additional summer peaking due to climate warming (Ritsema, 1993).

In general, policy objectives for GCC impact analysis on energy include: (1) Providing estimates of the magnitude and range of net impacts on energy as a component of overall impacts (damages or benefits) on society; these are needed for benefit/cost evaluation of policy choices for response strategies. (2) Identifying policies and RD&D programs that could reduce the vulnerability and potential impacts on energy systems through fostering beneficial adaptation.

An immediate focus for analysis is the Second Assessment Report of the Intergovernmental Panel on Climate Change (IPCC) in 1995. This report will emphasize four major areas, including three relevant to this workshop:

- regional climate predictions, climate change, and their impacts;
- natural and man-made causes of desertification and drought and their impacts;
- methodologies for impact analyses, also for purposes of the climate change convention.

Support of negotiations on the climate convention is a primary purpose for writing the IPCC report. Other analyses of impacts will be relevant to supporting development of U.S. policies and negotiating positions. The Department of Energy (DOE) needs to consider the second objective of identifying RD&D and policies for adaptation of energy supply and energy-using systems. Other possible applications of impact and adaptation analysis include input to the U.S. Country Studies program, which assists developing countries in studying their greenhouse gas emissions and needs for adaptation, and the ADEPT program of technology development and transfer.

OFFICE OF ENVIRONMENTAL ANALYSIS IMPACT ASSESSMENT PROGRAM

The DOE has been involved in assessment of climate change impacts in the energy sector for several years. In addition to staff contributions to the 1990 IPCC assessment and the 1992 IPCC supplement, early projects included: support for the MINK project, a multisectoral analysis of climate change impacts on the Missouri-Iowa-Nebraska-Kansas region; the Pacific Northwest regional assessment at Pacific Northwest Laboratory (PNL) with the U.S. Environmental Protection Agency (EPA); and analysis of sea-level rise, employing a geographic information system at Oak Ridge National Laboratory (ORNL). Based on the conclusions and recommendations in the 1992 IPCC paper, we have under way a limited impacts assessment program, with contracts at the national laboratories.

At PNL, we have a research project to examine impacts of climate change on energy use in the buildings sector. This work is examining the change in energy required by buildings in various regions, looking at changes in temperature, humidity, insolation and wind, along with regional differences in building types and space-conditioning equipment.

We have two projects under way at ORNL. The first, in conjunction with TVA, is an engineering study to examine the vulnerability of various technology systems associated with thermal power generation to different climate variables. The second project is an engineering analysis of the vulnerability of different renewable energy technologies to climate change.

At the National Renewable Energy Laboratory, we have research under way to look at total system impacts of climate change on renewable energy systems. This work will estimate climate change impacts on elaborate systems such as biomass-to-energy farms.

At Lawrence Berkeley National Laboratory a study is under way on climate impacts on the urban heat island effect, possible mitigation measures, and impacts on building energy use and air pollution.

IMPACTS ON ENERGY SUPPLY AND DEMAND— REGIONAL DIFFERENCES

Regionally oriented studies provide some important insights about GCC impacts on the energy sector. Studies have been conducted from both top-down (national and regional) and bottom-up viewpoints, for estimates of change in climate-sensitive energy demand. Other studies have examined particular technologies and vulnerabilities or reactions to changing climate. There is as yet no single, uniform framework to assemble the disparate methods, input assumptions, or projections. A few results are worth noting.

As part of the energy supply system, thermal power plants are generally considered to be quite robust; they are a well-developed, mature technology, suitable for a wide range of climate variability. The TVA has had a climate change assessment program since the late 1980s which has arrived at some interesting conclusions. There are, of course, the expected minor losses in operating efficiency in a high-temperature regime, from high intake air and cooling-water temperatures. Limiting factors associated with climate change often turn out to be the environmental constraints placed on the plants for discharge of heated effluent waters. This is also the case for the safety constraints for nuclear plants placed on the availability of sufficient cool water for emergencies (Miller et al., 1992). Analysis of the generating system reveals another interaction: plants are likely to be under the most operating stress, such as de-rating to cope with reduced cooling-water availability or operating efficiency, when the system demand for electricity is at its highest. Also, the TVA system has a number of thermal plants operating along the same river, where a number of plants may be affected simultaneously by reduced flow or heated waters.

On the energy demand side, a recent study performed for the Wisconsin Public Power Commission illustrates the breadth of interest in this subject (Ritsema, 1993). The analysis examines the impact of climate change on energy demand for the four utilities that supply Wisconsin with power. Even in this limited area, there is a great difference in impact on the utilities, depending on the makeup of customers, and the seasonal use of energy by the customers. The author concludes that this kind of analysis has many uses, such as ensuring that adequate reserve margin is left to meet summer peaking needs, as the potential effect of climate change on summer peaking needs is not included in planning.

Climate sensitivity in the energy sector and in energy use varies greatly by region. Hydropower is sensitive to temperature and precipitation; in some regions, even increased precipitation does not lead to increased stream flow due to evaporative losses (Scott, 1991). Thermal plants are only slightly sensitive to thermal stress, but water supply problems can lead to power interruption (Miller et al., 1992). Energy consumption for residential and commercial space conditioning is the most climate-sensitive aspect of demand. Areas where seasonal demand is highest in the winter may have significantly lowered total annual energy needs (Aittoniemi, 1991). In contrast, regions with high or increasing air-conditioning market saturation may result in significant increase in summer daily peaking, with high associated costs for additional peaking capacity and moderate additional energy costs. Researchers have found that although costs to plan for climate change may be high, costs induced by climate change without planning can be four times higher.

The importance of climate change in long-range utility planning is highly region-specific, as illustrated by these examples. Highly aggregated analyses assume that winter benefits are interchangeable with summer electricity needs. Regional analyses point out the differences in energy source (fuel or electricity) and load (peak vs. base) and associated costs. We believe these differences cannot be ignored. Analyses may need to include smaller spatial and temporal details in order to capture all effects and operational limitations.

IMPACTS ON RENEWABLE ENERGY

Renewable energy technologies are an excellent example of the need for more accurate, more detailed regional climate data for impact analysis and policy needs. Changes in physical parameters, such as insolation, precipitation (quantity and timing during the year), wind, humidity and evapotranspiration, can all have impacts on the operation of these systems and their performance in an energy system. Will wind power continue to be viable where it is currently deployed? Could probabilistic forecasts of climate change be used to guide siting and designs of wind farms in the near future? Currently there is much public debate about the future use of biomass as a fuel for generating electricity and providing transportation fuels. What impact will climate change have on biomass-to-energy systems? On siting energy plantations in the future? Will use of biomass as an energy resource compete with efforts to sequester carbon by reduced harvesting of trees?

Some impact analyses are available for geothermal energy. Currently, this technology is successfully deployed in various areas of the world, from California to South America, from the Philippines to Iceland. In New Zealand, for example, geothermal energy accounts for 7% of total electricity generation. At the Geysers in California there has been a loss in generation, believed due to overuse of the resource and resultant loss of natural recharging of the water. In the 1990 New Zealand national climate change impacts study, scenarios based on increases in precipitation and recharge rates were projected to maintain the viability of the geothermal industry (Mundy, 1990).

Climate records and geological history associated with the West African Continental Rift exhibit the effect of water recharge. Currently, the rift area has an arid climate, and geothermal activity is weak at the surface of the volcanoes, yet hot springs associated with deeply circulated ground water are present in the rift valleys. Former hot spring activity in evidence at high elevations on the volcanoes indicates the region was more geothermally active in the past. Geological evidence indicates that activity has declined due to water tables that are lower than in past humid climates, rather than because of waning heat sources (Sturchio et al., 1993).

IMPACTS ON DEVELOPING COUNTRIES

In the case of climate change impacts in developing countries, two key issues are: (1) potential increases in the ownership and use of air conditioning and how climate change might influence this; and (2) potential impact of climate change on availability of traditional biomass fuels and consequent changes in energy use.

Developing countries that are located in tropical and semitropical regions tend to have less need for space heating than those in temperate zones, but they have greater future potential for increasing the use of air conditioning and other appliances. Compared with developed countries, per capita electrical appliance ownership often is low, but it is increasing very rapidly in some countries. For example, household saturation of refrigerators in Beijing, Peoples Republic of China, increased from near zero to over 60% in just 6 years. Use of air conditioners also is increasing among upper-income groups in some countries, such as Malaysia (Meyers and Sathaye, 1985). Hence, the magnitude and patterns of energy use in many developing countries are likely to change rapidly in the future, perhaps posing a sharply increased demand for electricity.

Biomass energy includes traditional sources such as fuelwood, crop wastes, and dung used in cooking and heating; wood wastes and municipal wastes combusted for heat and power; and wood converted to charcoal. Potential climate impacts on biomass production for energy use should be similar to impacts on conventional agriculture and forestry except for differences in the scope of management practices. Many traditional biomass sources, such as gathering of fuelwood from unmanaged natural forests, are susceptible to shifts in climate that are unfavorable to the natural species and ecosystems. Intensively managed biomass systems using annual or short-rotation crops have wider latitude to adjust to climate change by changing plant species, irrigation, pest control, and other agricultural management practices, subject to limitations of economic viability and availability of land and water resources.

SUMMARY

Estimating the net impacts of GCC on energy systems is highly uncertain due to several factors: (1) regional patterns of future climate change are not reliably predictable at this time; (2) the nature of future energy supply and energy-using systems and their societal context at the time when impacts may occur is not well defined; and (3) the degree of adaptation by the private sector is uncertain. However, some important policy conclusions may be possible in spite of large uncertainties. The DOE policy objectives may be served, in part, to the extent that robust adaptation policies and technologies can be determined, even if the potential net damage due to GCC cannot be estimated accurately.

Since interactions among impacted sectors occur across various regional boundaries, analyses of responses to climate impacts may have to be treated in a hierarchical manner. This would require integral (strongly coupled interactions) analysis within a natural domain, such as a river basin or utility grid, with iterative (weakly coupled) interactions among river basins or other political/economic units.

If substantial climate change is several decades away, is there time to develop technological and institutional mechanisms for adapting energy systems to climate change and to incorporate those changes in a portion of the future capital stock of facilities? Stocks of buildings and space-conditioning systems may change slowly, but they can still be modified at some cost. However, unknown variations in climate are possible in the near term as a result of natural variations and past anthropogenic influences, while future anthropogenic changes are not yet fully predictable. Hence there may be a case for incorporating greater climate robustness into energy systems and buildings, depending on the costs. In any case, RD&D programs for new energy systems should consider vulnerability to climate change.

A key question for DOE is: What policies and actions are needed in the near future to help adjust to potential future climate change? That is, what can the government do to help the private sector and government agencies to adopt optimal adaptation strategies? The other key question is: What are the ranges and likely values of damages due to climate changes for given levels of greenhouse gas emissions? Characterizing potential impacts is a first step toward answering both of those questions, provided the appropriate parameters and variations are treated.

REFERENCES

Aittoniemi, P. **1991**. Influences of climate change on the Finnish energy economy. In: *Energy and Environment, 1991*, E Kainlauri, A Johanson, I Kurki-Suonio and M Geshwiler, eds. American Society of Heating, Refrigerating, and Air-Conditioning Engineers, Inc., New York.

Ball, RH and WS Breed. **1992**. Summary of Likely Impacts of Climate Change on the Energy Sector, submission to IPCC WG II for *Climate Change 1992: Supplementary Report to the IPCC Impacts Assessment.* (unpublished)

Cutter Information Corporation. **1993**. Feature Report in *Energy, Economics, and Climate Change*, August, 1993.

Energy Economist. **1988**. It is Not Just Farmers who Suffer from Drought." 81:7 July.

IPCC WG II. **1990**. *Climate Change: the IPCC Impacts Assessment*, McG. Tegart, WJ, GW Sheldon, and DC Griffiths, eds. Intergovernmental Panel on Climate Change, WMO/UNEP. Australian Government Publishing Service, Canberra, Australia.

IPCC WG II. **1992**. *Climate Change 1992: the Supplementary report to the IPCC Impacts Assessment*, McG. Tegart, WJ, and GW Sheldon, eds. Intergovernmental Panel on Climate Change, WMO/UNEP. Australian Government Publishing Service, Canberra, Australia.

Meyers, S and J Sathaye. **1985**. Energy use in cities in developing countries. Annu. Rev. Energy 10:109-133.

Miller, BA, et al. **1992**. *Impact of Incremental Changes in Meteorology on Thermal Compliance and Power System Operations* (interim report). Tennessee Valley Authority Engineering Laboratory, Norris, TN.

Mundy, C. **1990**. Energy. In: *Climate Change: Impacts on New Zealand*. New Zealand Ministry of the Environment.

Ritsema, MA. **1993**. Fire and ice, a method to assess climate change on peak demand. Public Utilities Fortnightly 130:45-47.

Scott, MJ. **1991**. *Effects of Climate Change on Pacific Northwest Water-Related Resources: Summary of Preliminary Findings*. U.S. Dept. of Energy (in review).

Sturchio, NC, PN Dunkley, and M Smith. **1993**. Climate-driven variations in geothermal activity in the northern Kenya Rift Valley. Nature 362:233-234.

Utility Environment Report. **1993**. *Global Warming Remains Low Priority for Electric Utilities*, Survey Finds.

CLIMATE UNCERTAINTY AND THE REGIONAL ECONOMIC IMPACTS OF GLOBAL CLIMATE CHANGE

M. J. Scott, A. M. Liebetrau, and C. A. Lopresti

Pacific Northwest Laboratory, Richland, Washington

ABSTRACT

During the period 1989 to 1992, the Department of Energy funded pioneering research analyzing the effects of climate change and carbon dioxide fertilization at the regional level in the Missouri-Iowa-Nebraska-Kansas (MINK) region. That study used existing process models of the agricultural, water, energy, and forestry sectors and of the regional economy to conduct a wide-ranging, integrated analysis of the regional economic effects of climate change.

However, neither the full integration of the models nor the full investigation of the uncertainties of climate and economic response to climate was accomplished, because (1) the process models were expensive to run, and (2) the MINK study used a single rich historical climate analog—the 1930s "Dust Bowl" period—to depict climate change. More analysis was required to develop ranges of climate influences over more extreme climate conditions (as well as probabilities of their occurrence). To do this, a more practical, inexpensive method was needed than was available in the MINK study.

In this paper, the MINK process models are integrated hierarchically in reduced form to efficiently explore uncertainty issues surrounding climate change effects. We summarize the essential features of the process models, using Multivariate Adaptive Regression Spline (MARS) to develop "response surfaces" (approximations to the underlying models). These response surfaces are integrated into a simple, reduced-form computer model, which is used to investigate the impacts of, and sensitivities to, alternative climate and technological and institutional response scenarios for the MINK region.

THE EFFECTS OF REGIONAL CLIMATE CHANGE ON SPACE CONDITIONING NEEDS AND THE ENERGY INDUSTRY IN THE GREAT LAKES REGION

M. E. Fernau,[1] E. D. Maloney,[2] and G. T. Bates[3]

[1]Sigma Research Corporation, Concord, Massachusetts

[2]Argonne National Laboratory, Argonne, Illinois

[3]National Center for Atmospheric Research, Boulder, Colorado

Key words: *(Regional) climate modeling, climate change, climate change effects, Great Lakes, utility industry*

ABSTRACT

To date, studies of the effects of potential climate change on energy use and demand have been done on a macro scale or with coarse model data. In reality, regional climate change effects will determine the actual behavior of energy users. The output from a 3-yr simulation of the coupled National Center for Atmospheric Research (NCAR) Community Climate model MM4 regional climate modeling system is used to examine changes in average temperature and temperature variability on a regional scale, the impacts of such change on the need for space conditioning in the Great Lakes region, and the subsequent changes in energy demand. The NCAR modeling system uses general circulation model results to drive a more highly resolved (60-km) mesoscale model to produce a detailed regional climate. A 3-yr run of both base case and doubled carbon dioxide climate for the United States has been produced. From these results, changes in heating and cooling degree days, and changes in consecutive days above or below various temperature thresholds were calculated. Heating and cooling energy use intensities that are representative of the residential building stock found in the region were used to convert climate data to energy demand. The implications for the energy industry are discussed. The model results indicate that the changed climate under doubled carbon dioxide conditions would have large impacts on energy use, although it is difficult to determine the balance between decreased heating needs and increased cooling needs. It was found that biases present in the temperature output of the modeling system are greater for the Great Lakes region than for the rest of the United States. This problem limits the usefulness of the current data set for determining the effects of climate change on energy use in that area. However, the model shows promise for regional studies. For example, recent successful coupling of a one-dimensional thermal eddy diffusion model to the NCAR modeling system to represent the Great Lakes suggests that the next iteration of climate change output from the NCAR system will yield important results when applied to effects studies.

INTRODUCTION

Policy-makers must be confident that they understand the likely effects of climate change on the regional scale that is of interest to their constituents before they will commit significant resources to mitigating potential climate change (Science and Policy Associates Inc., 1992). Effects studies that try to determine regional-scale impacts of climate change have been hampered by the poor spatial resolution (400 to 1000 km) of the general circulation models on which the climate change used in the studies is based (Smith and Tirpak, 1989; Takle and Zhong, 1991).

We use a new set of model results with much improved resolution to illustrate that output from a mesoscale model driven by general circulation model output can yield useful climate change information on a regional basis for use until improved computer technology makes possible running of general circulation models at high resolution. The National Center for Atmospheric Research (NCAR) has developed a modeling system in which a limited area mesoscale model known as MM4 is run for extended simulation periods (months to years) with a grid cell size of 100 km or less to yield a regional climate (Dickinson et al., 1989). MM4 has been modified to incorporate a GCM-compatible radiative transfer scheme, an improved boundary-layer parameterization, and a surface physics/soil hydrology package. The initial and boundary conditions for MM4 are provided either by gridded observation data or by output from a GCM. If the GCM output is from a future climate scenario, then climate change scenarios that are based on the large-scale circulation changes but that incorporate regional-scale features and effects can be generated with the modeling system. We use output from this modeling system to examine potential effects of climate change on energy demand in the Great Lakes region.

DESCRIPTION OF THE DATA SET

The NCAR regional climate modeling system is the first instance in which a mesoscale weather prediction model has been run without reinitialization for an extended simulation period of months or years. Results of an extensive comparison with the observations and the output from the GCM that are used to drive the model show that when driven by observations, the model system produces a realistic climate. When driven by GCM output, it retains the large-scale circulation information from the GCM but captures much more realistically the influences on climate of smaller-scale features such as complex terrain and water bodies (Giorgi et al., 1993a,b and additional references cited in these two papers).

These authors (Giorgi et al., 1993b) used the NCAR regional climate modeling system to produce two 3-yr simulations of the climate over the United States, at a 60-km resolution. One simulation used a present-day atmospheric carbon dioxide concentration of 340 ppmv, and the other used a value of 680 ppmv carbon dioxide to represent a doubled CO_2 scenario. Both cases were driven by the last 3 yr of 24-yr climate simulations of a version of the NCAR Community Climate Model (CCM) GCM known as GENESIS, described in Giorgi et al., 1993b. GENESIS is coupled to a 50-m mixed-layer ocean model, has a horizontal grid spacing of 4.5° × 7.5°, and has an enhanced surface physics package that has a 2° × 2° resolution. GENESIS represents the Great Lakes by three data points, while MM4 uses 64 grid cells.

Giorgi et al. (1993b) found that the average daily air temperature at 40 m for the current climate simulation using MM4 driven by GENESIS had an overall bias for the United States of +1.1°C (overprediction) in the five coldest months and –0.7°C (underprediction) for the five warmest months, but in the Great Lakes region the modeling system overpredicts temperatures by about 5°C. A similar bias is present in the GENESIS output. Apparently, the increased resolution in the Great Lakes area compared to that in the driving GCM is not sufficient to eliminate the bias that is present in the GCM. However, MM4 did moderate temperatures in the vicinity of the lakes, compared to GENESIS, and predicted about 1°C less warming in the Great Lakes under doubled CO_2 compared to GENESIS predictions. In the 1993b paper by Giorgi et al. the relatively short 3-yr, 40-m simulated temperatures are compared to observed, long-term, normal surface temperatures. It is possible that with a longer MM4 simulation a different bias might be obtained.

Recently, Bates et al. (1993) demonstrated a successful coupling of a one-dimensional thermal eddy diffusion model to the NCAR modeling system to represent the Great Lakes. The output gave more realistic behavior for lake temperatures. It is planned to generate new U.S. climate simulations for present-day and doubled carbon dioxide conditions using MM4 with the lake model incorporated. In addition, improvements have been made to the driving GCM. These simulations should yield a much reduced bias in the Great Lakes temperature field.

When this output is available, the analysis described in this paper will be repeated.

METHODS

We used the current-day (denoted 1 X CO_2) and the doubled carbon dioxide (2 X CO_2) MM4 simulations reported in Giorgi et al., 1993b, as the starting point for this analysis. The daily maximum and minimum temperatures at 40 m above ground level and in the air within the vegetation canopy foliage were extracted for a region encompassing the Great Lakes (Figure 1). They were converted to a surface air temperature using the following approximation (Dickinson et al., 1986; Giorgi et al., 1993a):

$$T_a = T_f + 0.55(T_m - T_f),$$

where T_a is the surface air temperature, T_f is the vegetation canopy air temperature, and T_m is the temperature at the lowest model. If no vegetation is present, the canopy temperature is set to the ground air temperature.

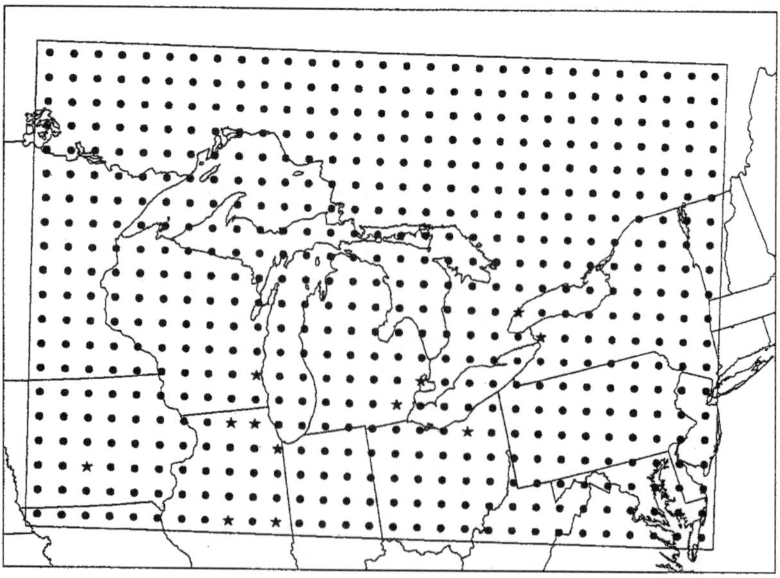

Figure 1. Grid point resolution of National Center for Climate Research regional climate modeling system used in study. Starred points correspond to points used to analyze climate change at major cities in Great Lakes region.

Using the derived surface air temperatures, the maximum and minimum daily surface temperatures were averaged to produce a daily mean temperature. This value was used to compute heating degree days, cooling degree days, and the difference in degree days between the control case and the doubled CO_2 case. The degree days were based on the common base of 65°F used by utilities, engineers, and the National Weather Service.

RESULTS

We selected 12 locations for analysis: Buffalo, Midway, and O'Hare airports in Chicago, Cleveland, Des Moines; City and Metro airports near Detroit, Milwaukee, and Toronto; and airports in Rockford, Springfield, and Urbana, Illinois. We obtained monthly temperature and degree-day data for these locations from National Weather Service data or other sources.

The model grid cell nearest to each location was determined using latitudes and longitudes and was used for comparison to observations. We examined several interpolation schemes but because of large differences in simulated temperature between the lakes and land in the vicinity of several locations these yielded poor results. For instance, observed annual mean heating degree-days differ by about 7% between Rockford and Chicago O'Hare, but when interpolating $1 \times CO_2$ case using the four closest grid cells, the simulated heating degree-days differ by over 20%. This is because the modeled Rockford value is based on four land grid cells, while the Chicago value is interpolated from both land and water cells. Therefore, we used the simulated values from the nearest grid cell for comparison to observations. The exception was at Chicago O'Hare, where the second closest grid point was used because the closest cell was located over the lake and compared poorly to observations. Figure 1 shows which grid points were used for each location.

For the $1 \times CO_2$ case, the simulated daily maxima and minima were used to obtain monthly mean maximum, minimum, and average temperature for each of the available 36 months. From these, the monthly and annual 3-yr averages were calculated, then compared to the observed monthly averages at each location. The temperatures were converted to degrees Fahrenheit because that is the unit for which heating degree-days and cooling degree-days are calculated in the United States.

The effect of the lakes is evident in the simulated monthly temperature averages. Figures 2 and 3 show the average temperatures for December through February and June through August for the 1 × CO_2 run. The contours follow the lakes closely.

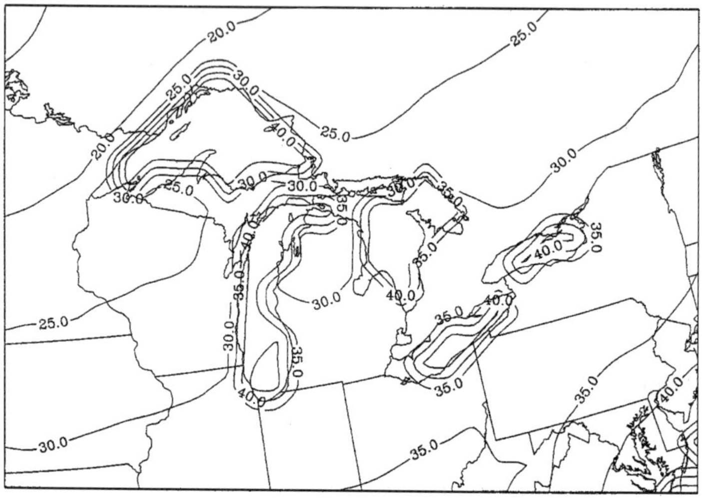

Figure 2. Average temperatures in degrees Fahrenheit for winter (December, January and February) for the 3-yr control simulation using MM4 driven by a general circulation model. Influence of the lakes is clearly evident in the contour pattern.

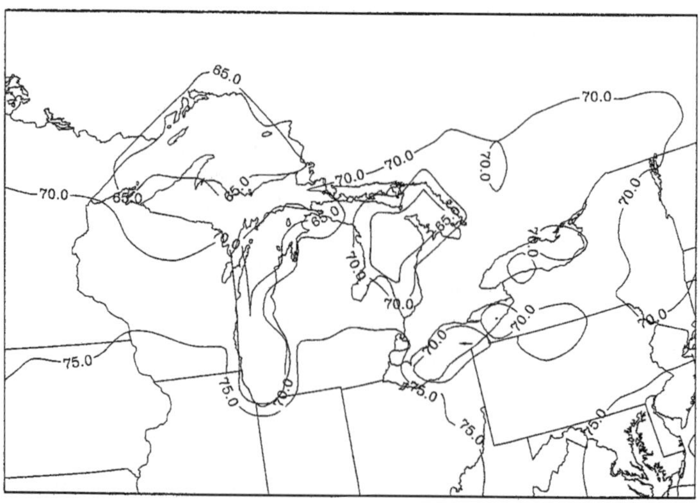

Figure 3. As in Figure 2, for summer (June, July, and August).

The bias discussed by Giorgi et al. (1993b) for the Great Lakes region was obvious in the 1 × CO_2 model results. The bias in the maximum, minimum, and average annual and monthly mean temperatures was calculated by comparing the model results with the observations. Over the 12 locations, the modeled maximum temperature averaged 4.7°F above observations, the minima averaged 6.7°F too warm, and the average temperature was 5.7°F too warm. The largest biases appeared in the summer months for maximum temperatures. The least bias, at times even negative, appeared in the spring months for minimum temperatures. Figure 4 shows the monthly mean observed and simulated temperatures for Midway airport in Chicago to illustrate the bias. This bias leads to annual heating and cooling degree-days for the 1 × CO_2 simulation that also are biased. For instance, the heating degree-days for Midway are too low by 15% and the cooling degree-days are too high by 50%.

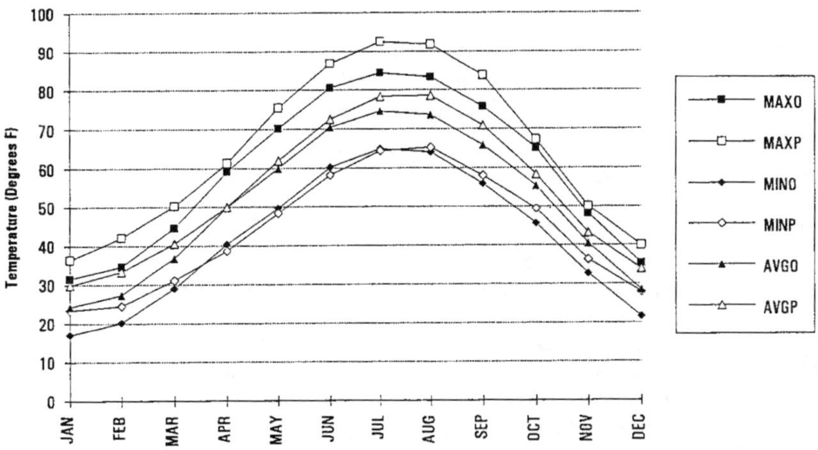

Figure 4. Monthly mean maximum (MAX), minimum (MIN) and average (AVG) temperatures for observed temperatures (solid shapes) and simulated base case temperatures (open shapes) at Chicago Midway Airport. Warm bias in simulated temperatures is evident.

Because of the large bias in the Great Lakes simulated temperatures, the absolute numbers for the two MM4 runs cannot be used for the energy analysis. We subtracted the monthly biases in the maximum and minimum temperature at each location from the simulated daily maxima and minima before calculating daily degree-days and the changes in degree-days between scenarios. For Toronto, the 2 months surrounding the midseason month were assigned the bias corresponding to the midseason month. Application of this method removed the bias from the 1 × CO_2 temperatures, leading to adjusted monthly mean temperatures and degree-days at each location that were almost identical to the observed means.

The changes in annual heating and cooling degree-days at each location that result after the biases are removed from both the 1 × CO_2 and 2 × CO_2 data sets are shown in Figures 5 and 6, respectively. Heating degree-day changes were very consistent from site to site, with the decrease varying from 31 to 34%. Under 2 × CO_2 conditions, heating degree-days disappeared entirely during July and August and dwindled to almost none in June and September. Cooling degree-days increased substantially in the 2 × CO_2 simulation. All locations but Des Moines, Springfield, and Urbana increase by 100% or more, with Buffalo and Milwaukee increasing by over 160%. Toronto almost triples, but this is probably an artifact of the way that the bias was assigned for that location. The spring and autumn biases at Toronto are very large compared to those at other sites, and these are assigned to all 3 months. The transition months of April, May, September, and October have substantially more cooling degree-days at each site under doubled CO_2 conditions.

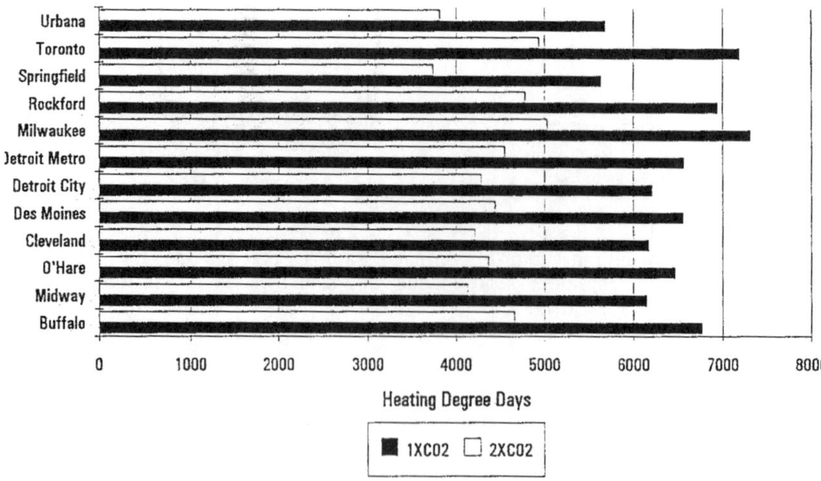

Figure 5. Change in 3-yr annual average heating degree-days at selected locations in the Great Lakes region between doubled CO_2 conditions (2 × CO_2) and present-day conditions (1 × CO_2), as simulated by MM4.

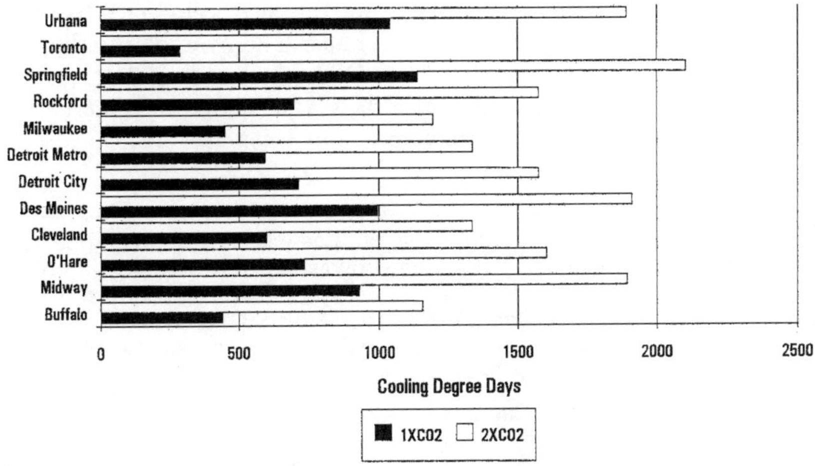

Figure 6. Change in 3-yr annual average cooling degree-days at selected locations in the Great Lakes region between doubled CO_2 conditions ($2 \times CO_2$) and present-day conditions ($1 \times CO_2$), as simulated by MM4.

After adjusting for the model bias, we examined the occurrence and frequency of temperature extremes. The dates of first and last freezes, defined as the last day of winter or spring and the first day of autumn or winter with a minimum below 32°F, were determined for the various locations (all but Detroit City and Chicago Midway) under the two scenarios. The effect of doubling CO_2 was to extend the frost-free period for almost all years and sites. The changes ranged from shrinking by 1 day to extending by 104 days the number of annual days between freezes. Urbana averaged an 83-day extension for the 3 yr, while Milwaukee only averaged 17. Most sites averaged 1 to 2 months more frost-free days. There was substantial year to year variation; Milwaukee's average of 17 consisted of a shrinkage of 1 day and extensions of 6 and 45 days; the extension at Chicago O'Hare varied from 7 to 57 days.

We examined changes in extremely warm days at three sites (Table 1). The period between the first and last occurrence of temperatures 90°F and above increased by an average of 3 to 5 weeks, with a range from a decrease of 11 days to an increase of 7 weeks, for the 3 simulated yr with a doubling of CO_2. Under $2 \times CO_2$ conditions, the first occurrence of temperatures at 95°F or above for the three sites shifted into late

May or early to mid-June from July, and the last occurrence usually shifted from July or August into September. The average number of days at or above 90°F increased by 32 at O'Hare, 38 at Des Moines, and 32 at Urbana. The average number of days at or above 95°F increased by 15 at O'Hare, 20 at Des Moines, and 14 at Urbana.

Table 1. Changes in extremely warm days at three locations between doubled CO_2 conditions ($2 \times CO_2$) and present-day conditions ($1 \times CO_2$), as simulated by MM4. Change Range 90 denotes the change in the length of the period between the first and last occurrence of temperatures 90°F and above, in days. Change Days 90 and Change Days 95 denote the change in the total number of days at or above 90°F and 95°F, respectively.

	Change Range 90	Change Days 90	Change Days 95
O'Hare Year 1	30	50	17
O'Hare Year 2	37	31	13
O'Hare Year 3	6	15	14
Des Moines Year 1	49	37	12
Des Moines Year 2	19	48	22
Des Moines Year 3	-11	30	26
Urbana Year 1	47	37	13
Urbana Year 2	23	35	18
Urbana Year 3	31	22	11

Consecutive day runs of extreme temperatures increased substantially at the three sites. For the control run, O'Hare only had one instance of 2 consecutive days at or above 95°F. Under $2 \times CO_2$, there were four instances of 2 consecutive days, two instances of 3 days, one instance of 4 days, two instances of 5 days and one instance of 6 consecutive days at or above 95°F. Ninety °F temperatures at O'Hare never occurred more than 5 days in a row under $1 \times CO_2$. There were seven instances of at least 5 consecutive days of 90°F or above under $2 \times CO_2$, one of which lasted for 25 consecutive days! Similar results were found for Des Moines and Urbana, with the number of long temperature runs increasing substantially. Under $2 \times CO_2$, Des Moines had one run of 10 consecutive days at or above 95°F and runs of 16, 13 and 21 days in a row at or above 90°F. Urbana went from no runs of 3 or more days of 95°F to eight runs at this threshold, the longest lasting 6 days.

DISCUSSION

We present a first use of simulated regional climate change from a limited area model driven by a general circulation model to examine changes in degree-days and temperature extremes on a regional scale. Because of uncertainties present in the driving GCM and the simple fashion in which the Great Lakes are portrayed in both models, the quantitative results presented here must be considered preliminary examples of the analysis that can be conducted with the coupled modeling approach as technical improvements reduce the model bias. Until GCM can achieve a resolution of less than 100 km, the coupled approach with improved GCM input and an improved treatment of the Great Lakes is a viable substitute for providing interim guidance to policy-makers who must grapple with decisions regarding the mitigation of potential climate change and its impacts on energy demand and other areas.

Studies (Smith and Tirpak, 1989; Linder et al., 1989) that used low-resolution output from GCM to drive energy demand models have suggested that trace-gas-induced climate changes could have major impacts on energy demand and peak capacity requirements of utilities. Increased air conditioning will require additional energy during the warmest periods. In more northerly areas this will be offset to a minor or major extent by decreased heating needs.

The qualitative trends described in this paper, derived from the more highly resolved coupled modeling system, support this conclusion and point out the potential magnitude of the problem over the next century. The 30% decrease in heating degree-days in the Great Lakes region with $2 \times CO_2$ implies a decrease in heating needs of similar magnitude (Blair, 1992). There is less agreement on the relationship between cooling degree-days and fuel use, because that depends also on humidity and the degree to which people will tolerate warmer weather. However, the doubling of cooling degree-days and the large increase in the number and persistence of very warm days under $2 \times CO_2$ suggest that growth in energy demand for summer air conditioning may be larger than the reduction in energy use that will result from lowered heating needs in the Great Lakes region and surrounding areas. The disparity in the magnitudes of these two trends suggests that this result likely will be upheld when results from the improved NCAR regional climate modeling system become available.

Sebold and Parris (1989) derived regression equations that relate residential end-use unit energy consumption to heating and cooling degree-days for conventional space heating and conventional central air conditioning, respectively. Application of these relationships to O'Hare and Urbana gives the result that unit energy consumption for space conditioning actually will decrease by about 2000 kwh per year per residential dwelling at these two locations under 2 × CO_2 conditions, contradicting the above intuitive result. However, these relationships contain considerable uncertainty, expressed as scatter about the regression line. Also, they are based on a limited sample and do not consider commercial changes. Thus, it is difficult to ascertain their accuracy.

Regardless of whether the impact on energy use is positive or negative, the NCAR regional climate modeling system predicts large changes in temperatures and, thus, degree-days under doubled CO_2 conditions. It is unlikely that energy demand will be unaffected if such conditions come to pass. The concept of using a mesoscale model driven by a general circulation model to simulate climate has been tested rigorously and yields useful insights into the potential impacts of climate change on a regional scale. These insights will be revisited and improved in further work as more sophisticated versions of the modeling system are implemented.

ACKNOWLEDGMENT

Work supported by the U.S. Department of Energy under contract W-31-109-Eng-38.

REFERENCES

Bates, GT, F Giorgi, and SW Hostetler. **1993**. Toward the simulation of the effects of the Great Lakes on regional climate. Mon. Weather Rev. 121:1373-1387.

Blair, FE, ed. **1992**. *The Weather Almanac*. Gale Research, Detroit, MI.

Dickinson, RE, PJ Kennedy, A Henderson-Sellers, and M Wilson. **1986**. *Biosphere-Atmosphere Transfer Scheme (BATS) for the NCAR*

Community Climate Model, NCAR/TN-275+STR. National Center for Atmospheric Research, Boulder, CO.

Dickinson, RE, RM Errico, F Giorgi, and GT Bates. **1989**. A regional climate model for the western United States. Clim. Change 15:383-422.

Giorgi, F, GT Bates, and SJ Nieman. **1993a**. The multiyear surface climatology of a regional atmospheric model over the western United States. J. Climate 6:75-95.

Giorgi, F, CS Brodeur, and GT Bates. **1993b**. Regional climate change scenarios over the United States produced with a nested regional climate model: spatial and seasonal characteristics. J. Climate, accepted for publication.

Linder, KP, MJ Gibbs, and MR Inglis. **1989**. *Potential Impacts of Climate Change on Electric Utilities*, Report No. EN-6249. Electric Power Research Institute, Palo Alto, CA. (Available from Electric Power Research Institute Distribution Center, Pleasant Hill, CA.)

Science and Policy Associates, Inc. **1992**. *Joint Climate Project to Address Decision Makers' Uncertainties, Report of Findings, May 1992*, Report No. TR-100772. Available from Electric Power Research Institute Distribution Center, Pleasant Hill, CA.

Sebold, FD and KM Parris. **1989**. *Residential End-Use Energy Consumption: A Survey of Conditional Demand Estimates*, Report No. CU-6487. Electric Power Research Institute, Palo Alto, CA. (Available from Electric Power Research Institute Distribution Center, Pleasant Hill, CA.)

Smith, JB and D Tirpak, eds. **1989**. *The Potential Effects of Global Climate Change on the United States: Report to Congress*, EPA-230-05-89-050. U.S. Environmental Protection Agency Office of Policy, Planning and Evaluation, Washington, DC.

Takle, ES and S Zhong. **1991**. Iowa's climate as projected by the global climate model of the Goddard Institute for Space Studies for a doubling of atmospheric carbon dioxide. J. Iowa Acad. Sci. 98:153-158.

THE HEALTH EFFECTS OF CLIMATE CHANGE

MIGRATION AND GLOBAL CHANGE: REGIONAL ISSUES IN HUMAN HEALTH AND WELFARE

J. Longstreth

Pacific Northwest Laboratory, Richland, Washington

ABSTRACT

Assessment of the human health and welfare impacts of global change represents a significant challenge to public health practitioners because of the complexity involved in humans' relationships with their environment. There are a few direct relationships, e.g., heat stress; however, by far the majority of interactions can be considered either secondary (one step removed) or tertiary (two or more steps removed). An example of a secondary effect would be changes in the distribution of disease to changes in the habitat of the vectors that carry them. Examples of tertiary effects would be an increase in infections in the host c ountry which receives refugees driven out of their homes by sea-level rise or drought-induced agricultural failure. As would be expected, estimating the impact of global change on human health and welfare increases in difficulty the less direct the relationship, because so many other factors need to be accounted for. In addition, all of the likely effects of global change on human health already exist in human populations, so that risk assessors have to contend with often significant background levels of a particular effect of interest.

I believe that a significant effect of global change will be an increase in human migration and that this increase will be accompanied by significant shifts in the distributions and incidences of a variety of infectious diseases. Believing this and establishing this are two entirely different things, however. This presentation will discuss our efforts to ascertain the relationship of emigrants to the importation of infectious diseases into the United States and current efforts to use migration out of the Great Plains during droughts of the "Dust Bowl" era and later as analogs for current migration from Mexico to the United States. Our goal in the latter study is to examine the impact of governmental intervention in the United States on drought-driven migration in order to explore policy options for modifying migration from Mexico using target agricultural support programs. Our overall hypothesis is that these two sorts of studies, coupled together, can help develop convincing evidence that although this is already a real problem today, under conditions of global change it is likely to get significantly worse.

THE DELIVERY OF CLIMATIC CHANGE PROJECTIONS: THE MECCA ANALYSIS PROJECT

C. Brescianini, W. Howe, and A. Henderson-Sellers

Climatic Impacts Centre, Macquarie University, Sydney, Australia

Key words: *Climate change, Greenhouse effect, Asthma, Impacts, Human health*

ABSTRACT

The Model Evaluation Consortium for Climate Assessment (MECCA) is a large research effort designed to quantify the probable range of future climatic change, provide policy makers with important information, and identify key topics of future research. To achieve these goals, MECCA's research plan includes two key elements: a set of carefully selected computer-based model experiments, and an analysis strategy that focuses on impact-orientated needs. One impact area of particular interest to the MECCA project is human heath. Here, we report on the delivery of the MECCA-sponsored climate-model information to the impact-modeling community. To give a demonstration of the use of these data we perform a simple analysis involving the potential distribution of asthma.

INTRODUCTION

The effectiveness of international health policies that are related to climatic change is significantly dependent on the reliability of climate-change forecasts. The world's climate is an immensely complicated system, and our predictions of the future climate must be made by simplified models of this system. These models vary from simple one-dimensional radiation balances of the earth to complex, fully three-dimensional computer simulations. The three-dimensional computer models on the upper end of this complexity range are commonly referred to as General Circulation Models, or GCM. These models represent the most advanced and sophisticated ever devised to simulate the world's climate. The models typically involve thousands of equations and hundreds of variables, and the task of solving these equations in a time-dependent fashion is enormous. Producing just one century-long climate forecast with a GCM can require months of processing time on a modern, state-of-the-art supercomputer.

The task of modeling the world's climate is so enormous that not even GCM are able to model with fine resolution all of the important physical processes that actually occur. Instead, numerous assumptions about these processes are made, and the grid resolution typically must be limited to the order of a few hundred kilometers. To further complicate matters there are also uncertainties in initial and boundary conditions, as well as in the observations used to test the climate models. As a result of these modeling assumptions, uncertainties are introduced into future-climate predictions.

Obviously, a health assessor, impact modeler, or policymaker would like know what the future climate will be like and, also, what the uncertainties are in these predictions. With the aim of providing accurate predictions of the world's future climate, as well as addressing the uncertainties in these predictions, the Model Evaluation Consortium for Climate Assessment (or MECCA) was established in 1991. MECCA is an international research effort that is sponsored by a consortium of public and private organizations. The original MECCA partners, and the partners added in 1992, are shown in Table 1. The MECCA partners proposed a 3-yr research plan with an estimated cost of $22 million. MECCA aimed to perform a variety of climatic predictions, using various state-of-the-art GCM from many leading research groups around the world. Central to the plan was the use of a dedicated supercomputing laboratory located at the National Center for Atmospheric Research (NCAR) in Boulder, Colorado, which would be used to perform the climate simulations and to store and analyze the extremely large data sets created by the GCM simulations.

Table 1. MECCA Partners.

University Corporation for Atmospheric Research (UCAR), USA
Electric Power Research Institute (EPRI), USA
Central Research Institute of Electric Power Industry (CRIEPI), Japan
Electricité de France (EdF), France
Italian Agency for New Technologies, Energy and Environment (ENEA), Italy
National Supercomputing Center for Energy and the Environment, USA
N.V. KEMA, The Netherlands
Southern California Edison Co., USA

The MECCA simulations were divided into Phase 1 and Phase 2 experiments, with 14 research groups taking part in phase 1 and 6 groups in phase 2. The phase 1 experiments were carried out on a time scale typically on the order of decades. A summary of the phase 1 experiments can be seen in Table 2. All of the phase 1 experiments have now been completed. The phase 2 experiments are currently under way and are focussing on a smaller number of high-resolution simulations where greenhouse-gas concentrations increase gradually over a long period of time (up to 100 yr).

Table 2. MECCA Phase 1 Experiments

1. Oglesby and Saltzman, GCM Sensitivity to Changing Atmospheric Carbon Dioxide Concentration
2. Henderson-Sellers, Tropical Deforestation: Local, Regional and Global Consequences
3. Washington and Meehl, Climate Sensitivity Experiments with Improved GCM
4. Williamson, Climate Model Sensitivity to Grid Resolution
5. Giorgi, Marinucci, Visconti, and De Canio, Regional Climate Change Scenarios for the United States and Europe
6. Thompson, Pollard, and Bergengren, GCM Surface Sensitivity Experiments
7. Krishnamurti, Climate Modelling with Emphasis on Tropical Rain
8. Wang, Assessment of the Climate Effect of Other Greenhouse Gases
9. McAvaney, Sensitivity of an Atmospheric GCM to Mass Flux Schemes
10. Meleshko, Sensitivity of Equilibrium Climate in a GCM to Cloud Parameterization Schemes
11. Holland and Pinardi, Sensitivity of Climate Simulations to Changes in Ocean Circulations
12. Schneider and Kinter, Internal Variability in Long-Term Climate Simulations
13. Chervin and Semtner, Sensitivity of Ocean Model Circulation to Atmospheric Forcing
14. Kittel, Regional Climate Change Impacts on Northern Great Plains Ecosystems

THE MECCA ANALYSIS TEAM

One of the special features of MECCA has been the establishment of a MECCA Analysis Team. The analysis team has as its overall objectives (1) the systematic examination of the results produced by the MECCA project, (2) the synthesis of MECCA assessments with results from other international projects, and (3) the effective dissemination of the MECCA results to the scientific, policymaking, and general communities.

The MECCA Analysis Team activities began in July, 1992, with the first analysis team workshop being held in January, 1993, at NCAR in Boulder. The team is currently based in the Climatic Impacts Centre at Macquarie University in Sydney, Australia. There are 10 team members in Sydney, consisting of researchers, computer officers, and students. The team also includes four co-opted researchers from other institutions around the world. Some of the important projects currently being undertaken by the analysis team include:

1. visualization of the GCM results, including color graphics and video;
2. assessment and intercomparison of GCM resuits using (1) hydrology, (2) radiative transfer, (3) vegetation schemes, and (4) present-day climate predictions and observations;
3. enhanced-greenhouse-gas impact assessments focussing on agriculture in Europe and severe events in Bangladesh;
4. creation of a homogenized data set of the GCM results in a user-accessible format.

THE MECCA PROTOCOL

One of the primary objectives of the analysis team is to ensure dissemination of the GCM results to the impact modeling community. In the past there has been hesitation on behalf of many GCM modelers to make their results available to impact modelers. This was mostly because they feared misapplication of their results by the impact modelers, since the latter are often unaware of the capabilities and limitations of the GCM. The GCM modelers were further frustrated by the shear volume of requests that they received for access to their data sets. From the impact modelers' point of view, GCM modelers were restricting their data. The impact modelers also complained that they could not obtain essential information on the GCM, such as the resolution, names of variables, and, importantly, where and how the data could be accessed.

The MECCA Analysis Team has been involved in bridging this gap between the two groups. This has been accomplished primarily by the creation of the MECCA Protocol, which consists of a Climate Model Output Warranty and an Impacts Requirements Specification. The Climate Model Output Warranty is essentially a questionnaire that is completed by the research group that performs the GCM simulation. By completing this form the details of the GCM and the type of experiment performed are gathered together. The impact modeler then completes the Requirements Specification. The completion of this form ensures that the impact modeler examines closely the validity of the GCM results to the proposed impact study. It also gives the impact modeler an opportunity to explain clearly to the GCM modeler what type of impact study will be performed. The analysis team can then examine the responses of both parties to the questions and make recommendations or suggestions. The GCM modelers who created the data then have the final decision as to whether or not the data are released to the impact modeler. Once permission has been given for data release, the MECCA analysis team will be able to supply all the data to the impact modelers. Any impact modeler who wishes to have access to the MECCA-sponsored GCM data should contact the MECCA Analysis Team Project Manager at Macquarie University. The applicant will be forwarded a Requirements Specification form, and the application can then proceed as set out above.

THE HOMOGENIZED DATA SET

The volume of data created from just one GCM simulation is enormous. In addition, the variable names, the grid resolutions, the data write intervals, the locations of the data, and the data format vary widely from one model to the next. This obviously makes any intercomparison of the GCM data an extremely difficuft proposition. The MECCA analysis team has attempted to alleviate this problem by creating a homogenized data set. All of the data in the set will be available in the same format and accessible in the same way. Because the volume of data is so large, the analysis team will initially restrict its data set to the phase 1 GCM simulations that included an ocean subscheme and offered daily data. (One GCM model has currently been included which offers only monthly-averaged data.) We believe this is a unique contribution to international climate assessment, since no other integrated-assessment group is currently archiving and evaluating on a daily basis. The availability of daily data should have considerable value to impact modelers and climate policy advisers who are concerned with climate variables. The homogenized data set will consist of horizontal slices (i.e., two-dimensional arrays) of the following variables:

1. Surface temperature (ground/ocean/sea ice), degrees kelvin
2. Total precipitation, millimetres per day
3. Mean sea-level pressure, pascals
4. Snow depth, millimeters of H_20 equivalent
5. Evaporation, watts per square metre
6. Lowest-level air temperature, kelvin
7. Total cloud amount, fraction.

From the MECCA phase 1 simulations the six groups listed in Table 3 will be used in the initial homogenized data set. Within each group there are two simulations that will be used in the archive. The first simulation is the control run, where the greenhouse gases are maintained at present-day concentrations; the second simulation uses increased concentrations of greenhouse gases. All of the increased-greenhouse-gas simulations have assumed an instantaneous increase in the greenhouse gas concentrations. (Transient increases in greenhouse gases are being studied in the MECCA phase 2 experiments.)

All of the data in the homogenized data set will be made available in Data Retrieval and Storage (DRS) format, a data storage system that has been developed by the Lawrence Livermore National Laboratory and is especially designed as a data storage system for GCM data. One of the biggest advantages of providing the data in DRS format is that the DRS files are portable between many computational platforms. For example, binary data can be transferred between Crays and Sun workstations without any explicit translation steps. (This is not usually the case for conventional binary data files.) The DRS system consists fundamentally of a set of FORTRAN library routines that enables users to create DRS files, and to read and write to existing DRS files. A full description of DRS is available in Drach and Mobley (1991a). DRS is also available with a utility program known as DRSED, which is

Table 3. General Circulation Model (GCM) Simulations used in Homogenized Data Set.

Principal Investigator	Name of GCM	Resolution[a]	Notes	Run	Length of Run (yr)
Henderson-Sellers	CCM1OZ	R15/L12	After equilibration Model includes BATS[b]	$1 \times CO_2$ $2 \times CO_2$	25 10
McAvaney	BMRC	R21/L9	Includes equilibration	$1 \times CO_2$ $2 \times CO_2$	15 15
Oglesby & Saltzman	CCM1	R15/L12	460 ppmv case included	330 ppmv CO_2 460 ppmv CO_2	135 95
Thompson & Pollard	GENESIS	R15/L12 atmosphere 2° × 2° surface	After equilibration Monthly averaged data	$1 \times CO_2$ $2 \times CO_2$	5 5
Wang	CCM1	R15/L12	Includes CH_4, N_2O, CFCs. Includes equilibration	year 1990 year 2050[e]	100[d] 107
Washington & Meehl	CCMOA	R15/L9	Includes equilibration Includes cirrus-albedo feedback.	$1 \times CO_2$ $2 \times CO_2$	30 30

[a]R = Rhomboidal truncation in spectral space, R15 is approximately 7.50 × 4.50. R21 is approximately 5.60 × 3.20. L = no. of vertical layers.
[b]BATS = Biosphere Atmosphere Transfer Scheme.
[c]Final 10 yr were conducted using non-MECCA funding. Results made available courtesy of Dr. McAvaney.
[d]Only 16.5 yr immediately available.
[e]Using IPCC scenario A.

described in Drach and Mobley (1991b). DRSED enables a user to easily browse through the contents of DRS files, and to print out any values of interest into a separate ASCII data file.

A strong motivation for using DRS is the availability of a special graphics package developed by PCMDI (Meyer et al., 1991), which is designed to use DRS data files. With this graphics package it is very easy to obtain a graphical display of the DRS data files. The PCMDI graphics is currently available only on SUN platforms, although versions for other platforms are expected to be available shortly. The DRS data files used in the homogenized data set are also easily convertible into GrADS (Doty, 1992) data files. GrADS is another excellent package for viewing and manipulating the data.

DRS FILES

A DRS file typically consists of two component files linked together by having the same file-name but a different file-name extension. The first file, known as the Data File (with a dat extension) contains the actual data, while the second file, the Dictionary File (with a dic extension) contains information that describes the data. When creating a DRS data file, the DRS libraries take the GCM output (which is usually in Cray 64-bit binary) and convert it into IEEE 32-bit binary by default. The data are then stored in binary form, which has the significant advantage of being compressed in size compared to equivalent ASCII files. The data may be stored in other forms if desired (such as Cray binary or ASCII) by setting certain options in the DRS libraries.

The DRS Dictionary File typically contains the following information:

1. Source of data (e.g., the name of the GCM, the run number, and the principal investigator);
2. variable name (e.g., TS, or PRECT);
3. complete title of the variable (e.g., surface temperature, or total precipitation);
4. units (e.g., °K, or mm/day).

The data stored in the dictionary file and the data file can be easily retrieved by using the DRSED utility, or by using DRS library calls. The information contained in the dictionary file is also displayed when using the PCMDI graphics.

USING THE HOMOGENIZED DATA SET

To demonstrate how the homogenized data set could be used to perform a climate-change-related study of human health, we have performed a simple analysis based on asthma. The actual relationship between asthma and the weather is, in general, a complex and indirect one (see e.g., Beggs and Curson, in preparation). Certain climatic conditions contribute to the presence and prevalence of allergens, such as house dust mites, moulds, and pollen, which contribute to asthma. In addition, these allergens are affected significantly by the indoor climate, which may be significantly different to the outdoor one. The study of asthma and climate is very interesting since it is well known that asthma is highly prevalent in certain geographical areas. Our simple analysis of asthma is based on some of the findings of Derrick (1972), who performed a comprehensive study of asthma and climate in the city of Brisbane. Brisbane is located at a latitude 27.5° S, in a subtropical region along the eastern coastline of Australia. Derrick confirmed that the relation of asthma to the weather was indeed a very complex one. However, he did find that asthma was most prevalent (as recorded by attendances at the Royal Brisbane Hospital) when the mean ambient temperature was about 21°C and the dew point (the temperature at which moisture in the air condenses) was about 13.5°C. The association of high asthma prevalence with a mean temperature of around 21°C was also found in North Queensland hospitals and in a children's hospital in Perth, although the dew points differed (being about 17°C in North Queensland, and about 10.5°C in Perth).

Derrick also noted that asthma is highly prevalent in cities such as Brisbane, Durban, and New Orleans, and that all of these areas have a mean annual temperature of 20 to 22°C and an average annual rainfall of over 100 cm (an average of 2.74 mm/day). (Although asthma has been found to be highly prevalent at these conditions, this does not mean that asthma is eliminated outside these temperature and precipitation ranges.) For our simple demonstration we have decided to concentrate on this final interesting observation, since values of surface temperature and precipitation are immediately available from the homogenized data set. To this end we have constructed a series of global maps (making use of the PCMDI graphics package), using two different GCM. On these maps we have displayed the areas of the world where the mean annual temperature is within the range of 20 to 22°C, and where the mean annual rainfall is predicted to be >100 cm. We were interested, primarily, to see if the GCM could accurately predict the climatic conditions in the regions of Brisbane, Durban, and New Orleans. We were also interested to see how the climatic conditions changed with an increase in CO_2.

CCM1OZ

Figure 1 shows the results for a 1 × CO_2 simulation in a randomly selected year (GCM year "137") using the GCM known as CCM1OZ (Henderson-Sellers et al., 1993). Figure 1A displays the 20°C and 22°C contours of annually averaged temperature. It can be seen that the regions enclosed by these two contours do, indeed, come very close to the cities of Brisbane, Durban, and New Orleans. The annual precipitation is shown in Figure 1B, where the shaded regions represent a rainfall of less than 100 cm/year. The rainfall predictions are good in the regions of New Orleans and Durban (within the grid resolution of the model, which is approximately 7.5° × 4.5°); however, rainfall is underpredicted in the Brisbane area. From our GCM simulation of the world's climate we might have expected (correctly) that Durban and New Orleans have very high rates of asthma; however, because of the lower-than-reality rainfall in Brisbane, we might not have realized that the Brisbane area also has high rates of asthma.

Figure 2 shows the results for an annually averaged year (again, GCM year "137") using a 2 × CO_2 simulation. As expected, the increased greenhouse gases cause an increase in the overall temperature of the world. This increase can be seen by the slight shift of the 20 to 22°C temperature bands toward the north pole in the northern hemisphere, and toward the south pole in the southern hemisphere. As a result of this shift, cities like Sydney, in Australia, and many areas in southern Europe, now lie within the 20 to 22°C temperature bands. Figure 2B shows the corresponding rainfall pattern (as represented by the rainfall being less than or greater than 100 cm/year) is quite similar to the rainfall experienced in the control run (Figure 1B).

CCMOA

One way to possibly access uncertainty in the GCM data is to compare the results obtained using different GCM. Figures 3 and 4 display results of 1 × CO_2 and 2 × CO_2 simulations, respectively, obtained using the CCMOA model (Meehl and Washington, 1992). Again, a randomly selected year was chosen (GCM year "106"). The 1 × CO_2 simulations, shown in Figure 3A, place the 20 to 22°C band correctly over the city of New Orleans in the northern hemisphere. However, the

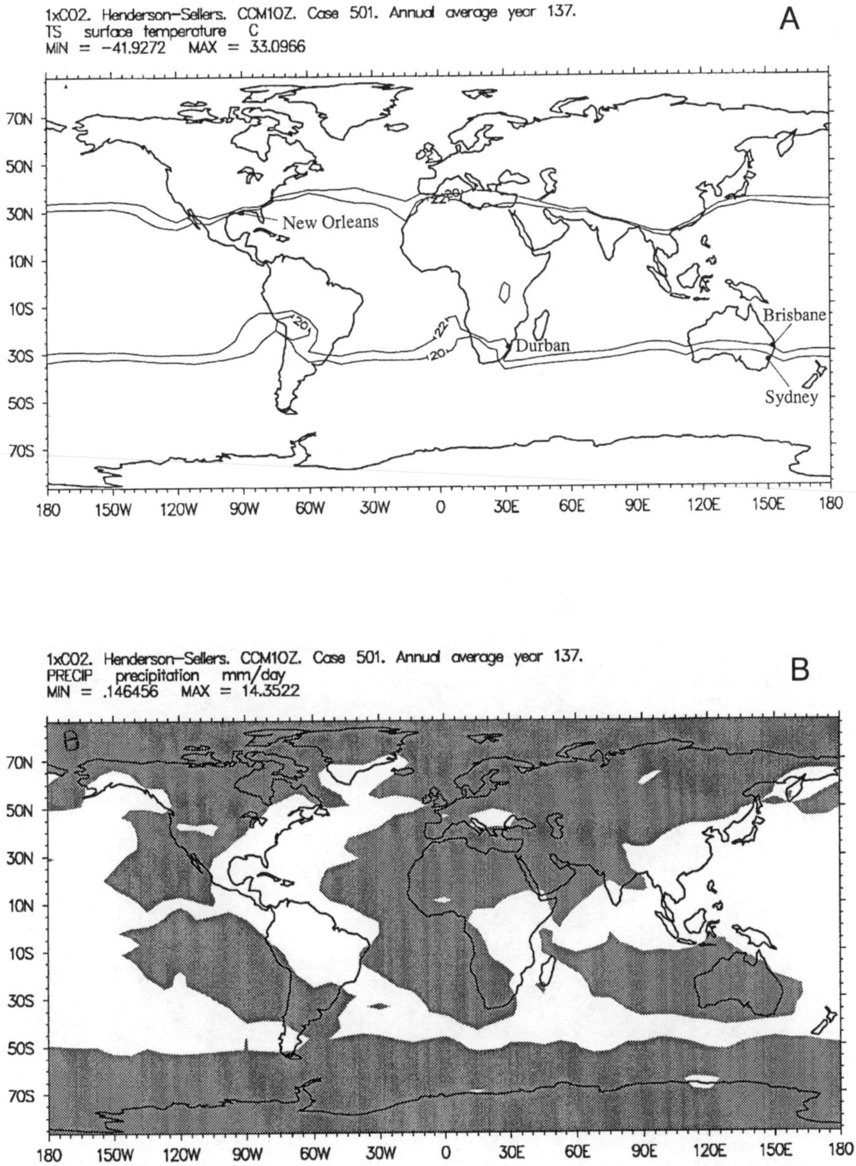

Figure 1. Control run for a 1 X CO_2 simulation using general circulation model CCMIOZ. A, Annually averaged surface temperature (20°C and 22°C contours); B, annually averaged total precipitation (shaded regions represent <100 cm/year).

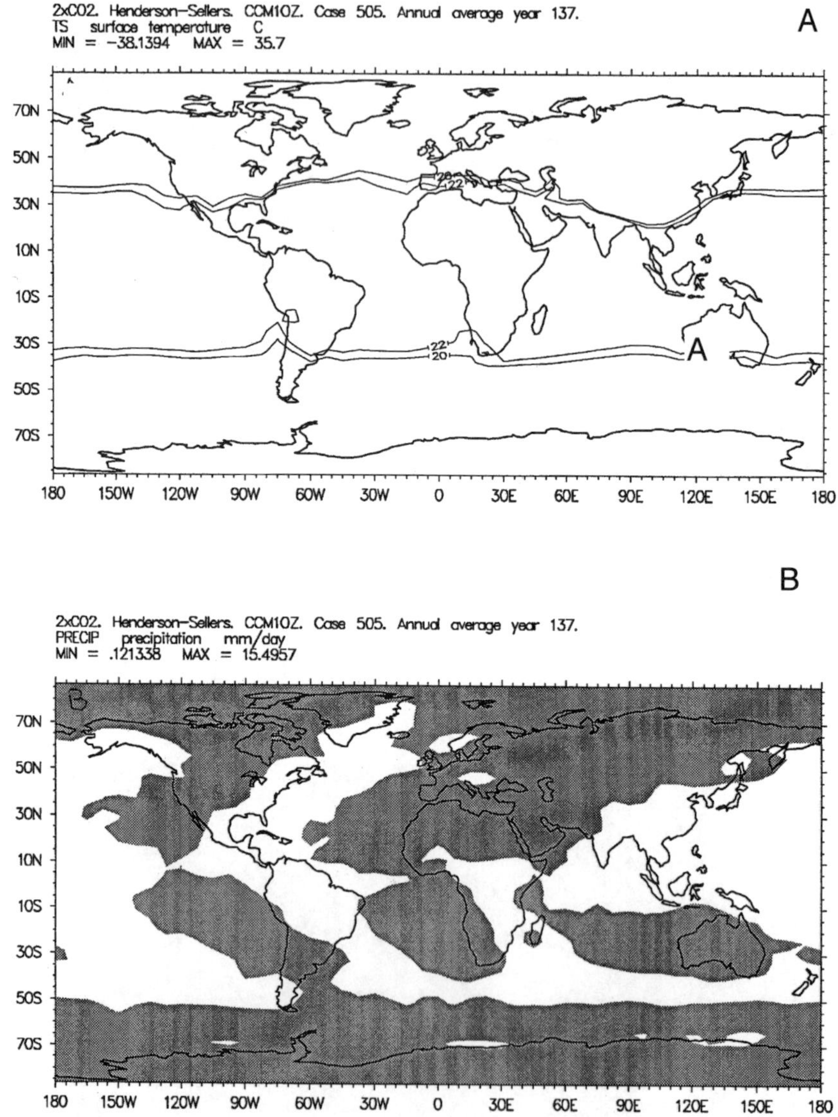

Figure 2. Control run for a 2 × CO_2 simulation using general circulation model CCM1OZ. A, annually averaged surface temperature (20°C and 22°C contours); B, annually averaged total precipitation (shaded regions represent <100 cm/year).

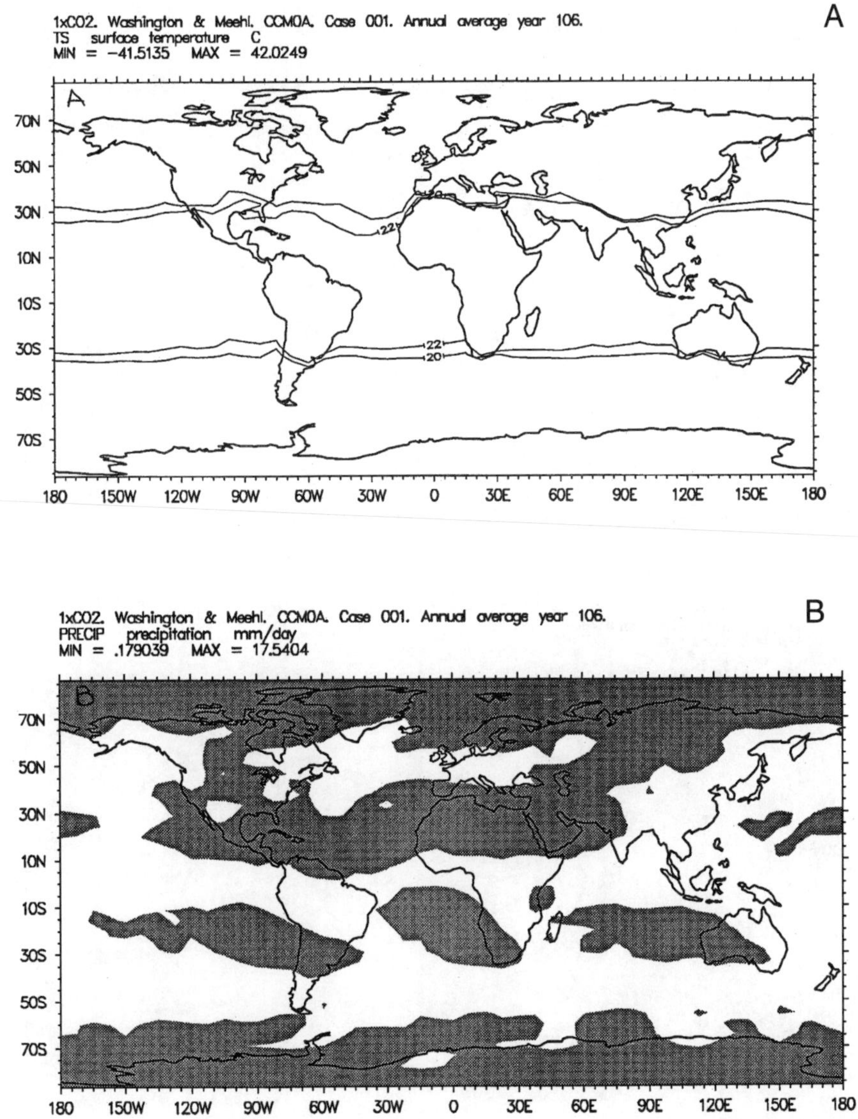

Figure 3. Control run for a 1 × CO_2 simulation using general circulation model CCMOA. A, annually averaged surface temperature (20°C and 22°C contours); B, annually averaged total precipitation (shaded regions represent <100 cm/year).

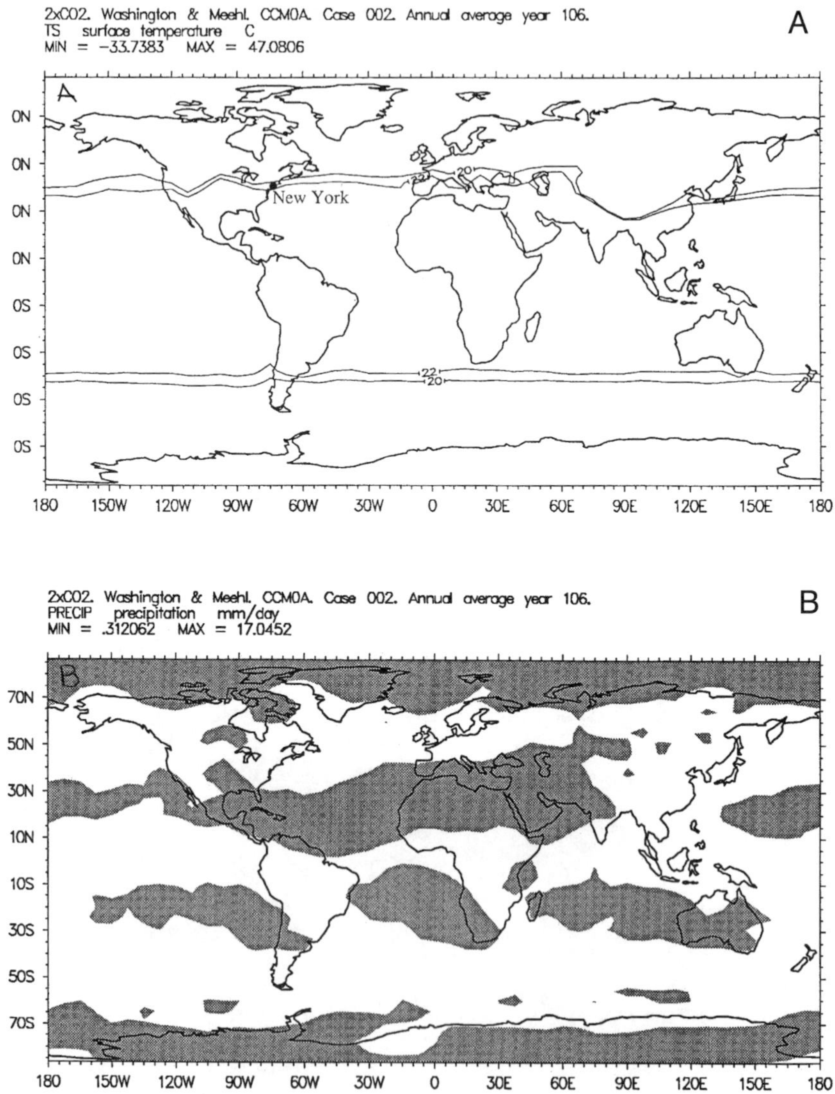

Figure 4. Control run for a 2 × CO_2 general circulation model CCMOA. A, annually averaged surface temperature (20°C and 22°C contours); B, annually averaged total precipitation (shaded regions represent <100 cm/year).

band is slightly too far south in the southern hemisphere, so it lies below Brisbane and Durban. Nevertheless, considering the grid resolution used within the model, the temperature predictions are still very good. The precipitation map, shown in Figure 3B, shows that the correct levels of rainfall have been predicted over Brisbane and Durban, however, the rainfall over New Orleans is less than in reality.

The 2 × CO_2 temperature map (Figure 4A) shows that the 20 to 22°C bands have shifted toward the poles, in agreement with the CCM1OZ simulations. The magnitude of the shift varies from region to region on the globe. Probably the most noticeable difference between the CCMOA and the CCM1OZ simulations is for the regions over the United States. The CCM1OZ simulations showed virtually no change in the location of the 20 to 22°C temperature band over the United States when the greenhouse gases were doubled. In comparison, the CCMOA simulation predicts a dramatic shift in this temperature band. In the 1 × CO_2 simulation of CCMOA this temperature band is around New Orleans and Florida, whereas in the 2 × CO_2 simulation the band is shifted to be near the Great Lakes and New York City. The precipitation map (Figure 4B) shows that a very significant part of this new 20 to 22°C band also has a precipitation greater than 100 cm/year. Hence, it is possible that we might see an increase in asthma in these areas.

DISCUSSION AND SUMMARY

To model possible human health impacts related to climate change, predictions of the future climate are required. MECCA has sponsored a series of climate-change predictions that have been performed by leading research scientists using state-of-the-art GCM. These results are to be made available to researchers who wish to perform climatic impacts studies such as those regarding health. Applications for the data are to be made to the MECCA analysis team at Macquarie University. Results of such studies are expected to be made public by the MECCA consortium.

We have given an example of the application of the MECCA results by calculating the regions of the globe where the climatic conditions match those experienced by the cities of Brisbane, Durban, and New Orleans. For the purposes of our demonstration these climatic predictions were generally found to be accurate. Simulations under an enhanced greenhouse-gas environment showed that the average temperatures currently experienced by these three cities moved toward the north pole in the northern hemisphere, and toward the south pole in the southern hemisphere. As a result, many regions of the world could experience an increased incidence of asthma.

Our study of asthma has only been for demonstration purposes. A thorough study of asthma and climate would require the development of an accurate model to describe the relationship between asthma and the climate. The variability in the climate (day to day, and year to year), the relationship between the indoor and the outdoor climate, and the relationship between allergens and the climate are all likely to be important factors.

ACKNOWLEDGMENT

We thank Paul Beggs, from the Climatic Impacts Centre, for many helpful discussions concerning the prevalence of asthma.

REFERENCES

Beggs, PJ and PH Curson. An integrated environmental asthma model. In preparation.

Derrick, EH. **1972**. Asthma and the Brisbane climate. Aust N Z J Med 13:235-246.

Doty, BE. **1992**. *Using the Grid Analysis and Display System (GrADS)*. Center for Ocean-Land-Atmosphere Interactions, Department of Meteorology, University of Maryland, College Park, MD.

Drach, R and R Mobley. **1991a**. *DRS Users Guide*, Program for Climate Model Diagnosis and Intercomparison, Lawrence Livermore National Laboratory, Livermore, CA.

Drach, R and R Mobley. **1991b**. *DRS Utilities Guide*, Program for Climate Model Diagnosis and Intercomparison. Lawrence Livermore National Laboratory, Livermore, CA.

Henderson-Sellers, A, RE Dickenson, TB Durbidge, PJ Kennedy, K McGuffie, and AJ Pitman. **1993**. Tropical deforestation: modelling local to regional-scale climate change. J. Geophys. Res. 98:7289-73

Meehl, G and W Washington. **1992**. Greenhouse sensitivity experiments with penetrative cumulus convection and tropical cirrus albedo effects. Climate Dynam. 8:211-223.

Meyer, MK, DN Williams, and RL Mobley. **1991**. *PCMDI Graphics Users Manual*, Program for Climate Model Diagnosis and Intercomparison. Lawrence Livermore National Laboratory, Livermore, CA.

GRAYING UNDER THE WEATHER: VULNERABILITY TO CLIMATE CHANGE IN AUSTRALIA

P. Curson

Climatic Impacts Centre, Macquarie University, Sydney, Australia

Key words: *Health, greenhouse warming*

ABSTRACT

It seems clear that climate change will not impact equally upon all social groups in developed countries. Some groups, by virtue of age and health status will be more susceptible to such changes.

This paper discusses recent trends in aging, chronic disease, disability and handicap in developed countries like Australia. Primarily, the paper is concerned with discussing how such changes in demographic and epidemiologic structure will heighten vulnerability to and risk from future climate change. In particular the paper addresses the relationship between particular sectors of the aged community (e.g., the "old old," the chronically ill old, the elderly disabled, the medications-dependent) and particular climatic factors (e.g., increasing temperatures, more frequent hot spells, higher humidity). The paper also explores the relationship between changing temperature and humidity and the effect on the medication-dependent elderly. In this respect the paper discusses the efficacy of some medications used in combination under certain climatic conditions.

INTRODUCTION

The population of Australia, like that of most other developed countries, is growing older. The elderly are rapidly increasing both in absolute numbers and as a proportion of the total population. Currently, almost 12% of the population are aged over 65, and this proportion is expected to increase to 20% within 35 or 40 yr. The largest increase during these years will be in the old-old group, people aged over 80. Population aging has been accompanied by an increase in the levels of chronic disease, disability, and handicap as well as in levels of drug dependence. In addition, current migration patterns among many of the elderly are directing them to rural towns and cities as well as to sunbelt coastal areas. In the context of climate change, and in particular warmer temperatures, there seems little doubt that such trends will increase vulnerability and greatly influence the geography of risk.

THE AUSTRALIAN CLIMATE CHANGE SCENARIO

While there remain many uncertainties about the nature, extent, and distribution of climate change in Australia, the latest predictions, given a 2 × CO_2 scenario, suggest the following (Whetton, 1993):

- for inland Australia, warming of 0.5–2.5°C
- for northern coastal areas, warming of 0.0–1.5°C;
- for southern coastal areas, warming of 0.5–2.0°C;
- changes in summer rainfall ranging from no change to +20% by 2030;
- changes in winter rainfall of ~10% with regional variations as follows:
  –probable decreases of 0 to 10% in inland southern Australia (including Adelaide),
  –probable change of -10% to +10% further south (including Perth),
  –probable increases of 0 to +10% further south (Hobart);
- general increase in rainfall intensity and heavy falls;
- possibility of longer dry spells and fewer rain days;
- possibility of more very hot days, fewer frosts, more floods.

POPULATION AGING

Australia's population has been gradually aging over the last 100 yr. At the beginning of this century, 35% of the population were aged under 15 yr and only 4% over 65 yr. By 1947, the proportion of elderly in the population had almost doubled, and by 1991 the aged comprised 11.4% of the total population, with those aged under 15 yr comprising less than 22%. This graying of the Australian population is likely to continue unabated over the next few decades, with the aged component, in particular, expected to grow very rapidly after the turn of the century, especially in the period after 2011. It can be confidently expected that the number of Australians aged over 65 will almost treble over the next 40 yr, from approximately 1.9 million in 1990 to 5.2 million or approximately 21% of the population, by 2031. This aging of the population owes much to the impact of the entrance of the baby-boom generations (i.e., those born between 1946 and 1961) into old age. In Australia this will commence about 2010 and continue into the mid 2020s. By 2020, the Australian population pyramid will have assumed a much more elderly structure, as illustrated in Figure 1.

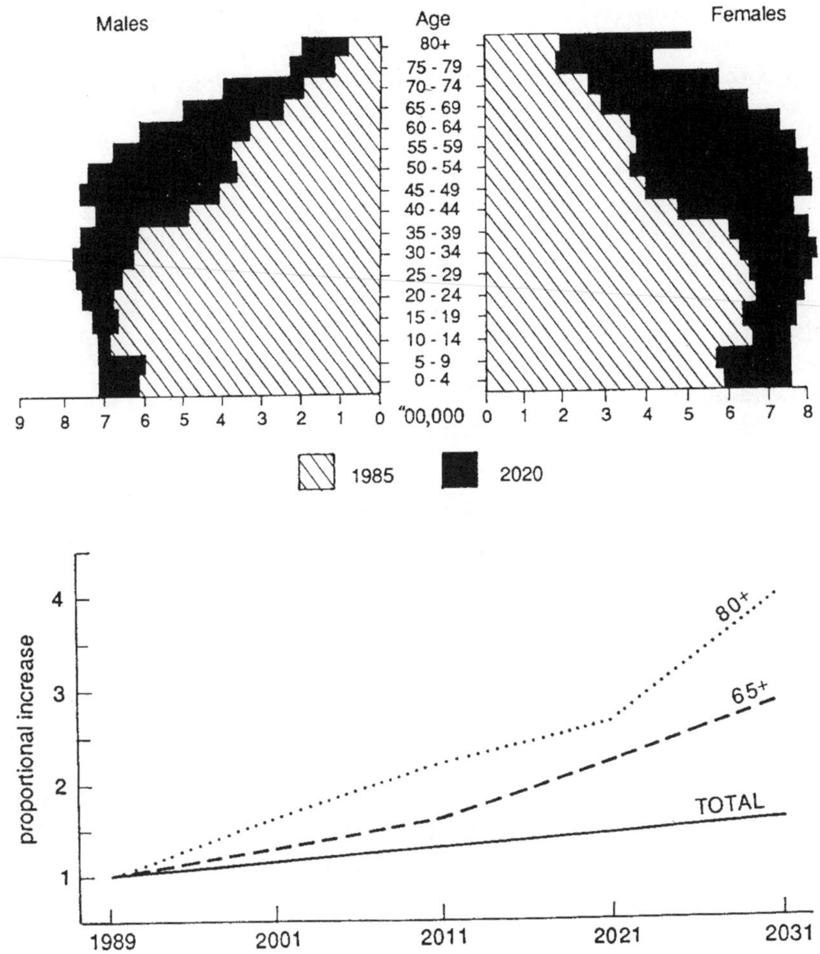

Figure 1. Changing Age/Sex Structure of the Australian Population (1985–2031).

Not only will there be a rapid and substantial increase in the number of aged in Australia, but within the aged group itself there will be important structural changes. The most important of these will be the rapid growth in the numbers of the *old-old,* or extreme aged (80+ yr). One hundred years ago, only 27% of males and 35% of females could expect to reach their 70th birthday. By the late 1980s, however, more than two-thirds of males and 81% of females survived to reach this age, and 37% and 58%, respectively, lived to the age of 80 yr (Rowland, 1991). As Figure 1 clearly indicates, the growth rate of the *old-old* group will outstrip both that of the aged population and the population as a whole. Between 1990 and 2031, while Australia's population is expected to increase by half (17 million to 26 million), the numbers of the 80+ age group are projected to quadruple. In 1991 there were approximately 395,300 persons aged over 80 in the Australian population (2.2%), compared to approximately 282,000 in 1984. By the year 2031 the number of people aged over 80 will exceed 1.4 million. The implications of such an increase are profound. This age group is at particularly high risk to episodes of severe illness as well as being subject to enduring chronic illness and disability. In addition, they are heavy users of a wide variety of drugs. Many of these problems have led to a heightened need for social and medical care and, often, institutionalization.

Another predictable change is that older women will continue to outnumber older men in Australia's future population. In 1992, there were approximately 131 females for every 100 males aged 65 yr and over; for those aged over 80 yr, there were 192 females for every 100 males. While sex differences may lessen over the next few decades, it is expected that females will continue to dominate the ranks of the elderly.

THE VULNERABILITY OF THE AGED TO CLIMATE WARMING

Of all the social groups in developed societies, the aged, and particularly the very old and chronically ill, would seem to be the most vulnerable to climate and climate change. This is largely because of a number of inter-related physiologic and social factors which, in combination and synergistically, serve to increase the elderly's vulnerability.

DIMENSIONS OF VULNERABILITY

Physiologic Vulnerability

There is strong evidence of the aged's sensitivity to temperature extremes (heat and cold waves). The effects of heat waves and high humidity are largely due to their impact on the body's thermoregulatory mechanism, the water and electrolyte balance, and on the increased demands made on the cardiovascular system by having to divert blood from central organs to peripheral body areas so as to augment heat loss and sweating (Ellis, 1972). Generally, the healthy elderly are capable of developing heat acclimatization when exposed to warmer conditions. For people with impaired cardiovascular systems, however, hotter temperatures and high humidity can substantially increase morbidity and mortality, particularly from ischemic heart disease. There is also some evidence that heat stress may provoke a mild hypercoagulable state which may aid the formation of clots in the blood vessels of elderly persons suffering from arteriosclerotic vascular disease, with implications for heart attacks, cerebral thrombosis/embolism, and stroke (Keatinge et al., 1986).

Those elderly with impaired respiratory systems, such as asthmatics, are often sensitive to a wide range of irritants and allergens such as dust mites, pollens, molds, and fungi, as well as concentrations of ozone, sulphur, and nitrogen dioxide, and particulates in the atmosphere. It seems clear that warmer, more humid weather will impact on such triggers with considerable implications for the health of such people (Curson, in press).

A cold environment also contributes to increased cardiovascular and cerebrovascular morbidity and mortality. Low temperatures may contribute to the development of coronary arteriosclerosis and its progression to myocardial infarction. Such exposure may also lead to the development of arrhythmias and increased cardiac insufficiency. In Australia, there is a marked seasonal variation in circulatory and cerebrovascular disease deaths, with a noticeable peak in the cold winter months.

While there is debate as to the actual mechanisms involved in this relationship, it is possible that respiratory disease may play a contributory role in enhancing or aggravating coronary heart disease. It is also possible that changes in lifestyle and behavior during winter months may contribute added stress. While these factors may contribute to increased morbidity and mortality, there would also seem to be direct effects associated with cold temperatures. Cold increases the myocardial oxygen demand by increasing sympathetic stimulation, systolic blood pressure, and cardiac diastolic pressure and volume. The myocardial oxygen supply may also be impeded by coronary vasoconstriction, especially in people suffering from arteriosclerosis. Cold temperatures may also actually promote arteriosclerosis by increasing blood pressure and blood concentration of cholesterol, catecholamine, and corticoid levels (Vuori, 1987). Given that winter temperatures may be warmer in the next 30 to 50 yr, it is likely that there will be a decline in winter morbidity and mortality rates among the elderly.

Chronic Illness, Disability and Handicap

Levels of chronic disease, disability, and handicap are currently high in Australia. At the last national survey, in 1988, 2.5 million or nearly 16% of all Australians were classified as being disabled, including 13% recorded as handicapped. (See notes to Table 2 for definition of *disabled* and *handicapped.*) Levels of chronic illness are also high. In 1989/1990, more than two-thirds of all Australians reported that they suffered from one or more chronic disorders, including 4.3 million, or one-quarter of the population, reporting a musculoskeletal problem (usually arthritis); 1.7 million reported a cardiovascular problem; and 1.2 million, hypertension. After New Zealand, Australia possesses the highest asthma morbidity and mortality rates in the world: possibly one-fifth of all elderly Australians suffer from the disease. While chronic disease and disability rates are generally high in Australia, rates increase markedly with age. For those over 65 yr of age, more than 94% reported at least one chronic condition in 1989/1990 (Table 1). The most commonly reported conditions were eyesight problems, arthritis, hypertension, hearing impairments, heart disease, and unspecified back trouble. Fifty-six percent of all those aged over 75 suffered from musculoskeletal disease, 48% from circulatory disease, and 64% from eyesight problems.

Not all major chronic diseases pose a threat to the quality of life or limit personal independence. In 1988, of the 1.8 million persons aged over 65, approximately 52% were classified as disabled, including 44% as handicapped. Like chronic disease, the risk of disability and handicap increases rapidly after age 65, as indicated in Table 2.

Survival curves derived from life tables modified to include disability and chronic disease states (see Mathers, 1991) provide an overall picture of the health expectancy of Australia's aged (Figure 2). These curves indicate, for males and females at various ages, the likelihood of a handicap, disability, or chronic-disease-free existence. Table 3 translates this material into life and health expectancy values. It is estimated, for example, that in 1988, males had a remaining life expectancy of 14.8 yr at age 65, roughly 45% of which would be spent disability-free and 54% handicap-free.

Table 1. Number of Persons (per 100 population) Aged Over 25 Years Experiencing Long-Term Illness: Selected Conditions, Australia, 1989-90. (Source: ABS, 1991.).

Age Groups:	24–44	45–64	65–74	75+
Diseases of the Circulatory System	84.7	252.6	432.8	482.7
Diseases of the Musculoskeletal System	269.6	430.8	526.5	560.4
Diseases of the Nervous System/ Sense Organs	317.6	721.8	787.8	790.0
Diseases of the Respiratory System	227.3	210.6	218.9	177.6
Diseases of the Digestive System	44.0	95.6	135.8	142.3
Skin Diseases	64.1	47.4	55.3	58.8
Mental Disorders	21.1	30.3	36.0	36.1
Total Reporting Long-Term Disease	675.5	896.5	949.2	958.3

Table 2. Proportion of Middle-Aged and Elderly Persons Experiencing Disability and Handicap, Australia, 1988. (Source ABS, 1989).

	Disability[(a)]		Handicap[(b)]		Mobility Handicap[(c)]	
Age Groups	Males	Females	Males	Females	Males	Females
45-59	21.1	18.7	17.5	16.3	12.6	12.7
60-64	43.5	27.9	37.6	24.5	27.2	19.6
65-69	47.2	35.9	35.5	28.6	29.9	25.8
70-74	50.8	45.9	40.4	28.4	33.4	33.0
75+	60.8	65.1	53.8	61.1	40.9	49.9
Total	16.0	15.2	13.0	13.0	8.8	9.5

[(a)]A person with one or more of a group of selected impairments and disabilities which had lasted or were likely to last for 6 months or more.

[(b)]A person reporting a disability which in some way limited their ability to perform certain self-care, mobility, communicative, school, or employment tasks.

[(c)]A person reporting difficulties in mobility (i.e., using public transport, movement within and without the house, walking up and down stairs.)

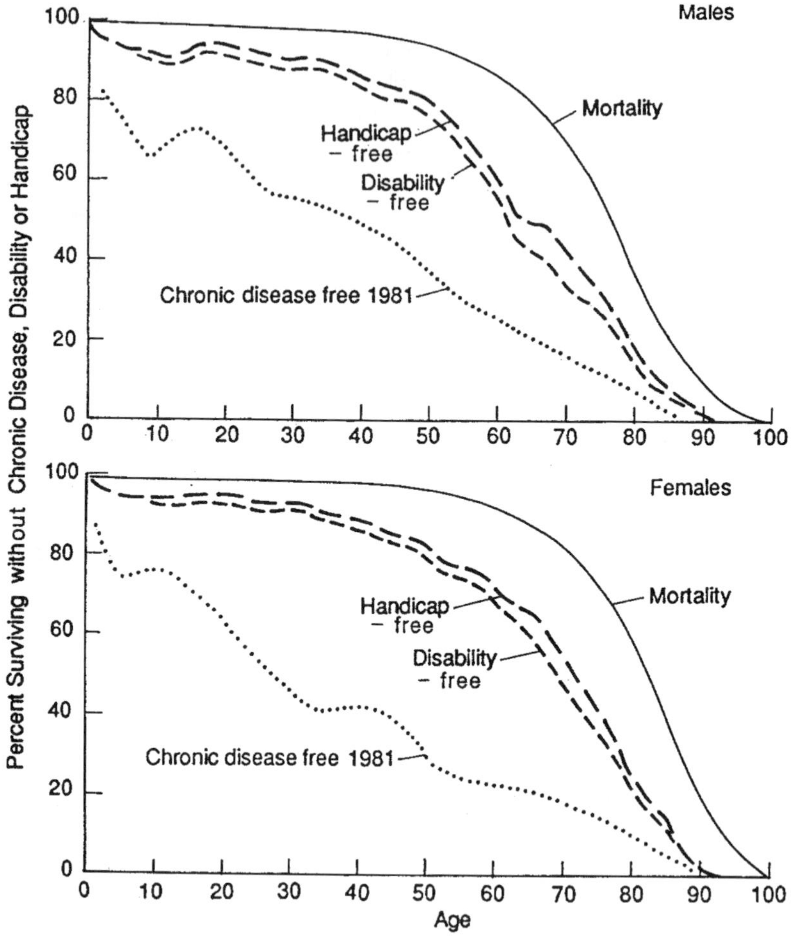

Figure 2. Survival and Health: State Curves, Australia, 1981–1988.

Australia's pattern of aging and health expectancy is broadly similar to that which has taken place in other developed countries. Evidence from the United States, Canada, England, and Wales suggests that the price paid for increasing life expectancy for the elderly over the last 20 yr has been higher levels of disability. It is now fairly widely accepted that in such countries, the substantial declines in mortality, particularly at older ages, have been more a product of ill people being saved than well

Table 3. Life and Health Expectancy of Older Persons, Australia, 1988. (Source: Mathers, 1991: Tables 4 and 5)

		Percent of Remaining Life	
	Life Expectancy (yr)	Disability-Free	Handicap-Free
Males At Age			
60	18.4	48.4	57.1
65	14.8	45.2	54.0
75	8.9	37.1	43.8
Females At Age			
60	22.8	51.3	56.6
65	18.7	46.0	57.3
75	11.4	32.4	36.8

people being rewarded by longer life (see Verbrugge, 1989). Australian evidence supports such a view and indicates an important increase in disability and handicap levels for elderly persons over the last decade (see Mathers, 1991). The reasons for such an increase are not completely clear but probably reflect the continued aging of the population, the exposure of *old-old* survivors to new illnesses (e.g., arthritis, deafness, and Alzheimer's disease) together with heightened public awareness about disease and improved medical diagnosis. It seems clear that chronic disease and disability rates will continue to increase as the population ages over the next 25 to 30 yr. In a warming climate, such changes are of considerable importance.

Drug Use

In most developed societies the elderly are major users of prescribed and nonprescribed drugs. Aging brings with it physiologic decline in many vital processes which, allied to various chronic diseases, compromises the elderly's capacity to adjust to, and cope with, social and environmental stress. In consequence, many aged become dependent on a wide variety of drugs, often taken in combination. In the 1989/1990 health survey, 91% of all Australians over the age of 75 yr were recorded as taking drugs, including 54% dependent on drugs for the control of heart, blood pressure, and fluid problems (ABS, 1991). In Australia, pensioner prescriptions under the Pharmacy Benefits Scheme account for 25% of all prescribed drugs and 55% of all psychotropic drugs (McCallum, 1990). In 1985, pensioners averaged more than six times the number of prescriptions issued to the nonpensioner population. It is also likely that nonprescription drug use is six- to seven-fold that of the general population. The annual increase in the elderly's use of drugs has considerably outstripped the growth in numbers of the elderly population. By the year 2031, it is likely that the elderly will utilize between 40 and 50% of total drug expenditure (Kurowski, 1989). The classes of drugs most commonly used by the elderly include cardiovascular drugs, analgesics, antimicrobials, hypoglycemic agents, psychotropic drugs, and antiarthritics. The widespread and increasing use of such drugs is of considerable importance, not only because older people are more vulnerable to adverse drug reactions, particularly those arising from multiple drug use, but also because there is an added risk stemming from drug interactions with climatic factors. Studies have repeatedly demonstrated not only a high prevalence of adverse drug reactions in the aged population but also that such people are poor compliers, as well as inaccurate and inappropriate drug consumers. Multiple drug use in the elderly is directly related to social isolation and the increased number of chronic disorders experienced. It would also seem that many doctors and patients believe that most disorders in the aged require drug intervention.

A number of factors influence adverse drug reaction in the aged, including increasing age, multiple drug use, allergic status, genetic and chronic disease status, and drug dose. Although little studied, climate would also appear to play an important role.

Some of the reasons for the increased incidence of adverse drug reactions concern changes in drug absorption, distribution, metabolism, and excretion rates. All of these are closely related to the physiologic changes of aging, such as modifications to the mucosal cells and surface area in the gastrointestinal tract, decreased gastrointestinal blood flow and rising pH levels of the stomach. Changes in body mass and composition, such as decreased lean body mass and decreased total body water, also play an important role by making it easier for some drugs to be stored for longer periods in fatty tissue. With respect to metabolism, most drugs are inactivated by enzyme activity in the liver. Liver size tends to decline significantly with advancing age, reducing the body's drug-metabolizing capacity. In many elderly, the elimination of drugs from the system is greatly decreased by changes in the renal system that influence blood flow, glomerular filtration, reabsorption and secretion.

Finally, it would seem that climatic factors such as thermal stress, ultraviolet radiation, air movement, and reduced partial oxygen pressure affect membrane permeability and the rate of drug absorption, metabolism, and excretion rates. The effects of drugs commonly used to treat musculoskeletal problems, hypertension, diabetes, heart and respiratory conditions vary markedly under different climate conditions. In particular, many of the medications used to treat these disorders behave differently in very warm to hot temperatures. Table 4 lists some potential drug/drug/climate interactions of importance for the elderly. Even excluding aspirin, which is probably the most commonly used drug among the elderly, nonsteroidal antiinflammatory drugs (NSAID) remain the most used, routinely deployed for the treatment of a variety of musculoskeletal problems. Possibly 10% of all prescriptions issued in Australia are for NSAID, and the consumption by the elderly of such drugs has risen steadily. In a study of Newcastle, Australia, more than 20% of all elderly respondents were found to be using NSAID (Henry, 1988). NSAID alone can produce side effects in the elderly; e.g., gastrointestinal upsets, fluid retention, edema, and heart failure. When NSAID are combined with other drugs, such as diuretics, in the elderly, the potential for adverse drug reactions is greatly increased. High temperatures and humidity and, particularly, heat waves may further increase the probability of adverse reactions by increasing body temperature, on occasion leading to heart stress, renal impairment, and heart failure (Table 4).

The toxicity of antipsychotics such as chlorpromazine, phenothiazine, and butyrophenone may also be increased under temperature extremes of hot and cold and, where such drugs are used in combination with anticholinergics, the risk of adverse reactions such as heat stress, hyperpyrexia, and stroke is greatly increased. Other significant drug/drug/climate interactions are beta-blockers such as Inderal, Trasicor, and Lopvesor which, when combined with calcium channel blockers such as verapamil or diltiazem can, independently of climatic conditions, lead to arrhythmias and congestive heart failure. Theophylline, a bronchodilator used to treat asthma, does not cohabit well when used in conjunction with antihypertension drugs in hot and

Table 4. Some Significant Drug Interactions Which May Cause Problems for the Elderly in a Changing Climate.

Drug	Potential Interacting Drug(s)	Climate Factor	Adverse Outcome
NSAIDS[a]	+Diuretics	High temperatures/ Heat waves High humidity	Heat stress Fluid Retention Renal Impairment Heart failure
Antipsychotics (e.g., phenothiazines, butyrophenones, thioxanthenes	+Anticholinergics	High temperatures/ Heat waves High humidity Low temperatures Cold waves	Heat stress Stroke Hyperpyrexia Hypothermia
Beta-blockers	+Calcium-channel blockers	Low temperatures	Bradycardia Heart failure
Beta-blockers	+Salbutamol	High temperatures Heat waves High humidity	Deterioration in respiratory function
Theophylline	+Diltiazem or Verapamil	High temperatures Heat waves High humidity	Tachycardia Nausea Hypotension Arrhythmia Heart attack
Benzodiazepines	+Tricylic anti-depressants	High temperatures Heat waves High humidity	Increased risk of side effects—dry skin, blurred vision, increased heart rate
Digoxin		High temperatures Heat waves	Each 1°C rise in temperature produces a significant increase in toxicity
Digoxin	+Calcium-channel blockers	High temperatures Heat waves	Increased serum levels of dioxin
Sympathomimetics (e.g., ephedrine, amphetamine, isoprenatine)		High temperatures Heat waves	Possible increased pyrexia due to vasoconstriction

[a]Nonsteroidal, anti-inflammatory drugs

humid conditions. Under such circumstances, fluid retention, nausea, and constipation may result. Other potential drug/drug/climatic interactions are given in Table 4.

Lack of Mobility

Lack of mobility as a result of pathological processes may render many elderly house- or bedbound. Immobility may, in itself, lead to diminished function of many vital processes, partly because of the loss of stimulus from movement, as in the case of the colon and bladder or the circulatory system, and partly because confinement in an unchanging environment has a substantial impact on receptor organs, which are not subject to a range of varying stimuli. Insufficient physical activity also tends to aggravate many musculoskeletal complaints. Immobility may also render many elderly more vulnerable to accidents, as well as making them more sensitive to climate and climate change, both in a social-adaptive and a physiologic way. There is evidence, for example, that the house- and bedbound elderly are more at risk from heat stroke than are the more active groups in the population. Inability to care for oneself has also been found to be a factor associated with risk of heat stroke (Kilbourne et al., 1982).

While no information exists on the number of elderly Australians who are house- and/or bedbound, the proportion of the population suffering from a mobility handicap varies from 26 to 50% for people over 60 yr of age (Table 2).

Changing Migration Patterns

Australians remain among the most mobile of societies, with approximately one person in six changing their place of residence annually. While old people move much less frequently than those of working age, retirement migration, particularly to coastal areas and from rural areas to towns and cities, is of considerable significance. Such mobility is driven by a variety of motives, ranging from the desire for independence and security and an improved life style (including a warmer climate), to more economic reasons of coping following retirement and/or the death of a partner. While it is difficult to see any change in existing migration patterns resulting from gradual climate change, there is little doubt that such migration patterns, if continued over the next 20 to 30 yr, while helping decrease the overall burden of winter mortality and morbidity, will increase the vulnerability of the aged to higher temperatures, as well as transport them into areas noted for vector-borne diseases like Ross River Fever and, possibly, dengue. In addition, many of these retirement areas currently do not possess the necessary health and social services infrastructure required for the support of large numbers of elderly persons.

THE GERIATRIC IMPERATIVE: CHALLENGES FOR HEALTH POLICY AND PLANNING IN A CHANGING CLIMATE

Concern about the social and environmental implications of an aging population and of climate change are relatively recent events in developed countries. Only since the 1960s have demographic projections of significant numbers of aging people captured public attention. In the case of climate change, public attention is an even more recent event. When projections of aging are matched with epidemiologic data about increasing chronic disease, disability, handicap, and drug use, it becomes clear that vulnerability to climate warming will be a critical issue over the next 30 to 50 yr. It is also clear that there will be a need to develop services and support structures which are sensitive to the needs of the elderly and, in particular, the very old, in the context of a changing climate. There will therefore be a need to develop special housing, rehabilitative services, in-home support services, medical care systems, and drug regimes sensitive to the management of chronic illness, disability, and drug dependence among the elderly in a warming climate.

Until now, climate and climate change have rarely been considered by those advocating social and health support programs to cope with an increasing aged population. It would seem clear that there is a need to develop risk profiles for the elderly as well as to develop specific geriatric climate change strategies designed to:

- appreciate the role of climate on the health and well-being of the aged,
- advance public education programs designed to draw attention to the vulnerability of the aged and chronically ill in a warming climate,
- develop appropriate long-term care and support structures sympathetic to the support of an aging population with high levels of chronic illness and drug dependence in a changing climate,

- consider specific adaptations to mitigate the effects of heat waves on the elderly and chronically ill, such as the provision of air-conditioned shelters,
- place less emphasis on drug therapy among the aged,
- develop drug management strategies that are appropriate to warmer/heat-wave temperatures and high humidity conditions,
- appreciate that under heat-wave/high humidity conditions the house-bound and bed-ridden, as well as those dependent on regular medications, are at higher risk,
- require more health-specific, long-range weather forecasts,
- appreciate that current migration patterns among the elderly may render them at higher risk,
- consider modifications to homes and retirement villages that are appropriate to the maintenance of the aged in a warmer climate.

REFERENCES

ABS. **1989**. *Survey of Disabled and Aged Persons 1988* (Preliminary Results). Australian Bureau of Statistics, Commonwealth Printer, Canberra, Australia.

ABS. **1991**. *National Health Survey 1989-90. Summary of Results. Australia*, Australian Bureau of Statistics, Commonwealth Government Printer, Canberra, Australia.

Curson, P. Climate and chronic respiratory disease in Sydney—The Case of Asthma. Clim. Change (In Press).

Ellis, FP. **1972**. Mortality from heat illness and heat-aggravated illness in the United States. Environ. Res. 5:1-58.

Henry, DA. **1988**. Side-effects of non-steroidal anti-inflammatory drugs. Balliere's Clin. Rheumatol. 2:425-454.

Keatinge, WR, et al. **1986**. Increased platelet and red cell counts, blood viscosity, and plasma cholesterol levels during heat stress, and mortality from coronary and cerebral thrombosis. Am. J. Med. 81:795-800.

Kilbourne, EM, et al. **1982**. Risk factors for heatstroke. JAMA 247:3332-3336.

Kurowski, W. **1989**. *Polypharmacy and Older People: A Background Paper,* Occasional Paper. Australian Council of Ageing, Melbourne, Australia.

Mathers, C. **1991**. *Health Expectancies in Australia in 1981 and 1988.* Australian Government Publishing Service, Canberra, Australia.

McCallum, J. **1990**. Health: The Quality of Survival in Older Age, pp. 195-240. In: *Australia's Health* 1990. Australian Government Publishing Service, Canberra, Australia.

Rowland, DT. **1991**. *Aging in Australia.* Longman Cheshire, Melbourne, Australia.

Verbrugge, LM. **1989**. Longer Life but Worsening Health? Trends in Health and mortality of Middle-Aged and Older Persons. Milbank Mem. Fund Q. 62:475-519.

Vuori, I. **1987**. The heart and the cold. Ann. Clin. Res. 19:156-162.

Whetton, P. **1993**. New climate change scenarios. Clim. Change Newsltr 5:7-8.

THE IMPACT OF CLIMATE CHANGE ON MALARIA

J. Porter

Climatic Impacts Centre, School of Earth Sciences, Macquarie University, Sydney, Australia

Key words: *Climate change, malaria, global distribution*

ABSTRACT

Malaria is considered to be the world's most serious and widspread vector-borne disease. This work represents a preliminary investigation of how climate change may affect the global distribution of malaria.

A simple bioclimatic model has been used to relate climate to overall mosquito distribution and from this infer the global pattern of malaria incidence. The model uses climate to determine the bioclimatic potential of a region indicating its suitability for malaria transmission. A geographical information system has been used to map the potential distribution of malaria on the basis of the current climate and a range of 2 × CO_2 scenarios obtained from five different general circulation models. The impact of changes in climate on potential malaria distribution varies quite considerably depending on the scenario under consideration.

This work provides an initial assessment of the possible effects of climate change on global malaria distribution and identifies some areas which may be at risk from such changes. In addition, the work highlights the problems encountered when trying the assess the impacts of global changes in climate on a disease which is greatly influenced by local conditions. It emphasises the need for impact studies to be based on regional scale changes in climate from a range of different GCM.

INTRODUCTION

Malaria is currently regarded as a major public health problem in the tropics and a serious obstacle to human progress and socioeconomic development, especially in rural areas (WHO, 1991). In many parts of the world the malaria situation is worsening as a result of changing environmental conditions (e.g., intensified exploitation of forests, agricultural and industrial development projects) which have led to increased contact between humans and the mosquito vectors of the disease (WHO, 1992a). Furthermore, there is concern that changes in climate may alter the distribution of malaria and may increase the risk of outbreaks in some areas (Liehne, 1988; WHO, 1990; Pearce, 1992). This paper examines the possible effects of climate change on the global distribution of malaria. A simple bioclimatic model has been used to delimit the current potential distribution of malaria and to examine broadscale shifts in malaria distribution that may occur as a result of changes in climate.

Malaria Ecology

The malaria cycle is a three-factor complex involving the malaria parasite, a mosquito vector, and a human host (May, 1950). The disease is caused by four species of parasite from the genus *Plasmodium* and is spread by around 60 species of mosquitoes from the genus *Anopheles*. Climate influences malaria transmission through its effects on both the vectors and the parasites. Each vector species has a specific range of climatic conditions suitable for its development. In addition, the development of the *Plasmodium* parasites within the mosquito vector is influenced by temperature (May, 1961). This means that the geographical distribution of both the mosquito and the parasite is restricted to those areas where climate, and other environmental conditions, are suitable. Furthermore, climate also has an indirect effect on malaria through its influence on suitable vegetation and vector breeding sites. Any change in climate is therefore likely to affect transmission and the overall geographical distribution of malaria. However, climate is only one of a large number of factors affecting malaria transmission. Malaria is a focal disease, and its transmission depends on the dynamics among humans, vectors, parasites, and the environment. More importantly, it depends on the effectiveness of control measures, socioeconomic development, and political stability (Nájera et al., 1992). The malaria situation is thus very complex.

The Current Situation

Approximately 40% of the worlds's population is still exposed to varying degrees of risk from malaria (WHO, 1992a). It is estimated that around 300 million people are infected with the malaria parasite, and approximately 120 million clinical cases of malaria occur each year, with around 1 million deaths (WHO, 1992a). Malaria is currently found in 99 countries or areas in the world (WHO, 1991). The situation is worst in Africa, where it is estimated that around 80% of the clinical cases occur, and around 90% of parasite carriers are living (WHO, 1992a). The majority of cases occurring outside Africa occur in the following countries in decreasing order of the magnitude of the disease: India, Brazil, Afghanistan, Sri Lanka, Thailand, Indonesia, Vietnam, Cambodia, and China (WHO, 1992b). In recent years malaria has extended to some originally malaria-free regions and other areas where it had been controlled in the past.

ASSESSING THE IMPACTS OF CLIMATE CHANGE ON POTENTIAL MALARIA DISTRIBUTION

General circulation models (GCM) are considered to provide the most credible estimates of greenhouse-gas-induced changes in climate. One of the major difficulties in using GCM-derived estimates to assess the effects of climate change on malaria is associated with differences in spatial scale. The coarse resolution of GCM, typically around 300 km, means that, in general, only large-scale estimates of climate change are produced. Methods of producing regional-scale changes in climate do exist (see, for example, Giorgi, 1990), however, these can be costly and time-consuming. In contrast to this, the transmission of malaria within an area is dependent on local conditions. Further difficulties arise because it is extremely difficult to model the distribution of malaria, particularly at the global scale. The complexity of the interaction among the number of mosquito and *Plasmodium* species involved, their climatic and environmental requirements, and the other sociocultural, economic, and political factors which affect the disease, makes it impossible to include all the variables and accurately model malaria distribution. For these reasons it was considered useful to examine the broadscale effects of changes in climate on malaria distribution before undertaking a more detailed study. In this study a simple bioclimatic model has been used to relate overall mosquito distribution to climate and, from this, the global pattern of malaria incidence is inferred.

The Bioclimatic Model

The bioclimatic model used in this work is based on the results of a study of the ecology of malaria mosquitoes and malaria transmission (Dutta and Dutt, 1978). The study determined a range of temperature, precipitation, and altitude limits that could be used to map the global incidence of malaria. Slight modifications to these limits have been made, and the resulting simple bioclimatic model has been used in this study to delimit areas with the bioclimatic potential for malaria transmission. The bioclimatic potential for malaria transmission can be calculated using the following simple criteria:

$$22°C \leq \text{mean } [T_{Jan} \text{ or } T_{Jul}] \leq 30°C + \text{Precip}_{annual} \geq 750 \text{ mm.}$$

Thus, areas where the mean monthly temperature in January or July is between 22°C and 30°C and annual precipitation is at least 750 mm are considered to be climatically suitable for malaria transmission to occur.

Mapping the Effects of Climate Change on the Potential Distribution of Malaria

One way of examining the impacts of climate change on potential malaria distribution is to map the present distribution under current climatic conditions, then perturb the climate to simulate a change, and remap the distribution of malaria. In order to do this, a geographical information system (GIS) called the Geographical Resource Analysis Support System (GRASS) has been used. GRASS is a public-domain GIS provided by the US Army Corps of Engineers, Construction Engineering Research Laboratory (USACERL, 1993). Information on the current climate was obtained from a global air temperature and precipitation dataset which is interpolated to a $0.5° \times 0.5°$ resolution (Legates and Wilmot, 1990a,b). The current potential distribution of malaria has been mapped using these data, and the resulting map has been used as a baseline from which to examine the impacts of climate change.

Estimates of possible changes in climate were obtained from $1 \times CO_2$ and $2 \times CO_2$ scenarios derived from five different GCM. Given the uncertainties surrounding GCM-derived projections of climate change, it is useful to include estimates from a number of different GCM in the assessment in order to examine the impacts of a range of possible changes in climate on malaria. To simulate a change in climate the change between the $2 \times CO_2$ and $1 \times CO_2$ model simulations was used to perturb the baseline. For temperature changes the difference between the two simulations was added to the baseline; for changes in precipitation, the baseline was multiplied by the ratio of the two simulations. The five GCM used in this study include those of the Australian Bureau of Meteorology Research Centre (BMRC) (Hart et al., 1990); Canadian Climate Centre (CCC) (Boer et al., 1984a,b), U.S. Goddard Institute for Space Studies (GISS) (Hansen et al., 1984; 1986), U.S.

Geophysical Fluid Dynamics Laboratory (GFDL) (Manabe and Wetherald, 1987; the more recent R30 Q-flux version of the model is used here); and U.K. Meteorological Office (UKMO) (Wilson and Mitchell, 1987). The disadvantage with most of these data is that they are from some of the earlier simulation experiments and so tend to estimate more extreme changes in climate than more recent simulations. However, the main reason for choosing these data is that all (except the BMRC model) are made easily available through the National Center for Atmospheric Research, Boulder, Colorado, USA. The BMRC data were supplied by Dr B McAvaney, BMRC, Melbourne, Australia.

RESULTS

Potential Distribution Under the Baseline Climate

Figure 1 shows the current potential distribution of malaria estimated by the bioclimatic model using the baseline climate. These limits represent the potential distribution of malaria based on mean climate alone. However, because malaria is a focal disease and its transmission is determined by local conditions, there will be many areas within these broad limits where the disease is not found. Furthermore, there are areas not included within this potential distribution where malaria is known to occur because microclimatic conditions and other factors are suitable for its development and transmission. For example, the model does not indicate the potential for malaria to occur in the Middle East or parts of east Africa, yet it is known to occur in these areas (WHO, 1991). The model also indicates the potential for malaria in some areas, such as Australia, eastern Brazil, parts of southern Europe, Korea, Japan, and the southeastern United States, where it is no longer found, although it was endemic in these areas until comparatively recently (Haworth, 1988). Indeed, the potential for malaria incidence based on suitable conditions for vector development and disease transmission still exists in many areas which are now considered malaria-free (Black, 1982). The model highlights a number of possible areas where such a situation may exist.

Potential Distribution Under a 2 × CO_2 Climate

Intermodel differences in the way each GCM models the climate system mean that there are differences in the estimates of changes in climate (see Mitchell *et al.*, 1990; Schlesinger and Mitchell, 1985 for further

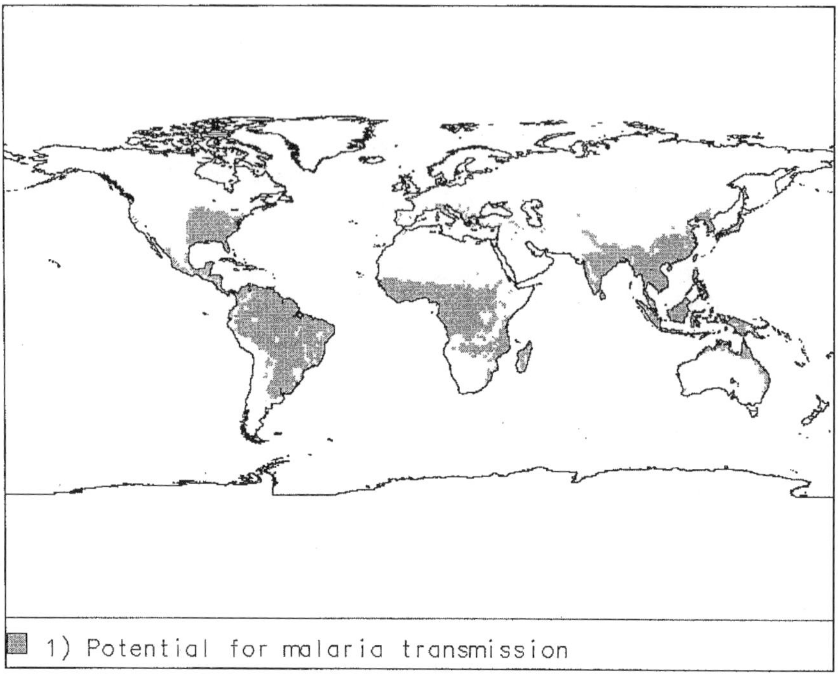

Figure 1. Baseline potential distribution of malaria, estimated using a simple bioclimatic model and based on current climatic conditions.

details). These differences in climate change are reflected in the variations in impact on potential malaria distribution estimated by each of the GCM. Table 1 shows the change in area of potential malaria occurrence estimated to occur as a result of the climate changes projected by the five GCM scenarios. The projected changes in climate result in both areas of increased and areas of reduced potential malaria occurrence. However, an overall increase in global potential malaria distribution is indicated for all projected climate changes.

Examination of Table 1 shows that the largest increase in potential malaria distribution would occur under the climate changes projected by the GFDL 2 × CO_2 scenario with an overall increase in area of 6,160,291 million km^2 (15.7%). However, the greatest overall *change* in malaria distribution would occur as a result of the climate changes estimated by the UKMO 2 × CO_2 scenario. The bioclimatic model suggests

Table 1. Changes in potential malaria distribution estimated to occur as a result of the climate changes projected by each of the five general circulation models (GCM) used in this study. Values (in km^2) represent the areas of increase and decrease in potential from the baseline, the area having the same potential as the baseline, and the total area of potential under each GCM. Figures in parentheses show percentage change in area from baseline area of 39,287,226 km^2.

	Area of potential for malaria transmission (km^2)			
GCM	Area of increase	Area of decrease	Same as baseline	Total area
BMRC	5 736 288 (14.6%)	2 745 006 (7.0%)	36 542 220	42 278 508 (+7.6%)
CCC	8 420 765 (21.4%)	5 621 945 (14.3%)	33 665 281	42 086 046 (+7.1%)
GFDL	11 325 374 (28.8%)	5 165 083 (13.1%)	34 122 144	45 447 517 (+15.7%)
GISS	10 261 994 (26.1%)	7 732 705 (19.7%)	31 554 522	41 816 515 (+6.4%)
UKMO	15 122 898 (38.5%)	9 997 755 (25.4%)	29 289 472	44 412 370 (+13.0%)

that, with such changes in climate, an area of 9,997,755 million km^2 would become climatically unsuitable for malaria occurrence and, at the same time, an area of 15,122,898 million km^2 would become suitable. This results in some dramatic shifts in potential malaria distribution, with an overall increase of 5,125,143 million km^2 (13.1%) in the area suitable for malaria transmission (Figure 2). Temperature increases of between 3 and 8°C projected by the UKMO 2 × CO_2 scenario mean that temperature would no longer be a limiting factor for malaria transmission in parts of the northeast United States, continental Europe, South America, and Africa, and hence there would be an increase in potential in these areas. Similar increases in temperature combined with increased precipitation would increase the potential for malaria transmission in Australia, northeast China, and eastern Europe. In contrast, in the southeast United States, northern India, Nepal, southeast China, Republic of Korea, parts of South America, Africa, and Indonesia the increase in temperature results in a decrease in potential for malaria and, consequently, a reduction in distribution. This decrease in potential reflects the fact that the temperature changes estimated by the GCM result in mean monthly temperatures in January or July in excess of the 30°C upper temperature threshold in the bioclimatic model. Therefore, according to the bioclimatic model, the area becomes unsuitable for malaria transmission.

In contrast to the large changes in malaria distribution estimated by the UKMO 2 × CO_2 scenario, the changes produced by BMRC 2 × CO_2 scenario are much smaller (Figure 3). With the climate changes

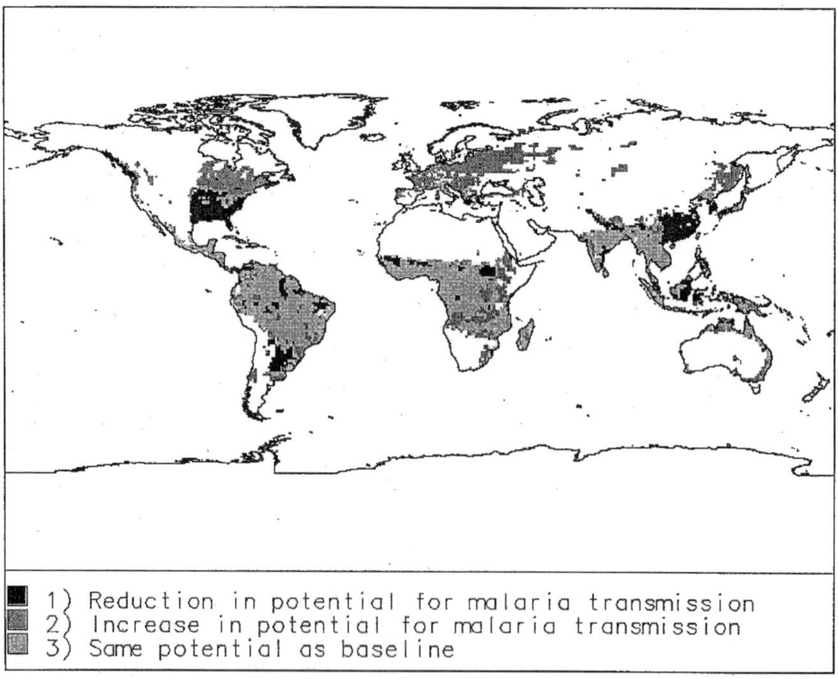

Figure 2. Change in potential distribution of malaria under climate changes estimated by the UKMO 2 × CO_2 scenario.

estimated by the BMRC 2 × CO_2 scenario, the bioclimatic model estimates that 2,745,006 million km^2 (7%) of the area currently considered suitable for malaria transmission would become unsuitable. This is combined with an expansion in area of 5,736,288 million km (14.6%) as the changes in climate increase the potential for malaria transmission in some areas. This results in an overall increase in global potential malaria distribution of 2,991,282 million km^2 (7.6%). The smaller impact on potential malaria distribution reflects the smaller temperature and precipitation changes projected by the BMRC 2 × CO_2 scenario. In general, temperature increases of 0°C to 6°C are estimated. The results from the BMRC 2 × CO_2 scenario used here are more recent and the simulation was run at a higher resolution than the UKMO 2 × CO_2 scenario simulation.

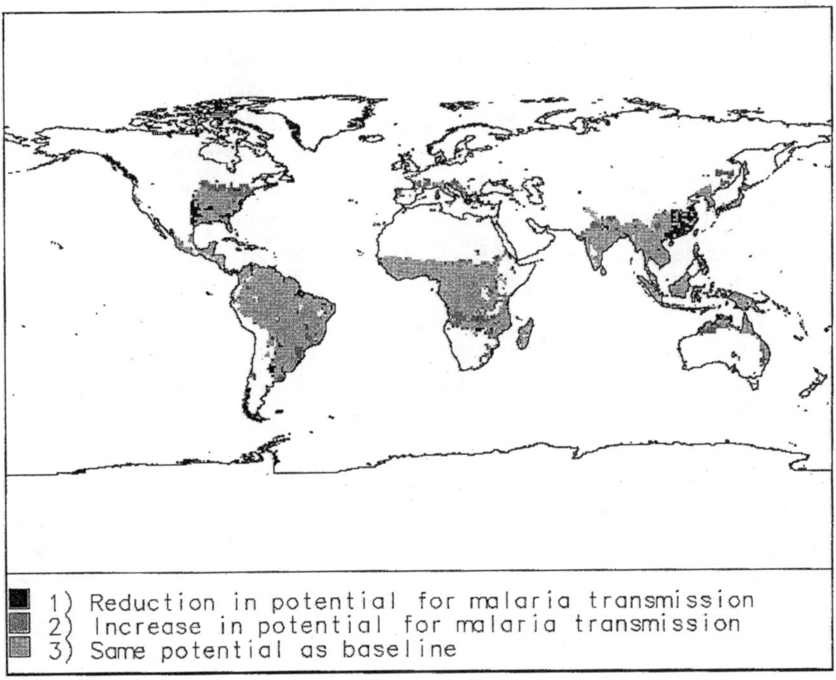

Figure 3. Change in potential distribution of malaria under climate changes estimated by BMRC 2 × CO_2 scenario.

DISCUSSION

The World Health Organization (WHO) considers that it is highly likely that changes in climate would result in a change in the distribution of malaria (WHO, 1990). The work outlined in this paper supports this view to the extent that it indicates changes in the *potential* distribution of malaria; i.e., potential shifts in the area considered to be climatically suitable for vector development and malaria transmission. That the extent of change in malaria distribution will depend on the magnitude of climate change (WHO, 1990) can clearly be seen by the range of changes in potential malaria distribution that are estimated to occur as a result of the climate changes projected by the five GCM used in this study (compare Figures 2 and 3). However, it is important to emphasize that the changes in potential distribution of malaria shown in this paper are not predictions of the impact of climate change on malaria, they are merely illustrative of some of the impacts that may occur.

There are a number of reasons why these results should be considered as preliminary and why further, more detailed, work is required. Firstly, there are the difficulties associated with modelling the occurrence of malaria. The simple model used in this study is based on only three climate variables: mean monthly temperature in January, in July, and annual precipitation. A much finer temporal resolution (probably daily values) and the inclusion of additional climate variables such as evaporation, would be necessary to make a detailed assessment of the possibility of malaria transmission. This is particularly true in epidemic areas, where malaria outbreaks are generally related to the seasonality of climate. As discussed above, there are a number of areas in which the bioclimatic model used in this study indicates there would be a reduction in potential for malaria occurrence and, consequently, a reduction in distribution. However, given the wide range of climatic conditions under which malaria currently occurs, it seems highly unlikely that any future change in climate would, by itself, result in a reduction in the distribution of malaria. Instead, the model may be highlighting areas in which a change in the length or timing of the malaria transmission season could occur as a result of high summer temperatures producing unfavorable conditions for mosquito and *Plasmodium* development. Another limitation of the model is that it can only be used to examine the impacts of climate change on malaria distribution. It makes no assessment of changes in the incidence of malaria in areas where the disease is currently present. Such changes in disease incidence and, consequently human, morbidity and mortality may be of greater overall significance than changes in malaria distribution.

As previously mentioned, additional problems in examining the effects of climate change on malaria relate to the spatial scale of the assessment. The fact that malaria is a focal disease means that a more detailed regional or local-scale examination of the effects of climate change on specific vectors is required before any useful prevention or control strategies can be considered. There are currently many uncertainties in the projections of climate change, particularly concerning the rate, magnitude, and regional pattern of changes. Such detailed examinations are likely to require an increased knowledge of the interrelationships between climate, vectors, and the parasite. Even in the absence of any change in climate, such knowledge could provide important information for use in malaria control activities. Finally, climate change is only one of a range of environmental, socio-economic, technical and political changes that will have an impact on the distribution and transmission of malaria. A great deal of work will be required before the full implications of these changes can be assessed.

CONCLUSIONS

The work outlined in this paper provides a simple first assessment of the possible impacts of climate change on the global distribution of malaria. The work highlights a number of problems encountered when trying to assess the impacts of global changes in climate on a disease which is greatly influenced by local conditions. Despite this, it can clearly be seen that the global malaria situation is influenced by climate, so any change in climate is likely to have an impact on the incidence of malaria. Above all, the work highlights a need to examine the impacts of climate change on a disease which is already extremely costly, both economically and in terms of quality of life.

ACKNOWLEDGMENTS

The author wishes to thank the following: A/Professor Russell Blong, A/Professor Bob Fagan, and particularly A/Professor Peter Curson, who generously sponsored the author's trip to the 32nd Hanford Symposium, Dr. Peter Love for his comments on this paper, and Dr. Philippe Martin (ISPRA) for his contribution to the early stages of the project.

REFERENCES

Black, RH. **1982**. *Malaria in Australia*, 1981, Topical Medical Technical Paper No. 8. Commonwealth Institute of Health, University of Sydney, Australian Government Publishing Service, Canberra.

Boer, GJ, R McFarlane, and R Laprise. **1984a**. The climatology of the Canadian Climate Centre general circulation model as obtained from a five-year simulation. Atmos.-Ocean 22:430-473.

Boer, GJ, R McFarlane, R Laprise, JD Henderson and JP Blanchet. **1984b**. The Canadian Climate Centre spectral atmospheric general circulation model. Atmos.-Ocean 22:397-429.

Dutta, HM and AK Dutt. **1978**. Malaria ecology: a global perspective. Soc. Sci. Med 12:9-84.

Giorgi, F. **1990**. Simulation of regional climate using a limited area model nested in a GCM. J. Climate 3:941-963.

Hansen, J, A Lacis, D Rind, G Russell, P Stone, I Fung, R Ruedy, and J Lerner. **1984**. Climate sensitivity: analysis of feedback mechanisms, pp. 130-163. In: *Climate Processes and Climate Sensitivity,* JE Hansen and T Takahashi (eds.). Maurice Ewing Series 5, American Geophysical Association, Washington D.C.

Hansen, J, A Lacis, D Rind, G Russell, I Fung, P Ashcroft, S Lebedeff, R Ruedy, and P Stone. **1986**. The greenhouse effect: projections of global climate change, pp. 199-218. In: *Effects of Changes in Stratospheric Ozone and Global Climate, Volume 1: Overview,* JG Titus (ed.). Environmental Protection Agency, United Nations Environment Program. WHO, Geneva.

Hart, TL, W Bourke, BJ McAvaney, and BW Forgan. **1990**. Atmospheric general circulation simulations with the BMRC global spectral model: the impact of revised physical parameterizations. J. Climatol. 34:436-459.

Haworth, J. **1988**. The global distribution of malaria and the present control effort, pp 1379-1420. In: *Malaria: Principles and Practice of Malariology*, WH Wernsdorfer and I McGregor (eds.). Churchill Livingstone, Edinburgh and London.

Legates, DR and CJ Wilmot. **1990a**. Mean seasonal and spatial variability in global surface air temperature. Theoret. Appl. Climatol. 41:11-21.

Legates, DR and CJ Wilmot. **1990b**. Mean seasonal and spatial variability in gauge-corrected, global precipitation. Int. J. Climatol. 10:111-127.

Liehne, PFS. **1988**. Climatic influences on mosquito-borne diseases in Australia, pp. 624-637. In: *Greenhouse: Planning for Climate Change*, G Pearman (ed.). CSIRO Publications, Melbourne.

Manabe, S. and RT Wetherald. **1987**. Large scale changes in soil wetness induced by an increase in carbon dioxide. J. Atmos. Sci. 44:1211-1235.

May, JM. **1950**. Medical Geography; its methods and objectives. Geograph. Rev. 40:9-40.

May, JM. **1961**. The ecology of malaria, pp. 161-230. In: *Studies in Disease Ecology*, JM May (ed.). Hafner, New York.

Mitchell, JFB, S Manabe, T Tokioka, and V Meleshko. **1990**. Equilibrium climate change, pp. 131-172. In: *IPCC Climate Change: The IPCC Scientific Assessment*, JT Houghton, GJ Jenkins and JJ Ephraums (eds.). Cambridge University Press, Cambridge, UK.

Nájera, JA, BH Liese, and J Hammer. **1992**. *Malaria: New Patterns and Perspectives,* World Bank Technical Paper No. 183. The World Bank, Washington DC.

Pearce, F. **1992**. A plague on global warming. New Sci. 136:12-13.

Schlesinger, ME and JFB Mitchell. **1985**. Model projections of the equilibrium climatic response to increased CO_2, pp. 81-147. In: *The Potential Climatic Effects of Increasing Carbon Dioxide,* MC MacCracken and FM Luther (eds.), DOE/ER-0237. U.S. Department of Energy, Washington DC.

USACERL. **1993**. *Grass 4.1 User's Reference Manual*, U.S. Army Corps of Engineers Construction Engineering Research Laboratories.

WHO. **1990**. *Potential Health Effects of Climatic Change*, WHO/PEP/90/10. World Health Organization, Geneva.

WHO. **1991**. Malaria, *World Health*, Sept-Oct, 1991.

WHO. **1992a**. World Malaria Situation in 1990. Weekly Epidemiol. Rec. 22:161-167; 23:169-174.

WHO. **1992b**. Press Release: Ministerial Conference on Malaria, Amsterdam, Netherlands, 26-27 October, 1992. WHO, Geneva.

Wilson, CA and JFB Mitchell. **1987**. A doubled CO_2 climate sensitivity experiment with a GCM including a simple ocean. J. Geophys. Res. 92:13315-13343.

REGIONAL HEALTH IMPACTS OF GLOBAL CLIMATE CHANGE: THE CASE OF ASTHMA

P. J. Beggs

Climatic Impacts Centre, School of Earth Sciences,
Macquarie University, Syndey, Australia

Key words: *Asthma, climate change, allergens, irritants, synergism*

ABSTRACT

Asthma is a major public health problem in Australia. It is often environmentally triggered and so is likely to be affected by climate change. A worsening of the status of this disease in Australia will undoubtedly have significant social and economic consequences. To date, reviews of climate change and health, with respect to asthma, have largely considered the impacts of atmospheric triggers such as urban air quality and, to a lesser extent, allergens and the direct impacts of climate and meteorological phenomena.

The objective of this study is to identify and examine some of the problems and difficulties associated with gauging the impacts of global climate change on asthma. These include difficulties in predicting regional and local climate change, assessing human and architectural influences on asthma triggers, and the problems associated with clarifying the effects of individual factors in a multifactorial disease such as this.

There appear to be many problems associated with gauging the effects of global climate change on asthma, and therefore the prospects for doing so appear limited. As in the recent past, our ability in the near future to gauge potential effects will at best be qualitative. A quantitative assessment would require significant improvements in both prediction of climate change at the regional and local scales which are more relevant to asthma occurrence and, perhaps, more importantly, improvements in our understanding of the complex pathogenesis of asthma and of the interactions between climate and natural systems.

INTRODUCTION

This paper reviews the existing literature relating to climate change and asthma. Also discussed are some of the difficulties associated with assessing possible trends in asthma associated with climate change in Australia. Emphasis will be placed on difficulties faced even before the difficulties of climate change prediction, and of translating such predictions to the regional, local, and "micro" scales.

While there is increasing discussion about the problems and difficulties associated with applying predicted climate change to impacts on regional issues such as health, there has been little mention of the problems that have existed for some time in determining the relationships between climate and disease and the mechanisms causing disease. This paper focuses on such problems.

ASTHMA DEFINED

Asthma is " . . . a disease characterized by an increased responsiveness of the airways to various stimuli and manifested by slowing of forced expiration (associated with widespread narrowing of the airways) which changes in severity either spontaneously or as a result of therapy" (ACCP-ATS, 1975). This airway-narrowing comes about through excessive contraction of the smooth muscle lining the bronchi, swelling of the lining of the bronchi, and hypersecretion of mucus in the bronchial tree. In many cases, asthma is clearly related to allergy—the individual is allergic to a number of substances known as allergens.

CAUSES VERSUS TRIGGERS

It is important to note the distinction between causes of asthma and triggers of asthma. Although several hypotheses partly explain asthma causation, much remains unknown. One hypothesis is that exposure to particular allergens in early childhood, perhaps in the first few months of life, can lead to sensitization to such allergens. Another hypothesis is that the occurrence of viral respiratory tract infections in the first year of life may lead to asthma. Far more is known about the triggers of asthma, the factors which provoke acute attacks and symptoms of the disease. The triggers of asthma in the atmospheric environment can be divided into three groups: climate/weather, allergens, and irritants (air pollutants). Climatic and meteorologic triggers include rapid temperature drops associated with the passage of a cold front. Examples of atmospheric allergens include mold/fungal spores, plant pollens, house dust mite feces, and cockroaches. The main atmospheric irritants triggering asthma are tropospheric ozone (O_3), nitrogen dioxide (NO_2), sulfur dioxide (SO_2), and particulates. Reviews of these three groups of atmospheric triggers of asthma include: for climate/weather (as well as irritants), Lopez and Salvaggio (1983); for allergens, Solomon and Mathews (1983), Baldo and Tovey (1984), and Bryant (1984); and for irritants, Whittemore (1981), and Abramson and Voigt

(1991). It is these triggering factors which may be affected by climate change, thereby leading to changes in the severity and frequency of asthma attacks.

COST AND MAGNITUDE OF ASTHMA IN AUSTRALIA AND NEW SOUTH WALES

Many people suffer from asthma but few die from it. About 800 to 900 deaths a year in Australia are due to asthma, a rate of 5.7 per 100,000. In New South Wales, approximately 320 people die from asthma each year. Estimates of morbidity in the Australian community suggest that 1 in 10 adults and 1 in 5 children, aged 5 to 9 years, suffer from asthma. Another indicator of the significance of asthma morbidity in Australia is hospital admissions. It is the most common cause of admission to pediatric hospitals (Mellis, 1988) and one of the primary causes for adult admissions to hospital (Australian Institute of Health, 1988).

A recent estimate of the cost of health care for one asthmatic over a 50-yr time span (based on doctor visits, drugs, and occasional treatment) is $A90,000 (1990 dollars; Ewan et al., 1991a). An estimate of the cost of asthma (based on hospital services, medical practitioner consultations, drugs and appliances, and loss of productivity) in 1989 to the community of New South Wales is $A209 million (Mellis et al., 1991).

ASSESSMENTS OF THE IMPACTS OF CLIMATE CHANGE ON ASTHMA

Climate change is generally expected to lead to increases in morbidity and mortality from asthma. Many of the effects of climate change on human disease, including asthma, will be indirect, or secondary, as well as direct, or primary (Goldstein and Reed, 1991; Haines and Fuchs, 1991; Doll, 1992). As will be discussed below, climate change will affect asthma indirectly through its impacts on atmospheric irritants and allergens and infections, and directly through its influence on the climate.

Air Quality (Irritants)

Longstreth (1991) recently suggested that ". . . respiratory diseases brought about by increased air pollution," e.g., asthma, would be ". . . one of the largest public health issues . . . " associated with global warming and stratospheric ozone depletion. Many studies have discussed the effects of atmospheric pollutants on asthma and how their occurrence may vary with climate change. An acceleration of photochemical reaction rates among the precursors of photochemical smog would be associated with increased temperatures and ultraviolet-B (UV-B) radiation (Leaf, 1989; McCally and Cassel, 1990; Rouviere et al., 1990; WHO, 1990; Ewan et al., 1991a; Doll, 1992). Thus, high levels of O_3 would be achieved earlier in the day and maintained longer and more frequently (Robinson, 1989; Longstreth, 1991). Such increases in ozone would be significant, as this irritant has been linked to summer peaks in the incidence of asthma (Haines and Fuchs, 1991). Grant (1990) has made two important observations regarding O_3 and asthma: (1) "The same magnitude of pulmonary decrement seen at a given O_3 level may represent a medically more significant effect for asthmatics . . . with already impaired respiratory functions than for normal, 'healthy' individuals." (2) "The synergistic potential between irritants and allergens, specifically that ozone `increases airway hyperresponsiveness to bronchoconstrictor agents such as histamine or other chemical mediators released in response to biological agents [causes] allergic reactions." Similar synergistic mechanisms between different irritants such as O_3 and SO_2 have been noted by Haines and Fuchs (1991).

Future emissions of NO_2 and SO_2 will rely primarily on human decisions and actions, being largely dependent on trends in fossil fuel consumption and, to a lesser extent, on the balance between the need for heating and the need for cooling (WHO, 1990). For example, higher winter temperature would lessen the demand for energy production for heating and thus would reduce acid aerosol exposure and reduce asthma morbidity (McCally and Cassel, 1990).

Regional (synoptic) scale conditions such as high-pressure systems (anticyclones) and local scale conditions such as katabatic (drainage) flows and sea breezes play a large role in air pollution production. Changes in the occurrence of such regional and local conditions will influence urban air quality (Ewan et al., 1991a).

Allergens

There is a clear relationship between many allergen-producing organisms, such as plants, mold, and house dust mites, and climatic factors such as humidity, temperature, rainfall, and sunshine. Increased levels of pollens and other biologic agents causing allergic respiratory system reactions in humans would result from higher temperatures coupled with elevated humidity. Last and Guidotti (1990/1991) and Last (in press) have suggested there would be significant changes in the distribution of vegetation, including many allergens. Similarly, Longstreth (1991) suggested " . . . changes in the prevalence or intensity of asthma . . . episodes in affected individuals . . ." could come about through likely " . . . quantitative and/or qualitative changes in the airborne concentration of allergens . . ." such as mold spores and pollens.

Climate

Wind plays a major role in the transport of pollens, dust, and other irritants (Davis et al., 1988). Waves of extreme heat and cold impose stress on asthmatic individuals (Longstreth, 1988). Ewan et al. (1991a,b) said that " . . . heat stress, combined with high humidity, acts as a trigger to attacks." Curson (1989) has suggested that ". . . drier winters and changed wind movements and cyclonic activity . . ." may lead to asthma increases in Australia.

Hennessy (1992) has estimated the change in frequency of extreme cold events in southeastern Australia (winter days when temperature falls below 0°C) associated with a 3°C warming (Figure 1). It is clear that such events would become far less frequent. It is likely that acute asthma symptoms triggered by sudden exposure to cold air would be reduced if such a climate change occurs in this region.

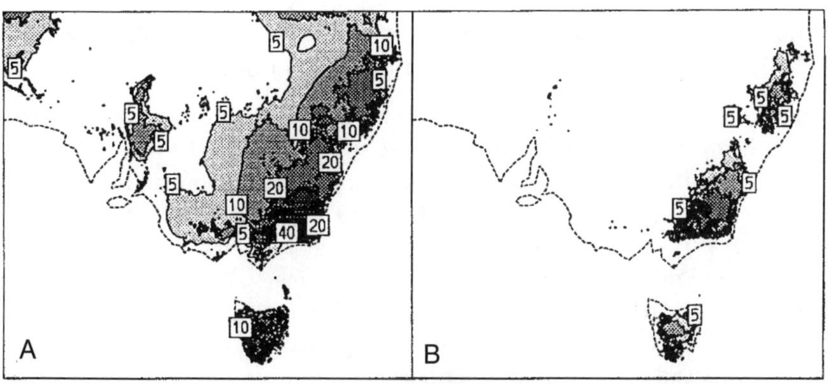

Figure 1. Frequency of extreme cold events in southeastern Australia (winter days when temperature fell below 0°C). A, current climate; B, estimated for warming of 3°C (from Hennessy, 1992).

Infection

In experimental studies, the resistance of animals to bacterial infection was lowered after exposure to low concentrations of NO_2 and O_3. Similar exposure could lead to increased rates of viral and microbial infections in humans. It has also been suggested that the immunosupressive effects of UV-B radiation could lead to increases in infection as depletion of the stratospheric O_3 layer continues (Davis et al., 1988; Last, in press). Ewan et al. (1991a) have stated that " . . . infections such as the common cold and influenza . . . " could increase in southeastern Australia and decrease in southwestern Australia due to predicted increases or decreases in rainfall. Such infections are commonly associated with asthma symptoms. These potential increases are significant since peaks in hospital admissions for asthma in autumn/winter have been related to viral infections (Haines, 1991; Haines and Fuchs, 1991).

PROBLEMS ASSOCIATED WITH GAUGING THE IMPACTS OF CLIMATE CHANGE ON ASTHMA IN SOUTHEASTERN AUSTRALIA

These recent assessments have been, rightly, conservative, at most indicating potential qualitative trends in asthma morbidity and mortality. To the author's knowledge, no concerted efforts to quantitatively and accurately predict the impacts of climate change on asthma have been made to date. This is perhaps due to a number of problems and difficulties (discussed below) associated with extending such predictions beyond their present elementary stage.

Problems Related to the Coarseness of General Circulation Model (GCM) Output and Applying This to Regional Forecasts

Longstreth (1991) has recognized one of the problems associated with determining the effects of global warming on human health: the ". . . vast uncertainties in the precise changes in weather anticipated with global warming." Haines and Fuchs (1991) suggested that we could talk generally about global warming but that ". . . regional changes in climate have not been modelled with confidence." Significant improvements in simulating regional climate have occurred as a result of nesting limited area models, with 70-km resolution, within GCM, with $4.5° \times 7.5°$ resolution. Using this approach, high-resolution simulations for the present have been conducted for Europe (Marinucci and Giorgi, 1992) and southeastern Australia (Pitman et al., 1991). High-resolution climate-change simulations have been conducted for Europe (Giorgi et al., 1992). Results obtained from these nested models have confirmed beliefs that simple interpolation of ". . . coarse resolution GCM output to obtain estimates of local changes in surface climatic variables for impact assessment . . ." (Giorgi et al., 1992) is inaccurate and inadequate. While nested limited-area models dramatically increase the resolution of climate-change predictions, there are problems with the approach. The physical nesting procedure is difficult; the final result may depend on how realistic the GCM simulation is. And the nested model must be able to simulate mesoscale circulation patterns within the region of interest for both the present and under conditions of climate change. While these are major research problems, they will be solved. Therefore, as the accuracy and availability of modeled regional change improve, so may our ability to predict impacts of future climate change on public health issues such as asthma.

Direct Effect of Climate/Weather

Most of the direct impact of climate/weather on asthma will be associated with extreme events. Indeed, there is a large body of literature describing the impacts of extreme events such as thunderstorms, cold fronts, and extremes of temperature that have been associated with asthma epidemics. Events of this scale are currently not very well modeled; thus, indications that their intensity will increase should be treated cautiously.

Haines and Fuchs (1991) noted that " . . . climate may affect the respiratory tract in three ways: seasonal effects, direct effects of specific weather conditions (such as thunderstorms and cold fronts), and combined effects of weather conditions and other environmental or topographical factors." They later noted that " . . . the mechanisms for some of these effects are still speculative." Haines and Fuchs (1991) state that " . . . the study of meteorological factors in human disease is complicated by a number of factors, including the interaction of different variables such as temperature and humidity . . ." and the varying responses by different subgroups according to local microenvironment. Further, Longstreth (1991) recognized that there are such complexities involved even in assessing the role normal weather plays in human disease that one can only speculate on the effects of global warming on human health.

Asthma, A Multifactorial Disease

Each asthmatic is unique in his sensitivity to factors that provoke acute symptoms. The composition of these factors is also individual. Therefore, it is difficult to generalize about the overall impacts that climate change will have on the status of asthma. Asthmatics rarely, if ever, react or are reactive to only one trigger. There is increasing evidence that triggers act synergistically or in combination, but such complex mechanisms are not fully understood. Indeed, Goldstein and Reed (1991), in their commentary on global atmospheric change and research needs in environmental health sciences, recognized the need to better understand both the individual and multiple effects of indoor and outdoor irritants and allergens on people with chronic respiratory diseases such as asthma. Thus, even if future climate can be predicted at appropriate scales and future major triggers can be accurately simulated, our lack of understanding about how symptoms really evolve means we cannot accurately predict the severity of asthma in the future.

Because asthma is a disease affected by a wide range of factors, it is possible that such factors will be suited to a variety of environmental conditions. Conditions close to the coast favor the house dust mite, while those inland favor grass pollen production (reflected in the status of the respective allergies present in the two populations). Different molds grow over very different temperature ranges and, no matter what the change in climate, there may be molds which can grow in that new climate.

Modeling Plant Distributions

As has been mentioned earlier, pollen produced by many species of plants can trigger acute attacks of asthma. Since most of the pollen of most windborne plant species is carried only a short distance (Solomon, 1984), the distribution of a particular plant species will have a large bearing on the population it will affect. The distribution of some plant species, based purely on climatic requirements or restraints, has been modeled by bioclimatologists.

Parietaria judaica is an example of such a species. Its potential bioclimatic distribution has been mapped (Bass and Bass, 1990) based on rainfall and temperature alone (Figure 2A), but this distribution bears little resemblance to the actual distribution of this species (Figure 2B). One of the factors leading to its deviation from this potential distribution is its preference for disturbed ground. Assuming future regional climates could be predicted accurately, the potential **future** bioclimatic distribution of this species may bear no resemblance to what it actually would be. In other words, it is not possible for us to accurately model the present distribution of *P. judaica* based on climate alone, therefore it is not valid to predict its future distribution based on climate alone.

Figure 2. A, Potential bioclimatic distribution of *P. judaica* in Australia (from Bass and Bass, 1990); B, actual distribution of *P. judaica* in Australia (from Parsons and Cuthbertson, 1992).

Another factor leading to the potential discrepancy between modeled and actual plant distributions is the human modification of local conditions (creating an artificial climate) by means of irrigation and fertilizer.

Variance of Indoor Microclimate to Outdoor Climate

We spend about 80 to 90% of our time indoors (Piver 1991). The major allergy in many parts of the world, including coastal southeastern Australia, is to house dust-mite allergens. The most abundant species of house dust mite in southeastern Australia is *Dermatophagoides pteronyssinus*. The house dust mite has rather specific climatic requirements, with optimum conditions occurring at 25°C and 75% relative humidity (Tovey, 1993). Although the indoor microclimate, particularly in beds and carpets where mites live, is not totally separate from the outdoor climate, and accounts for the commonly found decrease in mite numbers as one goes from coast to inland and low altitude to high altitude, there are often substantial differences. This makes it difficult to predict both present and future distribution and levels. Even if it were possible to predict climate change accurately for a region, such predictions would say little about the climate experienced in the home and even less about the microclimate within beds and in carpets.

Combined Architectural/Climatic Factors Influencing Allergens

Indoor mold is, to a certain extent, linked to the outdoor climate. Increases in rainfall and humidity tend to favor mold growth and subsequent spore release. Mold growth in homes often depends on sealing the house from the outside environment. Thus, the condition of the damp course and hence the age of housing in a region or local area will play some role in the response of indoor mold to the climate.

Reliance of Some Triggering Organisms Primarily on Factors Other Than Climate

Some triggers of acute asthma respond partially to climatic conditions but are mainly controlled by nonclimatic factors. Cockroaches produce allergens that affect many asthmatics. There are probably climatic requirements for cockroaches, such as temperature, but of equal or greater importance are factors such as human population density and housing age. Thus, predictions of their future distribution and abundance cannot be made on climate requirements alone.

Some major allergenic triggers of asthma attacks, such as cats, dogs, and horses, are not weather/climate-dependent. There is no reason to believe future climate change would affect their distribution.

Triggering Thresholds Are Currently Exceeded

The levels of house dust-mite allergen in most Sydney homes are already above that required to induce sensitization, that is, to develop immunoglobulin E (IgE) antibody and asthma (>2 μg of mite allergen Der pI/g of fine dust) and provoke symptoms in mite-allergic people (>10 μg/g) (Tovey, 1991). Specifically, 98.7% of Sydney homes have levels above 2 μg/g, and 88.3% have levels above 10 μg/g. Unlike some regions, no seasonal variation in dust-mite allergen levels has been observed in Sydney. Because the climate there is already conducive for year-round mite growth, climate change may not affect allergen levels. Even if it does, such a change may be insignificant in view of the already high levels.

Unexplained Geographic Distribution of Asthma Mortality

There is evidence to suggest that asthma mortality rates are not globally or regionally uniform. As we are unable to fully explain the present geographic distribution of asthma mortality, which is more than likely associated with environmental factors, it is doubtful whether attempts to predict future mortality distribution will be successful.

Perceptions of Health Risk

It has been recognized that ". . . health problems are caused as often by behavioral and social causes as by strictly biologic mechanisms . . ." (McCally and Cassel, 1990). As Gray (1991) rightly points out, humans can and do change their behavior in response to perceived health risk. It is possible that some of the potentially detrimental effects of climate change on asthmatics, such as increased house dust-mite allergen levels, will be mitigated, to a certain extent, by behavior modification; in this case, perhaps, by removal of carpets. Although such responses are difficult to predict, it is important that models used to gauge the impacts of climate change on asthma recognize this possibility.

CONCLUSIONS

It has been recognized that we have a poor understanding of the health consequences of global environmental change, and therefore associated research, and research into traditional environmental health, should be encouraged and expanded (McCally and Cassel, 1990). "Attempts to predict changes in respiratory disease as a result of climate change are difficult" (Haines and Fuchs, 1991).

It is imperative that researchers continue to tackle both ends of the problem of gauging health effects of climate change. It is clear that predicting the impacts of climate change on asthma with certainty is impossible, but if we can determine the impacts of climate change on some of the more certain triggers, such as O_3, house dust mites, and pollens, these predictions can be applied/translated to changes in triggers. This is true especially when it is possible to have accurate regional and local future climate change predictions, which will allow better predictions of their effects on triggering mechanisms. We must also continue trying to clarify the role of the atmospheric environment in asthma.

Future efforts to gauge the impacts of climate change on asthma must also attempt to consider and include human influences on natural and human systems.

REFERENCES

Abramson, M and T Voigt. **1991**. Ambient air pollution and respiratory disease. Med. J. Aust. 154:543-553.

ACCP-ATS Joint Committee on Pulmonary Nomenclature. **1975**. Pulmonary terms and symbols. *Chest 67*:583-593.

Australian Institute of Health. **1988**. *Australia's Health.* Australian Government Publishing Service, Canberra.

Baldo, BA and ER Tovey. **1984**. Allergy, allergens and allergen standardization. Med. J. Aust. 141(Special Supplement):S5-S8.

Bass, DA and DJ Bass. **1990**. *Parietaria judaica* L. A cause of allergic disease in Sydney. A study of habit and spread of the weed. Rev. Palaeobot. Palynol. 64:97-101.

Bryant, DH. **1984**. Allergies, mediators and asthma. Med. J. Aust. 141(Special Supplement):S2-S5.

Curson, P. **1989**. Human health in the greenhouse era, pp. 140-150. In: K McCracken, and R Blong, eds. *Planning for the Greenhouse, a Conference Designed for Planners, Engineers, Health Surveyors and Resource Managers,* July 17-18, 1989, Macquarie University, Sydney.

Davis, DL, V Miller, and JJ Reisa. **1988**. Potential public health consequences of global climate change, pp. 366-376. In: *Preparing for Climate Change,* Climate Institute, Government Institutes, Inc., Rockville, MD.

Doll, R. **1992**. Health and the environment in the 1990s. Am. J. Public Health 82:933-941.

Ewan, C, E Bryant, and D Calvert, eds. **1991a**. *Health Implications of Long Term Climatic Change.* Australian Government Publishing Service, Canberra.

Ewan, C, EA Bryant, GD Calvert, et al. **1991b**. Potential health effects of greenhouse effect and ozone layer depletion in Australia. Med. J. Aust. 154:554-559.

Giorgi, F, MR Marinucci, and G Visconti. **1992**. A 2XCO2 climate change scenario over Europe generated using a limited area model nested in a general circulation model; 2. Climate change scenario. J. Geophys. Res. 97:10011-10028.

Goldstein, BD and DJ Reed. **1991**. Global atmospheric change and research needs in environmental health sciences. Environ. Health Perspect. 96:193-196.

Grant, LD. **1990**. Respiratory effects associated with global climate change, pp. 81-88. In: *Global Atmospheric Change and Public Health,* JC White, ed. Elsevier Science Publishing Co. New York.

Gray, NJ. **1991**. Potential health effects of greenhouse effect and ozone layer depletion in Australia [letter]. Med. J. Aust. 155:207.

Haines, A. **1991**. Global warming and health. Br. Med. J. 302:669-670.

Haines, A and C Fuchs. **1991**. Potential impacts on health of atmospheric change. J. Public Health Med. 13:69-80.

Hennessy, KJ. **1992**. Regional Climate Scenario for South-Eastern Australia in 2030. In: *Agriculture and Greenhouse in South Eastern Australia,* IE Galbally and P Holper, eds. CSIRO Division of Atmospheric Research.

Last, JM. Global change: ozone depletion, greenhouse warming, and public health. Annu. Rev. Public Health 14:115-136.

Last, J and TL Guidotti. **1990/91**. Implications for human health of global ecological changes. Public Health Rev. 18:49-67.

Leaf, A. **1989**. Potential health effects of global climatic and environmental changes. N. Engl. J. Med. 321:1577-1583.

Longstreth, J. **1988**. Overview of the potential effects of climate change on human health, pp. 142-146. In: *Coping with Climate Change*, JC Topping, ed. Proceedings of the 2nd North American Conference on Preparing for Climate Change, December 1988.

Longstreth, J. **1991**. Anticipated public health consequences of global climate change. Environ. Health Perspect. 96:139-144.

Lopez, M and JE Salvaggio. **1983**. Climate-Weather-Air Pollution. pp. 1203-1214. In: *Allergy: Principles and Practice (Volume Two)*, E Middleton Jr., CE Reed, and EF Ellis, eds. The C. V. Mosby Company, St. Louis, MO.

Marinucci, MR and F Giorgi. **1992**. A 2XCO2 climate change scenario over Europe generated using a limited area model nested in a general circulation model; 1. Present-day seasonal climate simulation. J. Geophys. Res. 97:9989-10009.

McCally, M and CK Cassel. **1990**. Medical responsibility and global environmental change. Ann. Int. Med. 113:467-473.

Mellis, CM. **1988**. Important changes in the emergency management of acute asthma in children. Med. J. Aust. 148:215-217.

Mellis, CM, JK Peat, AE Bauman, et al. **1991**. The cost of asthma in New South Wales. Med. J. Aust. 155:522-528.

Parsons, WT and EG Cuthbertson. **1992**. Family Urticaceae, pp. 622-624. In: *Noxious Weeds of Australia.* Inkata Press, Melbourne.

Pitman, AJ, F Giorgi, and A Henderson-Sellers. **1991**. Southeast Australia's wintertime precipitation: sensitivity of climate predictions to model resolution. Aust. Meteorol. Mag. 39:21-35.

Piver, WT. **1991**. Global atmospheric changes. Environ. Health Perspect. 96:131-137.

Robinson, P. **1989**. The effects of climate change, pp. 23-41. In: *Global Climate Change Linkages: Acid Rain, Air Quality, and Stratospheric Ozone,* JC White, ed. Elsevier Science Publishing Co. New York.

Rouviere, C, T Williams, R Ball, et al. **1990**. Human Settlement; the Energy, Transport and Industrial Sectors; Human Health; Air Quality; and Changes in Ultraviolet-B Radiation. In: *Climate Change, The IPCC Impacts Assessment*, WJ McG Tegart, GW Sheldon, and DC Griffiths, eds. Australian Government Publishing Service, Canberra.

Solomon, WR. **1984**. Aerobiology of pollinosis. J. Allergy Clin. Immunol. 74:449-461.

Solomon, WR and KP Mathews. **1983**. Aerobiology and inhalant allergens. pp. 1143-1202. In: *Allergy: Principles and Practice (Volume Two)*, E Middleton Jr, CE Reed, and EF Ellis, eds. The C. V. Mosby Company, St. Louis, MO.

Tovey, E. **1991**. Monitoring biological particulates in homes. J. Occup. Health Safety - Aust. N.Z. 7:154-158.

Tovey, E. **1993**. Mites and their allergens, pp. 3-8. In: *Mites, Asthma and Domestic Design: Proceedings of a Conference Held at the Powerhouse Museum, E Tovey and A Mahmic, eds. 15th March 1993.* The University Printing Service, University of Sydney.

Whittemore, AS. **1981**. Air pollution and respiratory disease. Annu. Rev. Public Health 2:397-429.

WHO. **1990**. *Potential Health Effects of Climate Change*. World Health Organization, Geneva.

CLIMATE CHANGE AND NUCLEAR WASTE

DEFINING POSSIBLE CLIMATE STATES FOR THE GREATER CONFINEMENT DISPOSAL FACILITY AT THE NEVADA TEST SITE[a,b]

F. A. Duran,[1] F. M. Phillips,[2] J. Emery,[1] and E. Webb[3]

[1]GRAM, Inc., Albuquerque, New Mexico

[2]New Mexico Institute of Mining & Technology, Socorro, New Mexico

[3]Sandia National Laboratories, Albuquerque, New Mexico

ABSTRACT

The Department of Energy's Nevada Operations Office has disposed of a small quantity of transuranic waste at the Greater Confinement Disposal (GCD) facility in Area 5 of the Nevada Test Site (NTS). To assess the likelihood of complying with the Environmental Protection Agency standards for the disposal of transuranic waste, high-level waste and spent fuel, Sandia National Laboratories has performed a preliminary performance assessment (PPA) of the site. The results of the PPA analyses indicated that climate change may impact the performance of the GCD facility. Additional site characterization activities to support subsequent iterations of the PPA include a climate change study to define a range of possible climate states for the site.

The climate change study for the GCD site is based on a time series analysis of long-term paleoclimatic data, including lake surface-area data from Searles Lake, California, and $\delta^{d}0$ data from Devil's Hole, Nevada. These data will be interpreted to develop estimates of temperature and available water, and then extrapolated to predict a range of possible future climatic conditions. Sources of uncertainty in the data and analysis will be identified, and the significance of these data to the desert southwest will be discussed. Additional work for this study will include a comparison of the long-term data from Searles Lake and Devil's Hole with other, shorter-term indicators of paleoclimatic conditions.

[a]Sandia Document Control Number SAND93-1295A.

[b]This work performed at Sandia National Laboratories is supported by the U.S. Department of Energy under contract DE-AC04-76DP00789.

NATURAL RESPONSES TO QUATERNARY CLIMATIC CHANGE WITHIN THE NEVADA TEST SITE REGION

J. D. Gibson

Sandia National Laboratories, Albuquerque, New Mexico

ABSTRACT

Migration of hazardous contaminants depends on natural processes. Climatic fluctuations can affect the magnitudes and rates of many of these processes. In any long-term environmental evaluation of natural processes, responses to climatic change must be considered.

Five generalized categories of natural responses to Quaternary climatic change are recognized for the Nevada Test Site region of southwestern Nevada and adjacent California: (1) atmospheric, (2) biologic, (3) geomorphic, (4) hydrologic (surface and subsurface), and (5) pedologic/diagenetic. These responses dictate the potential effects of climatic change on contaminant transport, which is being examined by existing and planned environmental-restoration and waste-management programs within the region. Many of these programs require long-term (>10,000-yr) waste isolation. Specific responses corresponding to these categories are: (1) during the last glacial maximum (LGM), the Laurentide ice sheet may have bifurcated the jet stream and changed dominant atmospheric circulation patterns in the western United States. (2) Packrat-midden data suggest that, at a given locality, plant communities have changed radically over the last 18,000 yr. (3) Alternating episodes of incision and aggradation of alluvial fans reflect glacial and interglacial periods, respectively. (4) Extensive pluvial lakes developed in southern Nevada and adjacent California during the LGM. Changes in 6180 values from vein carbonate that was deposited by ground water at Devil's Hole, Nevada, provide an uninterrupted 500,000-yr climate history against which other paleoclimatic records can be calibrated and competing climatic theories can be evaluated. (5) Finally, increased effective moisture during glacial periods caused accelerated rates for soil-forming processes.

CLIMATE CHANGE AND URANIUM MILL TAILINGS DISPOSAL IN THE FOUR CORNERS REGION

W. J. Waugh[1] and K. L. Petersen[2]

[1]Environmental Sciences Laboratory,* P.O. Box 14000, Grand Junction, Colorado

[2]Westinghouse Hanford Company, P.O. Box 1970, Richland, Washington

*Operated by RUST Geotech Inc. for the U.S. Department of Energy

Key words: *Engineered cover, long-term performance, paleoclimate, paleoecology, uranium mill tailings*

ABSTRACT

The U.S. Department of Energy (DOE) is designing earthen repositories for the disposal of low-level radioactive uranium mill tailings. The DOE has been mandated to demonstrate that the repositories will isolate the tailings from biological intrusion, erosion, and the underlying aquifer for up to 1,000 yr. There is new evidence about past climate change and model predictions suggesting that global climatic variation will likely exceed the historical record as the lower atmosphere warms in response to increasing concentrations of carbon dioxide and other greenhouse gases. DOE therefore recognizes the need to incorporate possible ranges of future climatic and ecological change in the repository design process. In the Four Corners region of the U.S. southwest, the center of uranium mill tailings disposal activities, archaeologists and environmental scientists have worked independently to reconstruct past vegetation and climate; to understand changes in prehistoric agriculture that led to the growth and decline of Anasazi populations; and to help set bounding conditions for possible future environmental stresses on radioactive waste-disposal facilities. Therefore, paleoclimatic and ecological data have been used not only as a window on the past but also on likely local manifestations of future global change. Records have been developed from annual tree rings, from remains of flora and fauna preserved in packrat middens, and from plant pollen buried in lake sediments, alluvium, peat bogs, and archaeological sites. These sources reflect greater-than-historic variation in the magnitude and rates of local climate change, which can be compared with the modern record and with model projections.

INTRODUCTION

The U.S. Department of Energy (DOE) is responsible for disposal of tailings from mills which, between 1942 and 1960, provided uranium to the U.S Army's Manhattan Engineering District and later to the U.S. Atomic Energy Commission, for nuclear weapons and nuclear power generation. The highly acidic or alkaline milling processes produced more than 20 million metric tons of low-level radioactive and, sometimes, chemically toxic tailings (Portillo, 1992). Many of these mills were located in western Colorado, northwestern New Mexico, northeastern Arizona, and eastern Utah, a mostly arid and semiarid area known as the Four Corners region.

Over time, tailings buried in disposal facilities, even in the relatively arid Four Corners region, may be vulnerable to transport via rainwater percolation, gas diffusion, erosion, and intrusion by plant roots and burrowing animals. The DOE has been mandated to design disposal facilities that will isolate tailings from these transport pathways for at least 200 yr and, to the extent achievable, for 1,000 yr (USEPA, 1983). Considering this unprecedented legislative demand that disposal facilities effectively persist indefinitely, the design approach should exceed standard engineering practices and incorporate possible long-term changes in site hydrology and ecology. Thus, climate, the principal driver of potential contaminant transport processes, must be included in rational assessments of the long-term performance of tailings disposal facilities. This paper reviews the role of climate change in evaluations of disposal facility designs. We then infer a reasonable range of future climate scenarios at facilities in the Four Corners region using paleoclimate records.

LONG-TERM PERFORMANCE OF TAILINGS DISPOSAL FACILITIES

Containment of tailings and other wastes buried in disposal facilities is largely dependent on the long-term performance of earthen engineered covers (Anderson et al., 1993; Nyhan et al., 1990; Wing and Gee, 1990). Two general types of covers have been designed or constructed at disposal facilities in the Four Corners region (Caldwell, 1992): low-permeability designs, and capillary barrier designs (Figure 1). Low-permeability designs rely on compacted fine-soil layers to slow radon escape and deep infiltration of rainwater. The rocky upper layer provides erosion protection, and a sand bedding layer underlying the rock is intended to divert rainwater laterally. The capillary barrier design

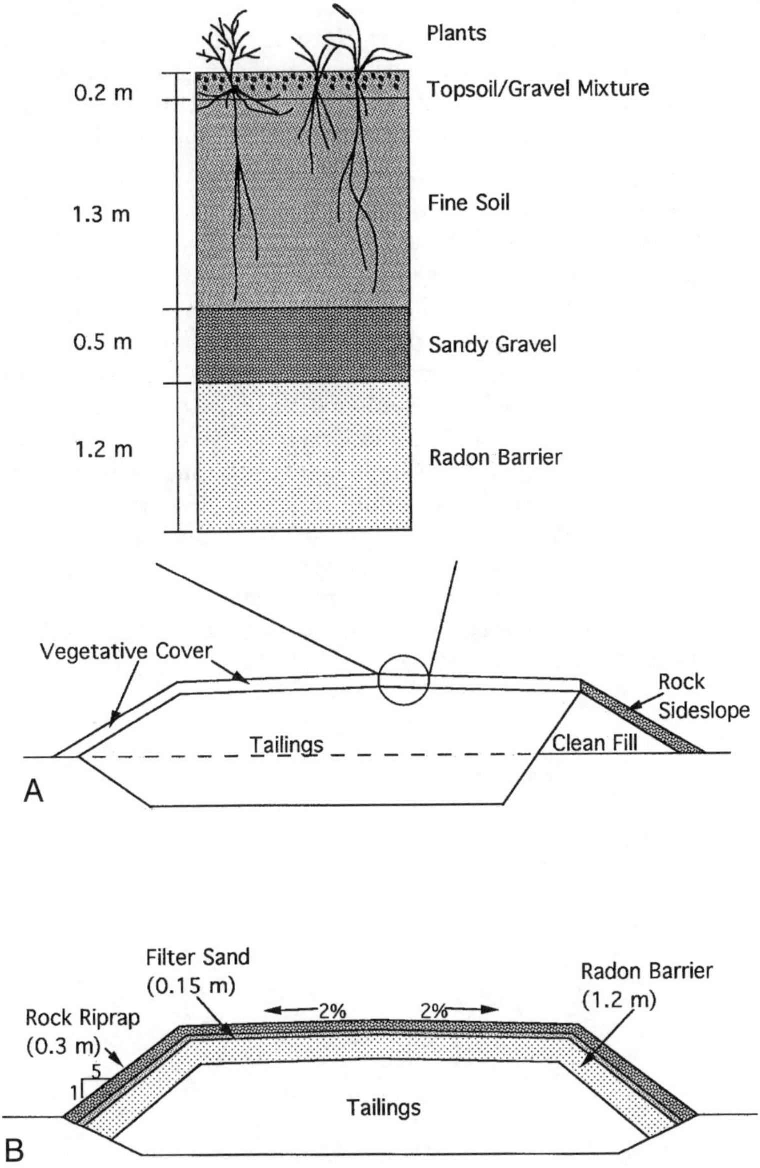

Figure 1. Typical (A) capillary barrier and (B) low-permeability designs for uranium mill tailings disposal facility covers. Rock side slopes are sometimes placed over clean fill to avoid excessive water infiltration.

adds a topsoil layer with favorable physical properties for storing rainwater close enough to the surface to be removed by evapotranspiration (Gee et al., 1992). With vegetation present, gravel mixed into the soil surface controls top-slope erosion, like a desert pavement, without negatively influencing the soil-water balance (Waugh et al., 1994a).

The triad approach recommended for assessing the long-term performance of covers (Figure 2) requires climatic data (Petersen et al., 1993). Models of soil-water movement (e.g., Fayer et al., 1992), erosion (e.g., Lane and Nearing, 1989), and radon diffusion (USNRC, 1984) engender an understanding of the complexity of environmental processes acting on engineered covers and can numerically simulate responses of designs to myriad conditions. Field tests (Figure 2), although expensive, provide the only real measures of performance and are needed to confirm model predictions (e.g., Gee et al., 1992; Waugh et al., 1991). Both modeling and field test results show that water infiltration of the cover, recharge, radon diffusion, biointrusion, and disposal cell stability are all sensitive to long-term changes in precipitation, temperature, and water extraction by plants. Natural analogs (Figure 2) provide clues from past environments as to the possible responses of engineered covers to long-term changes in these parameters (Waugh et al., 1994b).

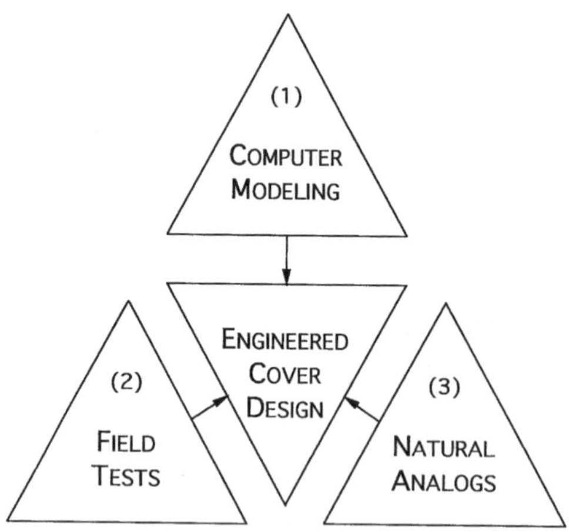

Figure 2. Integrated triad approach used to assess the long-term performance of engineered covers (Waugh et al., 1994b).

WHY USE PALEOCLIMATE ANALOGS?

Current cover design paradigms and practices tend to rely on meteorological records for evaluations of long-term performance (USDOE, 1989). Meteorological data should be used to model and test cover performance for present climatic conditions but not for possible future changes in climate. Climatologists generally agree that during the next decades and centuries, global climatic variation will exceed the historical record, as has occurred in the past (Houghton et al., 1990; Crowley and North, 1991) and/or as the lower atmosphere warms in response to increasing concentrations of anthropogenic carbon dioxide and other greenhouse gases (Hansen et al., 1988; Ramanathan, 1988).

Paleoclimate records provide not only a window on the past but also on possible local manifestations of future global change. By comparison, models of global and regional climate enhance our understanding of climate-change drivers but are inadequate for local climate projections. Nested regional models are continually improving (e.g., Giorgi et al., 1993), but available global climate models that could address the 1,000-yr design life of disposal sites are incapable of resolution on the spatial scales required (e.g., Hansen et al., 1988; Mitchell, 1989; Houghton et al., 1990). Ranges of possible future climate at Four Corners tailings disposal sites can best be inferred from natural proxy records of past climate change. A reasonable range of climatic conditions 1,000 yr into the future can be captured within a period spanning the Holocene and late Pleistocene. Proxy climate records covering this period have been developed for the Four Corners region from tree rings, plant and animal remains preserved in packrat middens, plant pollen buried in fluvial sediments, and Anasazi archaeology.

RECONSTRUCTING PALEOCLIMATES

The reconstruction of local paleoclimates of the Four Corners region using proxy climate records relies on an understanding of current climate drivers, of present-day relationships between plant distribution and climate, and on assumptions of cyclical climate and consistent responses of vegetation to climate change.

Climate Drivers

Winter (November to March) precipitation enters the region with migrating low-pressure systems following the westerly jet stream.

Troughs of low pressure typically follow a route north of a West Coast high-pressure ridge, entering the United States in the Pacific Northwest and tracking southward along the eastern side of the Rockies, often bringing partly cloudy skies and strong southwest winds to the Four Corners. An anomalous westward displacement of the winter high-pressure ridge routes the low-pressure trough south off the West Coast, entering the United States as far south as central California and tracking through the Great Basin and Colorado Plateau to the northeast. Once established, this pattern tends to recur, producing a succession of wet winter storms (Changnon et al., 1993). The pattern may be associated with El Niño-Southern Oscillation (ENSO) events (Redmond and Koch, 1991), during which low midlatitude storms draw moisture from the subtropical jet stream and bring heavy precipitation to the southwest (Cayan and Peterson, 1989). Tree-ring records suggest that ENSO events have caused anomalously wet southwestern winters for at least the past 300 yr (Swetnam and Betancourt, 1990).

Warm, relatively dry air masses governed by high-elevation anticyclones centered over the Mexican Highlands dominate the April, May, and June climate of the Four Corners region. In early July, a monsoon-like low pressure usually develops over the Colorado Plateau, drawing warm, moist air from the Gulf of Mexico and Gulf of California and spawning convective storms as a result of surface heating, convergence, or orographic lifting (Bryson and Lowry, 1955; Hales, 1974; Tang and Reiter, 1984). Development of a low-pressure trough in the Salton Sea region of southern California enhances air flow from the Gulf of California (Reyes and Cadet, 1988). The summer convective storms of the Four Corners region form in clusters many tens of kilometers across and, on average, persist for less than an hour (Barry and Chorley, 1970).

Interplay of Climate and Vegetation

The reconstruction of Four Corners paleoclimates assumes that (1) precipitation and temperature have an overriding influence on the elevational distribution of plants. Therefore, (2) past shifts in the elevational distribution of plants reflect past shifts in the elevational distribution of temperature and precipitation. In fact, plant distributions are delimited by many interacting factors that not only vary over time and space but also exhibit concomitant and synergistic effects. Elevational shifts occur when the complex of environmental conditions above or below a species' distribution changes so that conditions fall within its physiological tolerances. This may be a favorable change in temperature or effective moisture, but it may also be influenced by changes in species interactions that can diminish, augment, or otherwise offset climate-induced shifts. Further, considering the physiographic discontinuities in environmental gradients and the seasonality aspect of climate change, the elevational interplay of climate and vegetation should not be considered constant.

Recognizing this complexity and interdependence of factors that limit plant distribution, and considering the naiveté of thinking that the importance of single factors can be isolated, for the purposes of this study we proceeded with the above assumptions. After all, the Holocoentic concept of causal factors holds that factors are interactive, not that they have equal weight (Billings, 1952, 1978).

Figure 3 depicts a present-day elevational distribution of vegetation, temperature, and precipitation in the Four Corners region. Again, assuming some consistency of vegetation and climate interactions, evidence of past changes in the distribution of plants preserved at sites scattered along the elevational gradient (also shown in Figure 3) is the basis for reconstructing past climates. Temperature and precipitation curves were derived from meteorological data compiled by Petersen et al. (1987) for 12 sites located between 1,300 and 2,700 m elevation in southwestern Colorado and southeastern Utah. The generalized plant species distribution curves are based on numerous sources (e.g., Arno and Hammerly, 1984; Betancourt et al., 1990; Maher, 1961; Petersen, 1988). Lower limits of plants are generally considered to be set by moisture deficiencies, and upper limits by low temperatures. Table 1 presents the types of proxy records, dates, and references for the paleoclimate sites used in this study. Figure 4 is a map of paleosite locations.

DISPOSAL SITE CLIMATES

Records of Past Climate Change

The records in Table 1 indicate that plant distribution in the Four Corners region shifted hundreds of meters in elevation since the Late Pleistocene in response to fluctuations from arid to subhumid climates (Figure 5). Records from 18,000 B.P. to the present are well-constrained by radiocarbon dates and span the transition from the last full glacial maximum to the present interglacial period. These paleorecords reflect long-term shifts in average temperature and precipitation, not the high variability that exists in meteorological records (Nielson, 1986).

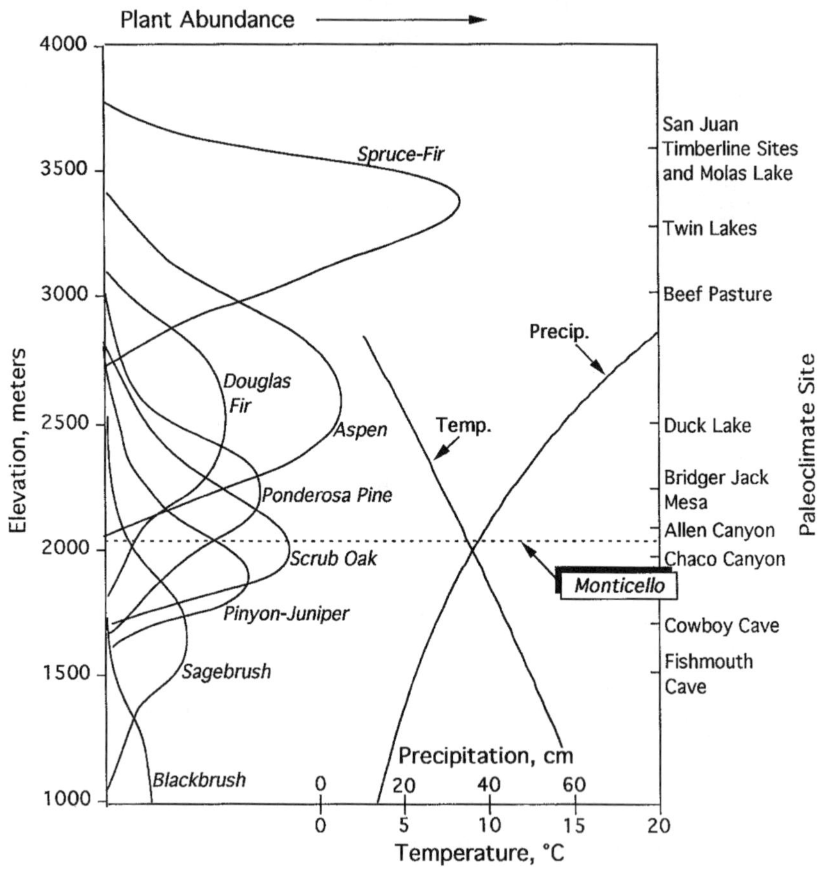

Figure 3. Elevational distribution of present-day vegetation, mean annual temperature ($r^2 = 0.97$), mean annual precipitation ($r^2 = 0.93$), and paleoclimate sites in the Four Corners region. The Monticello, Utah, tailings site is shown as an example.

Records of past climate change in the Four Corners region provide working limits for ranges of temperature and precipitation at tailings disposal sites. For example, the modern vegetation at Monticello, Utah, (Figure 4) is scrub oak, with pinyon-juniper on southern exposures and

Table 1. Types of records, dates, and references for paleoclimate sites used to reconstruct Four Corners region climates.

Paleoclimate Site	Record Type	Dates[a]	Reference
Fishmouth Cave	Packrat middens	13,800 B.P. to present	Betancourt et al., 1990
Cowboy Cave	Packrat middens, pollen, and plant macrofossils	13,000 B.P. to present	Jennings, 1980
Chaco Canyon	Packrat middens	10,600 B.P. to present	Betancourt et al., 1990
Allen Canyon Cave	Packrat middens	11,300 B.P. to present	Betancourt, 1984
Duck Lake	Pollen	Late Pleistocene	Betancourt and Biggard, 1985
Bridger Jack Mesa	*Pinus edulis* dendrochronology	500-yr record	Van Pelt, 1978
Beef Pasture	Pollen	6,000 B.P. to present	Petersen, 1988
Twin Lakes	Pollen	10,000 B.P. to present	Petersen, 1988
Molas Lake, San Juan Mountains	Pollen	16,000 B.P. to present	Maher, 1961
San Juan Mountains Timberline	Macrofossils and pollen in timberline bogs	10,000 B.P. to present	Carrara et al., 1991

[a] B.P. = before present.

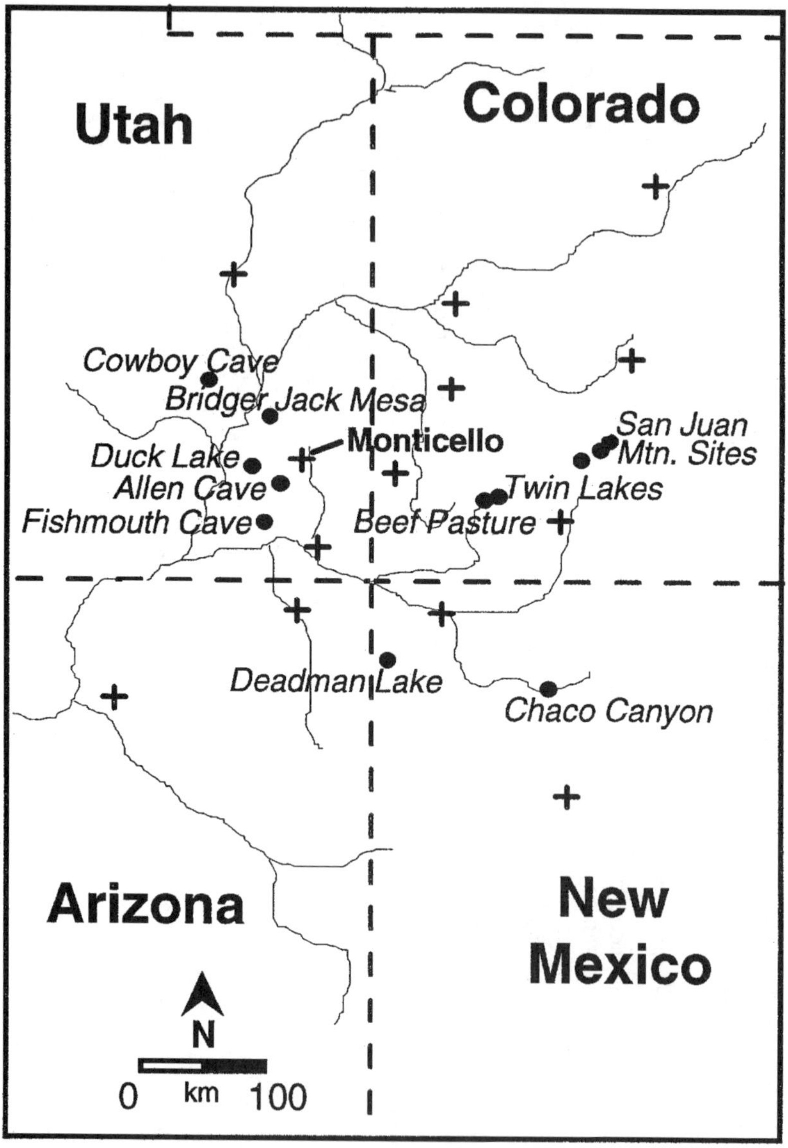

Figure 4. Map of paleoclimate sites (•) and uranium mill tailings sites (+) in the Four Corners region.

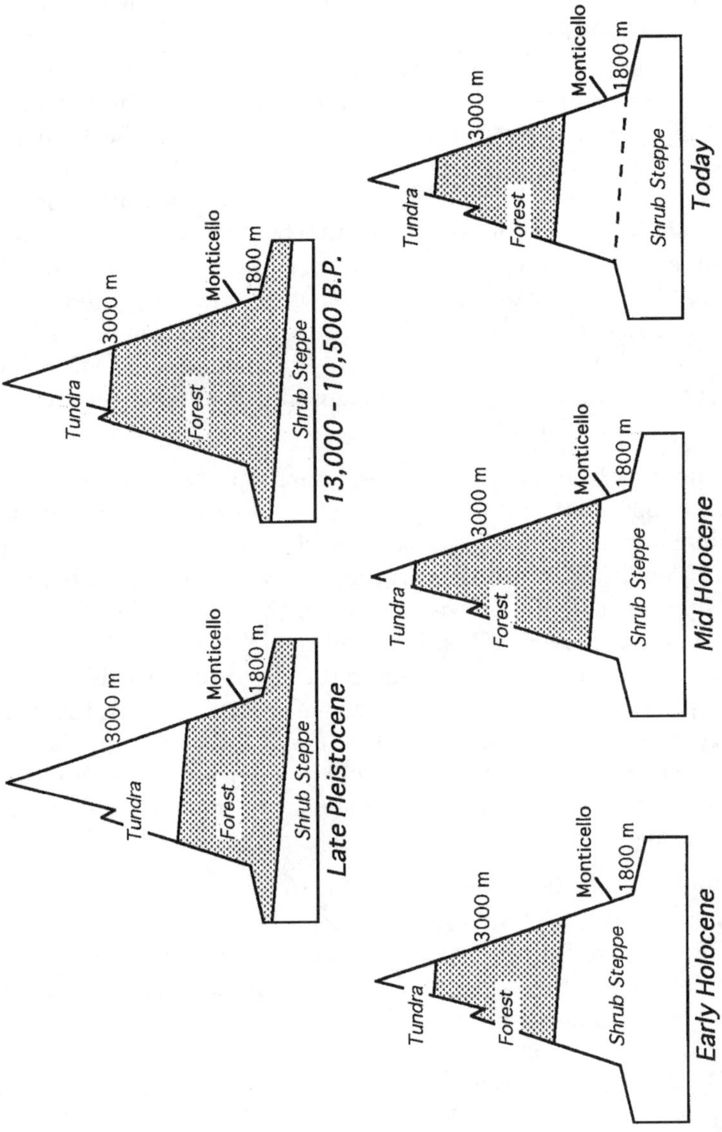

Figure 5. Generalized late Pleistocene and Holocene shifts in forest boundaries in the Four Corners region. Present lower extent of pinyon pine, a late Holocene newcomer, is marked with a dashed line.

ponderosa pine on northern exposures (Figure 3). Present mean annual temperature and precipitation are 8°C and 38 cm, respectively (Figure 3). During the Late Pleistocene, subalpine forests dominated vast areas now occupied by pinyon-juniper woodlands (Figure 5). This predominance of cold-tolerant species, the downward expansion of subalpine trees, and the apparent absence of warm-season species reflect a much wetter, colder climate than today (Lao and Benson, 1988; Thompson, 1990). The Laurentide ice sheet, thrust up into the lower atmosphere, may have spilt the jet stream into two branches. The southern branch was forced south across northern California, Nevada, Utah, and Colorado (COHMAP Members, 1988), resulting in long wet winters, cool summers, and an essentially nonexistent summer monsoon in the Four Corners. On the basis of a synthesis of Late Pleistocene paleorecords, we have set working-level values of mean annual temperature at 2°C and precipitation at 80 cm for Monticello.

Paleoclimate data from Table 1 sources record a postglaciation climate (13,000 to 10,500 B.P.) that continued wetter but warmer, with upper plant limits moving upslope but with little change in lower plant limits (Figure 5). Conditions of global warming increased the moisture-holding capacity of Pacific air masses; however, the jet stream continued tracking south of slowly retreating continental ice (Ruddiman and Wright, 1987), sustaining wet winters in the Great Basin and Colorado Plateau. By the early Holocene (approximately 10,000 B.P.), the Four Corners climate was significantly warmer and drier than previously, although seasonal patterns of wet winters and dry summers persisted. A shift toward a more modern summer monsoonal climate was delayed until the ice sheet retreated into northeastern Canada.

The middle Holocene (8,000 to 4,000 B.P.) was warmer and wetter than today: approximately 10°C and 60 cm precipitation at Monticello (Figure 5). The packrat midden record of vegetation change reflects cold and relatively dry winters, an increase in both summer and annual temperatures, a shift to monsoon-dominated summers, and greater effective moisture than at present (Betancourt, 1984; Betancourt and Biggard, 1985). The upper treeline was at least 80 m and as much as 140 m higher than at present (Carrara et al., 1991). This period, known as the Altithermal, has been viewed as an analog of future greenhouse warming, although the latter may develop more rapidly (within 50 to 100 yr) and have warmer winters (Houghton et al., 1990). By about 3500 B.P., the upper treeline had dropped to near its present elevation.

Petersen (1985, 1988) coupled pollen and archaeological records to infer fluctuations in the seasonality of climate in the Four Corners region for the last 2,000 yr. Changes in pollen influx for pinyon pine, a middle Holocene newcomer to the Four Corners region, may reflect changes in the northern latitudinal position of the summer monsoon boundary. The northern limit of Anasazi occupation also appears to have fluctuated in response to the northern position of the monsoon boundary. For example, a high pinyon pollen influx between A.D. 700 and 1200 corresponds to a period of increasing Anasazi population in the region. The Anasazi were subsistence farmers, who relied on summer monsoon rains to raise corn and other crops.

Historical climate change continued to influence human activities. Between the years 1250 to 1850, a period called the Little Ice Age, a colder climate displaced European agriculture to the south. In the Four Corners region, summer rains abated, growing seasons shortened, and the frequency of forest fires increased. With the end of the Little Ice Age, around 1840, pinyon pine began encroaching into deeper soils in woodland parks of the Four Corners (Van Pelt, 1978), reflecting an increase in summer rainfall (Petersen, 1988). Sustained by relatively high effective moisture, dry farming near Blanding, in southeastern Utah, began in 1905 but was abandoned in 1932 because of drought. Tree-ring records show that this was the wettest period in the Four Corners region during the last 300 yr (Stockton, 1976).

Future Climate Change

Global mean temperature may increase by 1.8 to 5.2°C in the next century; this is far beyond the limits of natural variability, in response to an industrial age buildup of carbon dioxide, methane, and other gases (Houghton et al., 1992). Model projections of the magnitude of warming vary, depending on whether factors such as CO_2 fertilization, feedback from stratospheric ozone depletion, and the radiative effects of sulfate aerosols are taken into account. Model projections of precipitation responses to greenhouse warming also are inconsistent (Houghton et al., 1990; Crowley and North, 1991; Washington and Meehl, 1984; Wilson and Mitchell, 1987; Schlesinger and Mitchell, 1987). Some regions may be effectively wetter and others drier, depending on the balance of the greater potential evaporation and the greater water-holding capacity of a warmer atmosphere.

Greenhouse warming will eventually be overwhelmed as the earth plunges into another ice age. Models of cyclic astronomical forcing of climate agree that, without anthropogenic disturbances, a long-term cooling trend that started about 6000 yr ago will continue, climaxing with a major glaciation in about 60,000 yr (Imbrie and Imbrie, 1980; Berger et al., 1991). In contrast, aperiodicity in the timing of past ice ages is evident in oxygen isotope records (Winograd et al., 1992). Other paleorecords suggest that certain feedback mechanisms have caused rapid and unpredictable transitions into ice ages (Berger and Labeyrie, 1987; Phillips et al., 1990).

Despite uncertainty about drivers of future climate change, climate extremes in the next 1,000 yr likely will not exceed those associated with the last glacial and interglacial periods (e.g., Spaulding, 1985). Therefore, paleorecords of full glacial and altithermal climates in the Four Corners region provide reasonable ranges of possible future climate and should be incorporated in assessments of the long-term performance of tailings disposal facilities. For the Monticello, Utah, example presented above, full glacial and altithermal climate reconstructions provide tentative working levels of 2 to 10°C mean annual temperature and 38 to 80 cm mean annual precipitation.

REFERENCES

Anderson, JE, RS Nowak, TD Ratzlaff, and OD Markham. **1993**. Managing soil moisture on waste burial sites in arid regions. J. Environ. Qual. 22:62-69.

Arno, SF, and RP Hammerly. **1984**. *Timberline: Mountain and Arctic Frontiers*. The Mountaineers, Seattle, WA.

Barry, RG, and RJ Chorley. **1970**. *Atmosphere, Weather, and Climate*. Holt, Rinehart, and Winston, New York.

Berger, WH, and LD Labeyrie (eds). **1987**. *Abrupt Climatic Change: Evidence and Implications*. Reidel, Dordrecht, The Netherlands.

Berger, A, H Gallee, and JL Melice. **1991**. The earth's future climate at the astronomical timescale, pp. 148-165. In: *Future Climate Change and Radioactive Waste Disposal: Proceedings of International Workshop*, GM Goodess and JP Palutikof (eds.). NSS/R257, U.K. Nirex Radioactive Waste Disposal, Ltd., Harwell, U.K.

Betancourt, JL. **1984**. Late Quaternary plant zonation and climate in southeastern Utah. Great Basin Nat. 44:1-35.

Betancourt, JL and N Biggard. **1985**. *Preliminary Assessment of Late Quaternary Vegetation and Climate of Southeastern Utah Based on Analyses of Packrat Middens*, BMI/ONWI-570. Battelle Memorial Institute, Columbus, OH.

Betancourt, JL, TR Van Devender, and PS Martin (eds.). **1990**. *Packrat Middens The Last 40,000 Years of Biotic Change*. University of Arizona Press, Tucson, AZ.

Billings, WD. **1952**. The environmental complex in relation to plant growth and distribution. Q. Rev. Biol. 27:251-265.

Billings, WD. **1978**. *Plants, Man, and the Ecosystem, 3rd ed.* Wadsworth, Belmont, CA.

Bryson, RA, and WP Lowry. **1955**. Synoptic climatology of the Arizona summer precipitation singularity. Bull. Am. Meteorol. Soc. 36:329-39.

Caldwell, JA. **1992**. Engineering perspectives for near-surface disposal, pp. 161-198. In: *Deserts as Dumps? The Disposal of Hazardous Materials in Arid Ecosystems*, CC Reith and BM Thompson (eds.). University of New Mexico Press, Albuquerque, NM.

Carrara, PE, DA Trible, and M Rubin. **1991**. Holocene timberline fluctuations in the northern San Juan Mountains, Colorado, U.S.A., as indicated by radiocarbon-dated conifer wood. Arct. Alp. Res. 23:233-246.

Cayan, DR, and DH Peterson. **1989**. The influence of north Pacific atmospheric circulation on streamflow in the west, pp. 375-397. In: *Aspects of Climate Variability in the Pacific and the Western Americas*, DH Peterson (ed.). Geophysical Monograph 55, American Geophysical Union, Washington, DC.

Changnon, D, TB McKee, and NJ Doesken. **1993**. Annual snowpack patterns across the Rockies: Long-term trends and associated 500-mb synoptic patterns. Mon. Weather Rev. 121:633-647.

COHMAP Members. **1988**. Climatic changes of the last 18,000 years: observations and model simulations. Science 241:1043-52.

Crowley, TJ and GR North. **1991**. *Paleoclimatology*. Oxford Monographs on Geology and Geophysics No. 16, Oxford University Press, New York.

Fayer, MJ, ML Rockhold, and MD Campbell. **1992**. Hydrological modeling of protective barriers: Comparison of field data and simulation results. Soil Sci. Soc. Am. J. 56:690-700.

Gee, GW, MJ Fayer, ML Rockhold, and MD Campbell. **1992**. Variations in recharge at the Hanford Site. Northwest Sci. 66:237-250.

Giorgi, F, GT Bates, and SJ Nieman. **1993**. The multiyear surface climatology of a regional atmospheric model over the western United States. J. Climate 6:75-95.

Hales, JE, Jr. **1974**. Southwestern United States summer monsoon source—Gulf of Mexico or Pacific Ocean? J. Appl. Meteorol. 13:331-42.

Hansen, J, I Fung, A Lacis, D Rind, S Lebedeff, R Ruedy, and G Russell. **1988**. Global climate changes and forecast by Goddard Institute of Space Studies three-dimensional model. J. Geophys. Res. 93:9341-9364.

Houghton, JT, GJ Jenkins, and JJ Ephraums (eds.). **1990**. *Climate Change: The IPCC Scientific Assessment*. Cambridge University Press, Cambridge, UK.

Houghton, JT, BA Callander, and SK Varney (eds.). **1992**. *Climate Change 1992: The Supplementary Report to the IPCC Scientific Assessment*. Cambridge University Press, Cambridge, UK.

Imbrie, J and JZ Imbrie. **1980**. Modeling the climatic response to orbital variations. Science 207:943-153.

Jennings, JB. **1980**. *Cowboy Cave*. University of Utah Anthropological Papers No. 104. University of Utah Press, Salt Lake City, UT.

Lane, LJ and MA Nearing. **1989**. *USDA - Water Erosion Prediction Project: Hillslope Profile Model Documentation*, NSERL Report No. 2. USDA-ARS National Soil Erosion Research Laboratory, West Lafayette, IN.

Lao, Y and L Benson. **1988**. Uranium-series age estimates and paleoclimatic significance of Pleistocene tufas from the Lahontan Basin, California and Nevada. Quat. Res. 30:165-76.

Maher, LJ, Jr. **1961**. *Pollen Analysis and Postglacial Vegetation History in the Animas Valley Region, Southern San Juan Mountains*, Ph.D. Dissertation. University of Minnesota, Minneapolis, MN.

Mitchell, JFB. **1989**. The "greenhouse effect" and climate change. Rev. Geophys. 27:115-139.

Nielson, RP. **1986**. High-resolution climatic analysis and Southwest biogeography. Science 232:27-34.

Nyhan, JW, TE Hakonson, and BJ Drennon. **1990**. A water balance study of two landfill cover designs for semiarid regions. J. Environ. Qual. 19:281-288.

Petersen, KL. **1985**. Palynology in Montezuma County, southwestern Colorado: The local history of pinyon pine (*Pinus edulis*), pp. 47-62. In: *Late Quaternary Palnology of the American Southwest*, B Jacobs, OK Davis, and P Fall, (eds.), ASSP Contribution Series 16. American Association of Stratigraphic Palynologists Foundation, Dallas, TX.

Petersen, KL. **1988**. *Climate and the Dolores River Anasazi*, University of Utah Anthropological Papers No. 113. University of Utah Press, Salt Lake City, UT.

Petersen, KL, JD Orcutt, and DA Breternitz. **1987**. *Dolores Archaeological Program: Supporting Studies: Settlement and Environment.* U.S. Department of Interior, Bureau of Reclamation, Engineering and Research Center, Denver, CO.

Petersen, KL, JC Chatters, and WJ Waugh. **1993**. *Long-Term Climate Change Assessment Study Plan for the Hanford Site Permanent Isolation Barrier Development Program*, WHC-EP-0569. Westinghouse Hanford Company, Richland, WA.

Phillips, FM, AC Campbell, R Roberts, and CV Kruger. **1990**. Abrupt interglacial/glacial transition in the 1.4 Ma water balance record from Searles Lake, California, pp. 28. In: *CANQUA/AMOUA 1990: Program and Abstracts*, First Joint Meeting, Canadian Quaternary Association and American Quaternary Association, University of Waterloo, Waterloo, Ontario, Canada.

Portillo, RS. **1992**. Mill Tailings Remediation: The UMTRA Project, pp. 281-302. In: *Deserts as Dumps? The Disposal of Hazardous Materials in Arid Ecosystems*, CC Reith and BM Thompson (eds.). University of New Mexico Press, Albuquerque, NM.

Ramanathan, V. **1988**. The Greenhouse theory of climate change: A test by an inadvertent global experiment. Science 240:293-299.

Redmond, KT and RW Koch. **1991**. Surface climate and streamflow variability in the Western United States and their relationship to large-scale circulation indices. Water Resour. Res. 27:2381-2399.

Reyes, S and DL Cadet. **1988**. The southwest branch of the North American monsoon during summer 1979. Mon. Weather Rev. 116:1175-1187.

Ruddiman, WF and HE Wright Jr. (eds.). **1987**. *North American and Adjacent Oceans During the Last Deglaciation*. The geology of North America, Volume K-3, Geological Society of America, Boulder, CO.

Schlesinger, ME and JFB Mitchell. **1987**. Climate model simulations of the equilibrium climatic response to increased carbon dioxide. Rev. Geophys. 25:760-798.

Spaulding, WG. **1985**. *Vegetation and climate of the last 45,000 years in the vicinity of the Nevada Test Site. south-central Nevada*, Professional Paper 1329. U.S. Geological Survey, U.S. Government Printing Office, Washington, DC.

Stockton, CW. **1976**. Long-term streamflow reconstruction in the upper Colorado River basin using tree rings, pp. 401-441. In: *Colorado River Basin Modeling Studies: Proceedings of a Seminar held at Utah State University. July 16-18, 1975*, CG Clyde, DH Falkenburg, and JP Riley (eds.). Water Research Laboratory, College of Engineering, Utah State University, Logan, UT.

Swetnam, TW, and JL Betancourt. **1990**. Fire-Southern oscillation relations in the Southwestern United States. Science 249:1017-1020.

Tang, M and ER Reiter. **1984**. Plateau monsoons of the Northern Hemisphere: a comparison between North America and Tibet. Mon. Weather Rev. 112:617-637.

Thompson, RS. **1990**. Late Quaternary vegetation and climate in the Great Basin, pp. 200-39. In: *Packrat Middens: The Last 40,000 Years of Biotic Change*, JL Betancourt, TR Van Devender, and PS Martin (eds.). University of Arizona Press, Tucson, AZ.

USDOE. **1989**. *Technical Approach Document, Revision II*, UMTRA-DOE/AL 050425.0002. Uranium Mill Tailings Remedial Action Program. U.S. Department of Energy, Albuquerque, NM.

USEPA. **1983**. *Standards for the Disposal of Uranium Mill Tailings*, 40 CFR 192. U.S. Environmental Protection Agency, Washington, DC.

USNRC. **1984**. *Radon Attenuation Handbook for Uranium Mill Tailings Cover design*, NUREG/CR-3533, prepared by Rogers and Associates Engineering, Salt Lake City, Utah. U.S. Nuclear Regulatory Commission, Washington, DC.

Van Pelt, NS. **1978**. *Woodland and Parks in Southeastern Utah*. Master's thesis, Department of Geography, University of Utah, Salt Lake City, UT.

Washington, WM and GA Meehl. **1984**. Seasonal cycle experiment on the climate sensitivity due to a doubling of $C0_2$ with an atmospheric general circulation model coupled to a simple mixed-layer ocean model. J. Geophys. Res. 89:9475-9503.

Waugh, WJ, ME Thiede, LL Cadwell, GW Gee, HD Freeman, MR Sackschewsky, and JF Relyea. **1991**. Small lysimeters for documenting arid site water balance, pp. 151-159. In: *Lysimeters for Evapotranspiration and Measurements*, RG Allen, TA Howell, W0 Pruitt, IA Walter, and ME Jensen (eds.). American Society of Civil Engineers (ASCE), New York.

Waugh, WJ, ME Thiede, DJ Bates, LL Cadwell, GW Gee, and CJ Kemp. **1994a**. Plant cover and water balance in gravel admixtures at an arid waste-burial site. J. Environ. Qual. 23:676-686.

Waugh, WJ, JC Chatters, GV Last, BN Bjornstad, and CR Hunter. **1994b**. *Barrier Analogs: Long-Term Performance Issues, Preliminary Studies, and Recommendations*, PNL-9004. Pacific Northwest Laboratory, Richland, WA.

Wilson, CA and JFB Mitchell. **1987**. A doubled $C0_2$ climate sensitivity experiment with a global climate model including a simple ocean. J. Geophys. Res. 92:13315-13343.

Wing, NR and GW Gee (eds.). **1990**. *Hanford Site Protective Barrier Development Program: Year 1989 Highlights*, WHC-EP-0318. Westinghouse Hanford Company, Richland, WA.

Winograd, IJ, TB Coplen, JM Landwehr, AC Riggs, KR Ludwig, BJ Szabo, PT Kolesar, and KM Revesz. **1992**. Continuous 500,000-year climate record from vein calcite in Devils Hole, Nevada, Science 258:255-260.

INCORPORATING LONG-TERM CLIMATE CHANGE IN PERFORMANCE ASSESSMENT FOR THE WASTE ISOLATION PILOT PLANT

P. N. Swift,[1] B. L. Baker,[2] K. Economy,[3] J. W. Garner,[4] J. C. Helton,[5] and D. K. Rudeen[6]

[1]WIPP Performance Assessment Department 6342, Sandia National Laboratories, Albuquerque, New Mexico

[2]Technadyne Engineering Consultants, Inc., Albuquerque, New Mexico

[3]Ecodynamics Research Associates, Albuquerque, New Mexico

[4]Applied Physics, Inc., Albuquerque, New Mexico

[5]Arizona State University, Tempe, Arizona

[6]New Mexico Engineering Research Institute, Albuquerque, New Mexico

Key words: *Climate change, nuclear waste disposal, performance assessment, Waste Isolation Pilot Plant*

ABSTRACT

The United States Department of Energy (DOE) is developing the Waste Isolation Pilot Plant (WIPP) in southeastern New Mexico for the disposal of transuranic wastes generated by defense programs. Applicable regulations (40 CFR 191) require the DOE to evaluate disposal-system performance for 10,000 yr. Climatic changes may affect performance by altering groundwater flow.

Paleoclimatic data from southeastern New Mexico and the surrounding area indicate that the wettest and coolest quaternary climate at the site can be represented by that at the last glacial maximum, when mean annual precipitation was approximately twice that of the present. The hottest and driest climates have been similar to that of the present. The regularity of global glacial cycles during the late Pleistocene confirms that the climate of the last glacial maximum is suitable for use as a cooler and wetter bound for variability during the next 10,000 yr. Climate variability is incorporated into groundwater-flow modeling for WIPP performance assessment by causing hydraulic head in a portion of the model-domain boundary to rise to the ground surface with hypothetical increases in precipitation during the next 10,000 yr. The variability in the modeling of disposal-system performance by allowing head values to vary over this range is insignificant compared to variability resulting from other causes, including incomplete understanding of transport processes. Preliminary performance assessments suggest that climate variability will not affect regulatory compliance.

INTRODUCTION

The Waste Isolation Pilot Plant (WIPP) is planned as a research and development facility to demonstrate the safe disposal of transuranic wastes generated by the U.S. Department of Energy (DOE). It is located in semiarid rangeland in southeastern New Mexico, 42 km east-southeast of the city of Carlsbad (Figure 1), at a surface elevation of 1040 m above mean sea level. Bedded halite of the late Permian Salado formation (approximately 255 million yr old) has been selected as the host stratum because of its extremely low permeability, long-term geologic stability, and creep properties that facilitate sealing. Excavation of the waste-emplacement panels 655 m below the ground surface is partially complete, and additional panels will be excavated in the future as needed.

Figure 1. Location of Waste Isolation Pilot Plant.

No transuranic waste is currently at the WIPP, and before disposing of waste in the facility the DOE must evaluate compliance with applicable long-term regulations of the Environmental Protection Agency (EPA). The regulation of interest here is Subpart B of 40 CFR 191 (Environmental Standards for the Management and Disposal of Spent Nuclear Fuel, High-Level and Transuranic Radioactive Wastes, Final Rule [USEPA, 1985]), which requires evaluation of the consequences of future inadvertent intrusion into the repository. As discussed below, climate change will affect performance only if the repository is breached. Compliance with long-term regulations for which human intrusion does not apply, including those implementing the Resource Conservation and Recovery Act (RCRA), will not be affected.

Sandia National Laboratories is performing iterative preliminary performance assessments (PA) to provide guidance to the WIPP Project while preparing for final evaluations of compliance with applicable long-term regulations. Preliminary PA for the WIPP have been performed in 1990 (Bertram-Howery et al., 1990; Rechard et al., 1990; Helton et al., 1991), 1991 (WIPP PA, 1991a,b,c; Helton et al., 1992), and 1992 (WIPP PA, 1992a,b; Sandia WIPP Project, 1992; WIPP PA, 1993a,b). As stipulated by Congress (Public Law 102-579, 1992), biennial preliminary PA will continue to be performed for the WIPP until the DOE is prepared to submit a final PA to the EPA (currently scheduled for 1998).

The Containment Requirements of 40 CFR 191 set probabilistic limits on the 10,000-yr cumulative releases of radionuclides at the "accessible environment" boundary, which is the ground surface and, for the WIPP, a vertical plane in the subsurface 2.4 km at its closest point from the waste (WIPP PA, 1992a, 1993a). The regulation requires estimating the probability of all releases, and allows larger releases at lower probabilities. "Inadvertent human intrusion" (e.g., by drilling during future exploration for natural resources) must be considered as a possible event.

Preliminary PA to date indicate that, without human intrusion, the repository will comply with 40 CFR 191 B without difficulty (WIPP PA, 1992a, 1993a). Significant quantities of radionuclides from the WIPP will be transported only as solutes in a liquid phase (brine), and all modeling to date indicates that brine will not migrate more than a few tens of meters from the waste in 10,000 yr if the waste-disposal panels are not breached by intrusion (WIPP PA, 1993a,b). However, if human intrusion occurs, radionuclides may reach the accessible environment by two paths. First, some material will be brought to the ground surface immediately during drilling. This release may be important for regulatory compliance (WIPP PA, 1992a) but will not be affected by climate change and is not discussed further here. Second, additional radionuclides may reach the subsurface boundary of the accessible environment, long after the intruding borehole is abandoned, by transport as solutes in brine that migrates up the borehole and laterally away from the repository in an overlying permeable unit (the Culebra Dolomite Member of the late Permian Rustler formation). Long-term changes in climate have the potential to affect disposal-system performance by altering flow and transport in this subsurface pathway.

LONG-TERM CLIMATE VARIABILITY

Present Climate

Mean annual precipitation at the WIPP has been estimated to be between 28 and 34 cm/yr (Hunter, 1985). Freshwater pan evaporation in the region is estimated to be 280 cm/yr (U.S. DOE, 1980). At Carlsbad (100 m lower than the WIPP surface elevation), 53 yr (1931-1983) annual means for precipitation and temperature, are, respectively, 32 cm/yr and 17.1°C (University of New Mexico, 1989). Annual precipitation is dominated by a late-summer monsoon, when solar warming of the continent creates an atmospheric pressure gradient that draws moist air inland from the Gulf of Mexico (Cole, 1975). Winters are generally cool and dry.

Paleoclimates and Climatic Variability

Geologic data from southeastern New Mexico and the surrounding region show repeated alternations of wetter and drier climates throughout the Pleistocene, corresponding to global cycles of glaciation and deglaciation. Data from plant and animal remains and paleo-lake levels permit quantitative climate reconstructions for the region only for the last glacial cycle; they confirm the interpretation that conditions were coolest and wettest during glacial maxima (Swift, 1993). The hottest and driest conditions since the last glaciation have been similar to those of the present. Modeling of global circulation patterns suggests that these changes resulted from the disruption and southward displacement of the winter jet stream by the ice sheet, causing an increase in the frequency and intensity of winter storms throughout the American southwest (COHMAP Members, 1988). Mean annual precipitation 22,000 to 18,000 yr ago, when the last North American ice sheet reached its southern limit (roughly, 1500 km north of the WIPP), was approximately twice that of the present (Figure 2). Mean annual temperatures may have been as much as 5°C colder than at present.

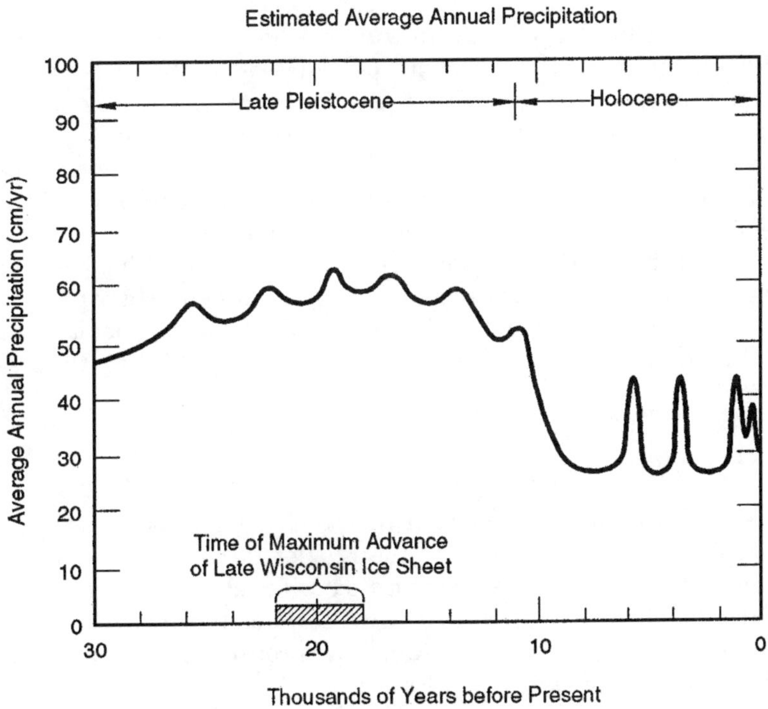

Figure 2. Estimated mean annual precipitation at the Waste Isolation Pilot Plant (WIPP) during the late Pleistocene and Holocene periods (Swift, 1993).

Glacial periodicities have been stable for the last 800,000 yr (Milankovitch, 1941; Hays et al., 1976; Imbrie et al., 1984; Imbrie, 1985). Barring anthropogenic changes in the earth's climate, relatively simple modeling of climatic response to orbital changes in insolation suggests that the next glacial maximum will occur in approximately 60,000 yr (Imbrie and Imbrie, 1980). The extent to which unprecedented anthropogenic climate changes may alter this conclusion is uncertain, but currently available models of climatic response to an enhanced greenhouse effect (e.g., Mitchell, 1989; Houghton et al., 1990) do not predict changes of a larger magnitude than those of the Pleistocene. Furthermore, published models do not suggest significant increases in precipitation in southeastern New Mexico following global warming (Washington and Meehl, 1984; Wilson and Mitchell, 1987; Schlesinger and Mitchell, 1987; Houghton et al., 1990). Even allowing for anthropogenic change, climate variability at the WIPP can be bounded by Pleistocene extremes (Swift, 1993).

Relatively shorter-term climatic fluctuations have occurred throughout the Pleistocene and Holocene, with periodicities on the scale of hundreds to thousands of years (Figure 2). The causes of these nonglacial fluctuations are, in general, unknown, but paleoclimatic data indicate that precipitation may have approached glacial highs at some times during the Holocene (Swift, 1993). Based on the past record, fluctuations of this sort are probable during the next 10,000 yr, and must be included in long-term assessments. The climate-variability model selected for the WIPP PA conceptually incorporates uncertainty in both glacial and nonglacial climatic fluctuations by allowing conditions to reach, at a maximum, glacial extremes three times during the next 10,000 yr.

HYDROLOGIC MODELING

The WIPP PA models groundwater flow and radionuclide transport in the Culebra Dolomite Member of the Rustler formation because it is the most transmissive water-saturated unit above the repository (WIPP PA, 1992b, 1993a). Present groundwater flow is substantially less in other units, and only the Culebra is considered to represent a possible pathway for radionuclide release.

The Culebra is a fractured dolomite, approximately 7 m thick, and is present throughout the region of interest at depths typically 200 m or greater. In most locations, it is bounded above and below by low-permeability mudstones and evaporites (Beauheim and Holt, 1990; Brinster, 1991). WIPP PA modeling treats the Culebra as a perfectly confined aquifer, with flow occurring in only two dimensions in the model domain (Figure 3) (WIPP PA, 1992b, 1993a).

No direct evidence exists for the location of either recharge to or discharge from the Culebra. Potentiometric-surface maps constructed from available well data imply inflow to the model domain from the north and outflow to the south. Mercer (1983) suggested that recharge probably occurs 15 to 30 km northwest of the WIPP, where the Rustler Formation crops out. Lambert (1991) and Lambert and Carter (1987) have speculated on the basis of isotopic evidence that little if any recharge may be occurring now and that present flow reflects long-term draining from recharge during Pleistocene glacial periods. Preliminary modeling indicates that long-term draining is not incompatible with observed hydraulic properties (Davies, 1989; Corbet and Wallace, in press). Three-dimensional regional flow modeling in progress will permit additional testing of this hypothesis and provide an improved model for the spatial and temporal variability in vertical flux (Corbet and Wallace, in press).

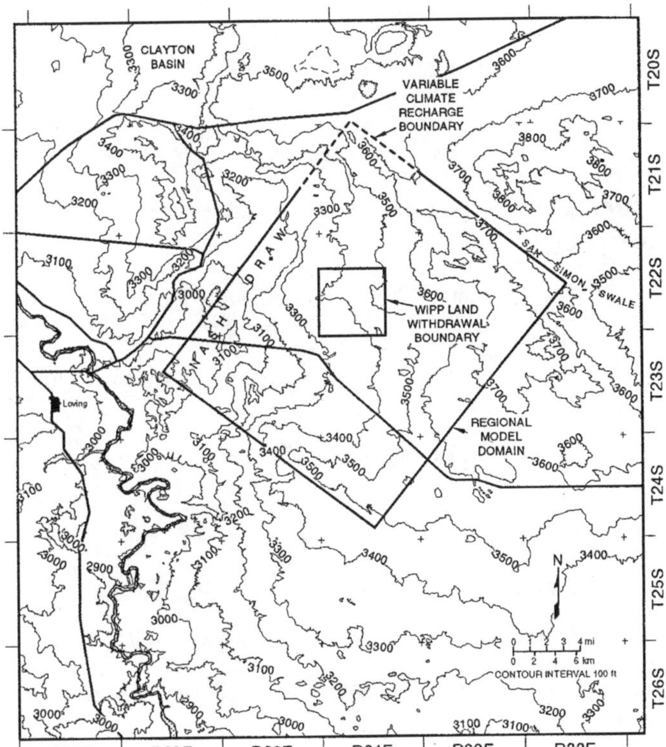

Figure 3. Topographic map of Waste Isolation Pilot Plant (WIPP) area showing the boundaries of the regional groundwater flow model used in 1991 performance assessment (modified from WIPP PA, 1991b).

For the purposes of PA modeling, the location of recharge is unspecified but is assumed to occur north of the model domain. The amount of present recharge is not specified, except through the assumption that present hydraulic head values within the model domain reflect steady-state conditions. Changes in recharge are not modeled explicitly because the assumed recharge area is outside the model domain. Instead, they are approximated by varying head values in a "recharge strip," through which most inflow occurs along the northern edges of the model domain (Figure 3). Heads are not varied along other boundaries, reflecting the belief that all recharge occurs north of the model domain. For the 1991 and 1992 preliminary PA, climatic variability in recharge to the Culebra has been approximated using the following relationship (Swift, 1991; WIPP PA, 1991b,c; Helton et al., 1992):

$$\frac{h_f(t)}{h_p} = \frac{3A+1}{4} - \frac{A-1}{2}\left(\cos\theta\, t + \frac{1}{2}\cos\phi\, t - \sin\frac{1}{2}\phi\, t\right). \qquad (1)$$

This function defines time-dependent head values in the "recharge strip," where

$h_f(t)$ = head (m) in selected boundary cells in the Culebra at time t,

h_p = estimated head (m) in selected boundary cells in the Culebra now,

A = recharge amplitude factor (dimensionless), as described below,

θ = frequency (Hz) for Pleistocene glaciations,

ϕ = frequency (Hz) for second-order climatic fluctuations, and

t = time (sec) after decommissioning of the WIPP.

Figure 4 shows values of the function at 1000-yr time intervals, as implemented in the 1991 WIPP PA.

This function is not used to predict future climates, but rather is designed to provide a simple way to examine the influence of possible climatic changes during the next 10,000 yr. Variable parameters permit examining sensitivity to both the frequency and amplitude of climatic change.

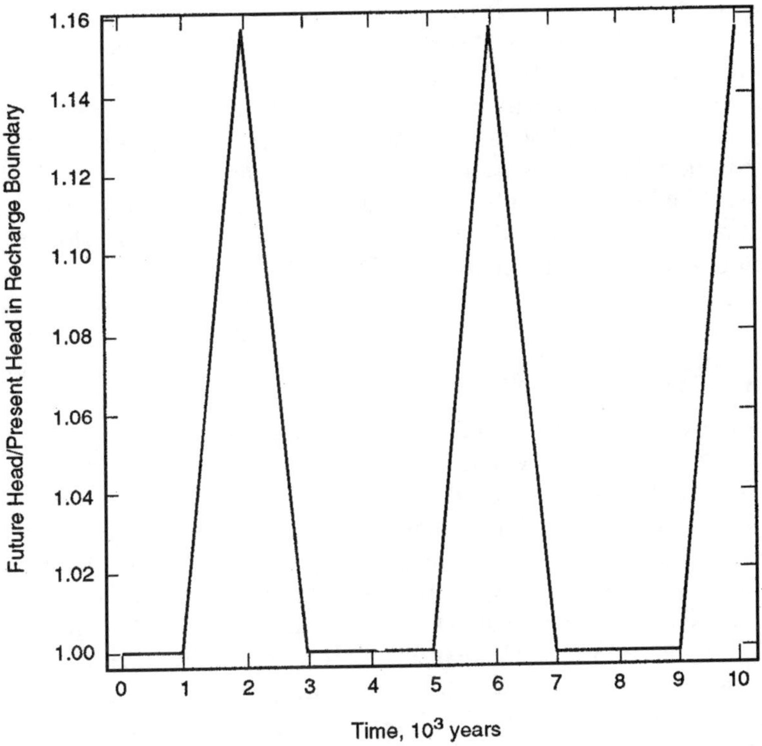

Figure 4. Boundary head function [equation (1)] as implemented in the 1991 WIPP performance assessment (WIPP PA, 1991b). For this figure, A = 1.16, θ = 1.7×10^{-12} Hz, and ϕ = 2×10^{-10} Hz.

Periodicity of the function is controlled by two terms, θ and ϕ, that can be adjusted to approximate the periodicities observed in the paleoclimatic record. In preliminary PA to date, fixed values have been used for these two parameters, yielding a glacial periodicity of 60,000 yr and a second-order periodicity of 3000 yr. If performance is believed to be sensitive to the frequency of climatic change, different values for these parameters can be used in future analyses.

Amplitude of the function is controlled by A, which can be scaled appropriately for the groundwater-flow model parameter to be varied. For the 1991 PA, this parameter was varied from 1 to 1.16. The minimum value, 1, results in no change in boundary head values in the "recharge strip" during the entire 10,000-yr period. The maximum value, 1.16, causes head values to rise from their initial (present) elevation (e.g., 880 m in the northernmost cell) to the elevation of the ground surface (1030 m in the northernmost cell) at the end of the 10,000-yr period (Figure 4). Geological evidence suggests that this increase in head may not be unrealistic: fossil spring deposits at lower elevations in the region indicate discharge from a water table at the ground surface during the late Pleistocene (Bachman, 1981, 1987). Relatively low topographic relief in the region precludes head rising significantly above the ground surface.

DISCUSSION

In keeping with the probabilistic requirements of 40 CFR 191, consequence modeling for the WIPP PA is performed using a Monte Carlo approach that relies on multiple realizations of system performance using deterministic models of physical processes (WIPP PA, 1992a,b). Values for uncertain parameters are selected using a Latin hypercube sampling strategy (McKay et al., 1979) from distributions based on available data, and each realization uses a separate input vector of sampled parameter values. The methodology is well-suited for conducting uncertainty and sensitivity analyses that provide quantitative and qualitative insights about the potential variability in model results caused by uncertainty in specific input data (Helton et al., 1991, 1992; Helton, 1993).

The recharge amplitude factor, A, defined above for equation (1), was one of 45 parameters sampled for use in 60 realizations in the 1991 preliminary PA (Helton et al., 1992) and one of 49 such parameters used in 70 realizations in the 1992 PA (WIPP PA, 1993a). Analyses were performed for scenarios involving a single intruding borehole and two intruding boreholes. In both 1991 and 1992, simulations were repeated using the full suite of realizations for each of several conceptual models for radionuclide transport in the Culebra. The choice of transport model for use in a final PA will be made after additional data are obtained (USDOE, 1993). Cases considered to date include transport in a single-porosity, fracture-only medium and transport in a dual-porosity medium which allowed diffusion into the pore volume of the dolomite matrix. Cases were considered both with and without chemical retardation of radionuclides by sorption. Computational modeling, including discussion of the computer codes used and the results of these analyses are described in detail elsewhere (WIPP PA, 1991b; Helton et al., 1992; WIPP PA, 1992b, 1993a).

Variability in the recharge amplitude factor contributed significantly to variability in total releases only for the conceptual model that included dual-porosity transport without chemical retardation (Helton et al., 1992). Even in this case, regression analysis shows variation in boundary head values to have been a minor contributor to overall variability in model outcomes, ranking below parameters used to describe radionuclide solubility in the source term, permeability of the borehole pathway from the repository to the Culebra, and fracture spacing in the Culebra.

Two analyses were conducted as part of the 1991 PA that specifically examined the importance of boundary head variations by using fixed minimum (1.00) and maximum (1.16) values for the recharge amplitude factor (Helton et al., 1992). The full suite of 60 realizations were repeated for each analysis with sampled values used for all other parameters, resulting in two sets of outcomes which were in all ways comparable except for the value used for A. Results are shown in Figure 5 for both single-porosity and dual-porosity transport with chemical retardation. For all but a few of the single-porosity realizations, using the maximum recharge factor has essentially no effect on total releases. This lack of sensitivity apparently occurs because single-porosity transport is rapid enough that most of the long-lived radionuclides (e.g., uranium) that enter the Culebra reach the accessible environment within 10,000 yr regardless of the head gradient. For the relatively slower dual-porosity transport, the maximum recharge factor increased releases for essentially all realizations in which subsurface releases occurred.

CONCLUSIONS

For preliminary comparison with requirements of 40 CFR 191, performance estimates are displayed as complementary cumulative distribution functions (CCDF) that indicate the probability of exceeding various levels of cumulative radionuclide releases to the accessible environment. Because the modeling system and data base are incomplete, no CCDF

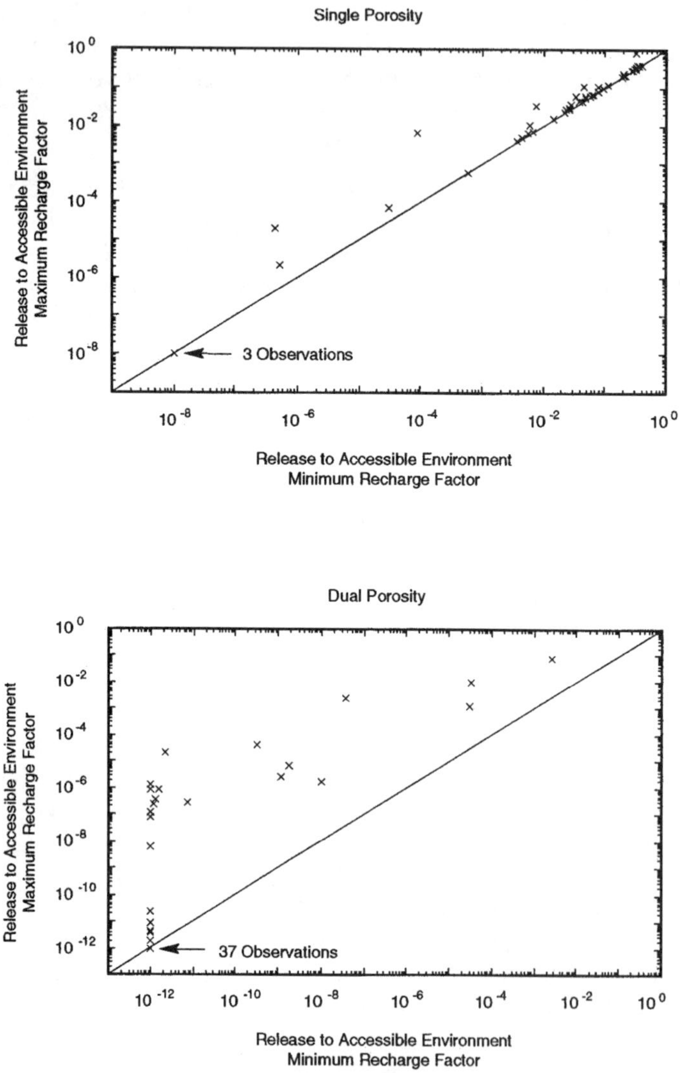

Figure 5. Scatterplots showing 10,000-yr cumulative radionuclide releases to subsurface boundary of accessible environment 60 for realizations using minimum and maximum values for the recharge amplitude factor (Helton et al., 1992). Releases are normalized to total inventory, as specified by 40 CFR 191. Releases are shown for single (top) and dual (bottom) porosity conceptual models for radionuclide transport, including chemical retardation. Both plots show releases from scenarios involving two intrusions into same panel 1000 yr after decommissioning [See WIPP PA, 1991b for discussion of scenario definitions.] Normalized releases below 10^{-8} (above) and 10^{-12} (below) are plotted at those values.

presented to date in WIPP PA are suitable for compliance evaluations. Regulatory limits are commonly displayed, however, on preliminary CCDF to provide guidance to the Project and to assist in identifying those areas in which uncertainty has the potential to affect compliance.

Figure 6 is a composite display of mean CCDF that shows the relative importance of the recharge factor in the 1991 PA (Helton et al., 1992). Estimated performance is shown for three alternative cases: transport in a single-porosity medium with chemical retardation and sampled values for the recharge factor, transport in a dual-porosity medium with chemical retardation and the maximum recharge factor used in all realizations, and transport in a dual-porosity medium with chemical retardation and the minimum (i.e., present) recharge factor used in all realizations. Except for those parameters used to describe the alternative transport cases, all parameter values were the same for each case. Results do not include releases at the ground surface during drilling, and therefore are an incomplete measure of overall performance. Limits specified by 40 CFR 191 B are given only for reference.

Single-porosity transport results in the largest releases and the CCDF closest to the EPA limits. Separate curves are not shown for the single-porosity extreme climate cases because, as shown in Figure 5, changing the recharge factor has essentially no effect on the largest releases that determine the location of the mean CCDF. Varying boundary head values within this range does not affect regulatory releases if transport occurs in a single-porosity medium.

The two mean CCDF shown for dual-porosity transport are significantly further from the EPA limits than the CCDF for the single-porosity case. Allowing boundary head values to rise in response to climatic change does result in an increase in estimated releases, but not sufficiently to affect compliance.

Although all results shown here are conditional on the assumptions used in the analyses and may change as flow and transport models are improved, the conclusion that climate change is unlikely to affect compliance appears robust. The uncertainty remaining about the correct conceptual models for both climatically varying recharge and radionuclide transport in the Culebra is substantial. However, as long as releases calculated with extreme head elevations and the least favorable transport model considered remain below the EPA limits, climate change alone will not lead to regulatory violations. Defensibility of this conclusion will depend, in part, on improved understanding of regional flow.

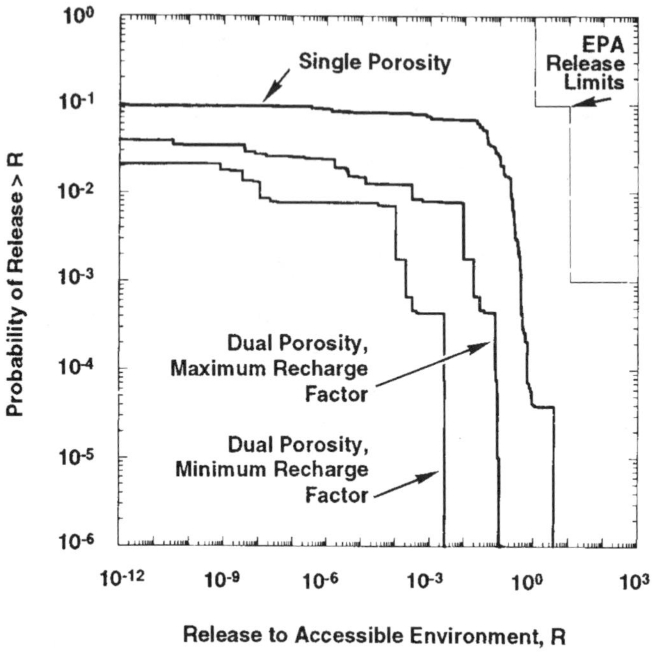

Figure 6. Mean complementary cumulative distribution functions (CCDF), showing estimated performance of the Waste Isolation Pilot Plant (WIPP) for subsurface releases only, following intrusions at 1000 yr after decommissioning (Helton et al., 1992). Releases normalized to total inventory, as specified by 40 CFR 191. Curves shown for three cases: single-porosity transport with sampled values for recharge factor, dual-porosity transport with maximum value for recharge factor, and dual-porosity transport with minimum value for recharge factor. All cases include chemical retardation. EPA limits are shown for reference only. Results here are not suitable for direct comparison to regulatory limits because they are preliminary (i.e., based on an incomplete modeling system and data base) and because they do not include releases at the ground surface during drilling.

ACKNOWLEDGMENTS

This work was performed at Sandia National Laboratories, Albuquerque, New Mexico, and was supported by the United States Department of Energy under Contract DE-AC04-76DP00789. The authors are members of the WIPP Performance Assessment Department (6342) at Sandia National Laboratories, and gratefully acknowledge contributions of approximately 40 colleagues in the department who have also worked on WIPP performance assessment under the direction of DR

Anderson. TF Corbet and CL Axness provided critical reviews of the manuscript. Illustrations were prepared by D Marchand of Tech Reps, Inc.

REFERENCES

Bachman, GO. **1981**. *Geology of Nash Draw, Eddy County, New Mexico*, Open-File Report 81-31. U.S. Geological Survey, Denver, CO.

Bachman, GO. **1987**. *Karst in Evaporites in Southeastern New Mexico*, SAND86-7078. Sandia National Laboratories, Albuquerque, NM.

Beauheim, RL and RM Holt. **1990**. Hydrogeology of the WIPP Site, pp. 131-179. In: *Geological and Hydrological Studies of Evaporites in the Northern Delaware Basin for the Waste Isolation Pilot Plant (WIPP), New Mexico*, Geological Society of America 1990 Annual Meeting Field Trip #14 Guidebook. Dallas Geological Society, Dallas, TX.

Bertram-Howery, SG, MG Marietta, RP Rechard, PN Swift, DR Anderson, BL Baker, JE Bean Jr, W Beyeler, KF Brinster, RV Guzowski, JC Helton, RD McCurley, DK Rudeen, JD Schreiber, and P Vaughn. **1990**. *Preliminary Comparison with 40 CFR Part 191, Subpart B for the Waste Isolation Pilot Plant, December 1990*, SAND90-2347. Sandia National Laboratories, Albuquerque, NM.

Brinster, KF. **1991**. *Preliminary Geohydrologic Conceptual Model of the Los Medaños Region Near the Waste Isolation Pilot Plant for the Purpose of Performance Assessment*, SAND89-7147. Sandia National Laboratories, Albuquerque, NM.

Cole, FW. **1975**. *Introduction to Meteorology*, 2nd edition. John Wiley and Sons, New York.

COHMAP (Cooperative Holocene Mapping Project) Members. **1988**. Climatic changes of the last 18,000 years: observations and model simulations. Science 241:1043-1052.

Corbet, TF and MG Wallace. Post Pleistocene patterns of shallow groundwater flow in the Delaware Basin, southeastern New Mexico and west Texas. In: *New Mexico Annual Field Conference Guidebook, Carlsbad Region, Permian Basin, New Mexico and West Texas*. JW Hawley and DW Love (eds.). (in press).

Davies, PB. 1989. *Variable-Density Ground-Water Flow and Paleohydrology in the Waste Isolation Pilot Plant (WIPP) Region. Southeastern*

New Mexico, Open-File report 88-490. U.S. Geological Survey, Albuquerque, NM.

Hays, JD, J Imbrie, and NJ Shackleton. **1976**. Variations in the earth's orbit: pacemaker of the Ice Ages. Science 194:1121-1132.

Helton, JC. **1993**. Risk, uncertainty in risk, and the EPA release limits for radioactive waste disposal, SAND91-1255J, Nucl. Technol. 101:18-39.

Helton, JC, JW Garner, RD McCurley, and DK Rudeen. **1991**. *Sensitivity Analysis Techniques and Results for Performance Assessment at the Waste Isolation Pilot Plant*, SAND90-7103. Sandia National Laboratories, Albuquerque, NM.

Helton, JC, JW Garner, RP Rechard, DK Rudeen, and PN Swift. **1992**. *Preliminary Comparison with 40 CFR Part 191, Subpart B for the Waste Isolation Pilot Plant, December 1991 — Volume 4: Uncertainty and Sensitivity Analysis Results*, SAND91/0893/4. Sandia National Laboratories, Albuquerque, NM.

Houghton, JT, GJ Jenkins, and JJ Ephraums. **1990**. *Climate Change: The IPCC Scientific Assessment*. Cambridge University Press, New York.

Hunter, RL. **1985**. *A Regional Water Balance for the Waste Isolation Pilot Plant (WIPP) Site and Surrounding Area*, SAND84-2233. Sandia National Laboratories, Albuquerque, NM.

Imbrie, J. **1985**. A theoretical framework for the Pleistocene Ice Ages. J. Geol. Soc. 142:417-432.

Imbrie, J and JZ Imbrie. **1980**. Modeling the climatic response to orbital variations. Science 207:943-953.

Imbrie, J, JD Hays, DG Martinson, A McIntyre, AC Mix, JJ Morley, NG Pisias, WL Prell, and JJ Shackleton. **1984**. The orbital theory of Pleistocene climate: support from a revised chronology of the marine $\delta^{18}O$ record, pp. 269-305, Part 1. In: *Milankovitch and Climate, Proceedings of the NATO Advanced Research Workshop on Milankovitch, Palisades, NY, November 30-December 4, 1982*. AL Berber, J Imbrie, J Hays, G Kukla, and B Saltzman (eds.). D. Reidel Publishing Co., Boston, MA.

Lambert, SJ. **1991**. Isotopic constraints on the Rustler and Dewey Lake groundwater systems, pp. 5-1 through 5-79. In: *Hydrogeochemi-*

cal Studies of the Rustler Formation and Related Rocks in the Waste Isolation Pilot Plant Area, Southeastern New Mexico, SAND88-0196. MD Siegel, SJ Lambert, and KL Robinson (eds.). Sandia National Laboratories, Albuquerque, NM.

Lambert, SJ and JA Carter. **1987**. *Uranium-Isotope Systematics in Groundwaters of the Rustler Formation, Northern Delaware Basin, Southeastern New Mexico, I: Principles and Preliminary Results*, SAND87-0388. Sandia National Laboratories, Albuquerque, NM.

McKay, MD, RJ Beckman, and WJ Conover. **1979**. A comparison of three methods for selecting values of input variables in the analysis of output from a computer code. Technometrics 21:239-245.

Mercer, JW. **1983**. *Geohydrology of the Proposed Waste Isolation Pilot Plant Site, Los Medaños Area, Southeastern New Mexico*, Water-Resources Investigations Report 83-4016. U.S. Geological Survey, Albuquerque, NM.

Milankovitch, M. **1941**. *Canon of Insolation and the Ice-Age Problem*. Königlich Serbische Akademie, Beograd. (English translation by the Israel Program for Scientific Translations; published the U.S. Department of Commerce and the National Science Foundation, Washington, DC).

Mitchell, JFB. **1989**. The "Greenhouse Effect" and climate change. Rev. Geophys. 27:115-139.

Public Law 102-579. **1992**. *Waste Isolation Pilot Plant Land Withdrawal Act*. U.S. Government Printing Office, Washington, DC.

Rechard, RP, H Iuzzolino, and JS Sandha. **1990**. *Data Used in Preliminary Performance Assessment of the Waste Isolation Pilot Plant (1990)*, SAND89-2408. Sandia National Laboratories, Albuquerque, NM.

Sandia WIPP Project. **1992**. *Preliminary Performance Assessment for the Waste Isolation Pilot Plant, December 1992 — Volume 3: Model Parameters*, SAND92-0700/3. Sandia National Laboratories, Albuquerque, NM.

Schlesinger, ME, and JFB Mitchell. **1987**. Climate model simulations of the equilibrium climatic response to increased carbon dioxide. Rev. Geophys. 25:760-798.

Swift, PN. **1991**. Climate and recharge variability parameters for the 1991 WIPP PA calculations, pp. A-107 through A-122. In: *Preliminary Comparison with 40 CFR Part 191, Subpart B for the Waste Isolation Pilot Plant, December 1991—Volume 3: Reference Data*, SAND91-0893/3. WIPP PA Division, Sandia National Laboratories, Albuquerque, NM.

Swift, PN. **1993**. Long-term climate variability at the Waste Isolation Pilot Plant, Southeastern New Mexico, USA, SAND91-7055J. Environ. Management 17:83-97.

University of New Mexico. **1989**. *New Mexico Statistical Abstract 1989*. Bureau of Business and Economic Research, University of New Mexico, Albuquerque, NM.

USDOE. **1980**. *Final Environmental Impact Statement: Waste Isolation Pilot Plant*, DOE/EIS-0026. U.S. Department of Energy, Washington, DC.

USDOE. **1993**. *Test Phase Plan for the Waste Isolation Pilot Plant*, DOE/WIPP 89-011, Rev. 1. U.S. Department of Energy WIPP Project Integration Office, Albuquerque, NM.

USEPA. **1985**. 40 CFR Part 191: environmental radiation protection standards for the management and disposal of spent nuclear fuel, high-level and transuranic radioactive wastes, final rule. Fed. Reg. 50:38066-38089.

Washington, WM and GA Meehl. **1984**. Seasonal cycle experiment on the climate sensitivity due to doubling of CO_2 with an atmospheric general circulation model coupled to a simple mixed-layer ocean model. J. Geophys. Res. 89:9475-9503.

Wilson, CA and JFB Mitchell. **1987**. A doubled CO_2 climate sensitivity experiment with a global climate model including a simple ocean. J. Geophys. Res. 92:13,315-343.

WIPP PA. **1991a**. *Preliminary Comparison with 40 CFR Part 191, Subpart B for the Waste Isolation Pilot Plan—Volume 1: Methodology and Results*, SAND91-0893/1. Performance Assessment Division, Sandia National Laboratories, Albuquerque, NM.

WIPP PA. **1991b**. *Preliminary Comparison with 40 CFR Part 191, Subpart B for the Waste Isolation Pilot Plant—Volume 2: Probability* *and Consequence Modeling*, SAND91-0893/2. Performance Assessment Division, Sandia National Laboratories, Albuquerque, NM.

WIPP PA. 1991c. *Preliminary Comparison with 40 CFR Part 191, Subpart B for the Waste Isolation Pilot Plant—Volume 3: Reference Data*, SAND91-0893/3. Performance Assessment Division, Sandia National Laboratories, Albuquerque, NM.

WIPP PA. 1992a. *Preliminary Performance Assessment for the Waste Isolation Pilot Plant, December 1992—Volume 1: Third Comparison with 40 CFR 191, Subpart B*, SAND92-0700/1. Performance Assessment Department, Sandia National Laboratories, Albuquerque, NM.

WIPP PA. 1992b. *Preliminary Performance Assessment for the Waste Isolation Pilot Plant, December 1992—Volume 2: Technical Basis*, SAND92-0700/2. Performance Assessment Department, Sandia National Laboratories, Albuquerque, NM.

WIPP PA. 1993a. *Preliminary Performance Assessment for the Waste Isolation Pilot Plant, December 1992—Volume 4: Uncertainty and Sensitivity Analyses for 40 CFR 191, Subpart B*, SAND92-0700/4. Performance Assessment Department, Sandia National Laboratories, Albuquerque, NM.

WIPP PA. 1993b. *Preliminary Performance Assessment for the Waste Isolation Pilot Plant, December 1992—Volume 5: Uncertainty and Sensitivity Analyses of Gas and Brine Migration for Undisturbed Performance*, SAND92-0700/5. Performance Assessment Department, Sandia National Laboratories, Albuquerque, NM.

PROJECTING THE RANGE OF POTENTIAL FUTURE CLIMATE CHANGE AS AN AID IN ASSESSING THE EFFECTIVENESS OF THE HANFORD SITE PERMANENT ISOLATION BARRIER

K. L. Petersen

Westinghouse Hanford Company, Richland, Washington

Key words: *Performance assessment, waste disposal, paleoclimate, future climate*

ABSTRACT

The Hanford Site Permanent Isolation Barrier Development Program was organized to develop an in-place disposal capability for low-level nuclear waste for the U.S. Department of Energy at the Hanford Site in southeastern Washington. Layered earthen and engineered barriers are being developed by Westinghouse Hanford Company and the Pacific Northwest Laboratory that will function, in what is presently a semiarid environment (annual precipitation, 150 mm), for at least 1,000 yr by limiting the infiltration of water through the waste. The Long-Term Climate Change Task is one of several key barrier tasks. Based on the recommendation of a panel of internationally recognized climate and modeling experts, climatic data for this task are being acquired in a stepwise and multidisciplinary manner. The specific research strategy includes literature review and specialized studies to obtain pollen-derived climatic reconstruction, documented historic weather patterns, and Global Circulation Model output of potential future climate changes related to both the greenhouse effect and the cycling into the next ice age. The specific goals of the task are to: (1) obtain defensible probabilistic projections of the long-term climate variability in the Hanford Site region at many different time scales; (2) develop several test-case climate scenarios that bracket the range of potential future climates; and (3) use the climate scenarios both to test and to model protective barrier performance.

INTRODUCTION

The Hanford Site Permanent Isolation Barrier Development Program was organized to develop the technology needed to provide an environmentally secure, in-place disposal capability for the Hanford Site (Adams and Wing, 1986; Cadwell, 1991; Wing and Gee, 1990). The goal of the Barrier Development Program is to provide defensible evidence that final barrier design(s) will adequately control water infiltration, plant and animal intrusion, and wind and water erosion for a minimum of 1000 yr; it will also isolate wastes from the accessible environment and warn inadvertent human intruders by using markers. Evidence for barrier performance will be obtained by conducting laboratory experiments, field tests, computer modeling, and other studies that establish confidence in the ability of the barrier(s) to meet a 1000-yr design life.

The Barrier Development Program system uses engineered layers of natural material to create an integrated structure with redundant protective features. The natural construction materials (e.g., fine soil, sand, gravel, riprap, clay, asphalt) have been selected to optimize barrier performance and longevity. Figure 1 shows a conceptual drawing of a typical barrier and warning marker system. Once developed, it is anticipated that the permanent isolation barrier design could be used at other arid to semiarid sites. Because of the time desired for waste isolation, long-term changes in the local climate could have a significant impact. Much evidence (Crowley and North, 1991) suggests that long-term climate change is cyclical in nature. Therefore, reconstructing climate changes over the past 10,000 to 125,000 yr should provide insight into the range of bounding conditions of potential future changes and increase confidence in the barrier design.

In 1989, a panel of technical experts in paleoclimatic reconstruction and future climate projection reviewed the Barrier Development Program's climate study program and recommended that emphasis should

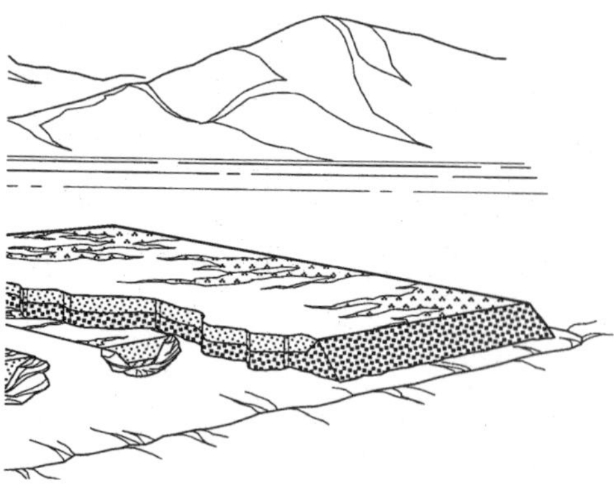

Figure 1. Layered earthen and engineered barrier being developed by Westinghouse Hanford Company and Pacific Northwest Laboratory to help isolate waste at Hanford Site.

be given to detecting climatic extremes that could make a difference in the barrier's performance for 1000 yr and longer. These climatic extremes include changes in temperature, precipitation, and vegetation related to "greenhouse warming" and the cycling of the climate into the next ice age.

The following were selected as examples of potential climatic extremes that could impact barrier performance if such extremes were to occur in the future:

- the last interglacial/glacial transition (beginning Wisconsin, nominally 115,000 yr ago);
- the last glacial maximum (nominally 18,000 yr ago);
- the early Holocene (9000 yr ago);
- the mid- to late Holocene (6000 to 3500 yr ago).

Understanding the underlying causes of the Pasco Basin's modern climatic pattern provides a basis for evaluating past and especially potential changes that might be important in any long-term performance assessment.

The current approach for obtaining needed climatic information for the Barrier Development Program is presented in a study plan (Petersen et al., 1993). The work described in this plan is consistent with the Protective Barrier and Warning Marker System Development Plan (Adams and Wing, 1986; Wing and Gee, 1990). Figure 2 shows how climatic data are being acquired in a stepwise and multidisciplinary manner.

MODERN CLIMATIC PATTERNS

The Hanford Site is located in the Pasco Basin in south-central Washington (Figure 3). The climate of this area can be classified as midlatitude semiarid or midlatitude desert, depending on the climatological classification scheme being used. Summers are warm and dry, with abundant sunshine. During this season, large diurnal temperature variations are common, resulting from intense solar heating and nighttime cooling. Daytime high temperatures in June, July, and August can exceed 40°C. Winters are cool, with occasional precipitation. Outbreaks of cold air associated with modified arctic air masses infrequently reach the area and seldom cause temperatures to drop below –18°C. Overcast skies and fog occur periodically in this season.

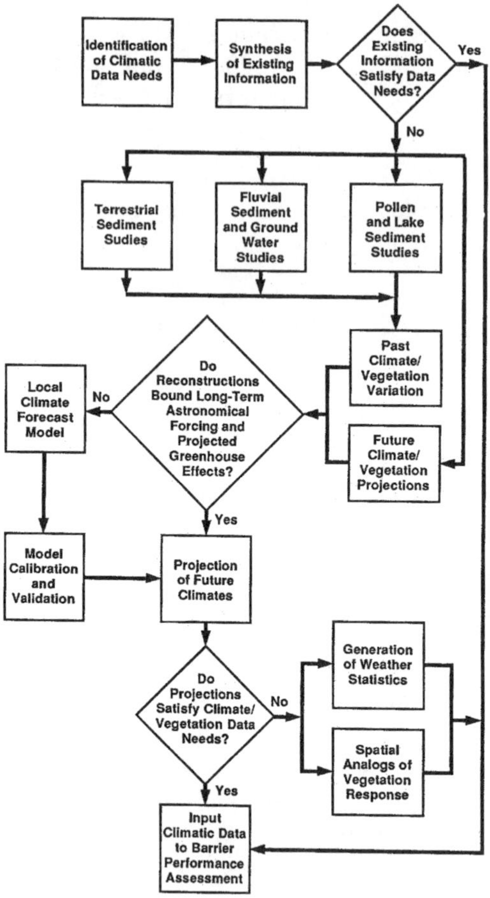

Figure 2. Flow chart depicting strategy for obtaining long-term climatic data to aid in future climatic prediction at Hanford Site.

The Pasco Basin is in the rain shadow of the Cascade Range. These mountains limit much of the maritime influence of the Pacific Ocean, producing a climate that is more continental in nature than if the mountains were not there. The local topographic features have significant effect on temperature, wind, and precipitation because all air masses that reach the Pasco Basin undergo some modification as a result of passing over the relatively complex topography of the area. The Cascade Range also serves as a source of cold air drainage, which contributes to the northwest component of the wind regime on the Hanford Site.

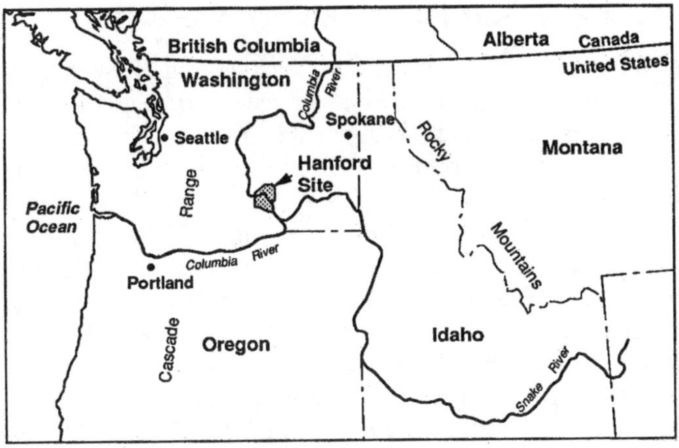

Figure 3. Location of Hanford Site, east of Cascade Range in south-central Washington.

Climatological data are available for the Hanford Meteorological Station (HMS), which is located between the 200-East and 200-West Areas situated on a plateau at about 215 m elevation (Stone et al., 1983). Average monthly temperature at the HMS is 12°C. The unusual cool nights that occur are a result of cool gravity winds originating from the Cascade Mountains. Temperatures at the Hanford Site are colder in the winter and warmer in the summer than would be the case without the Cascades blocking the more equitable maritime climate to the west. The Pasco Basin is warmer during the winter than a site at an equivalent latitude on the Great Plains in the midcontinent because mountain ranges to the north and east shield the area from many of the arctic surges that descend from Canada. Half of all winters are free of temperatures at or below –18°C.

Precipitation measurements have been recorded at the HMS since 1945; they were recorded at the nearby Hanford Town Site between 1912 and 1945. The long-term average annual precipitation at the HMS and the Hanford Town Site is 16 cm. The three months of November through January contribute 44% of this total, while the three months of July through September contribute only 13%. About 38% of all precipitation during the months of December through February is in the form of snow. Winter monthly average snowfall ranges from 0.8 cm in March to 13.5 cm in January. An accumulation of as much as 15.24 cm of snow is expected in only one winter in four. As a seasonal average, only four days accumulate 15.24 cm or more of snow on the ground.

HOLOCENE AND WISCONSIN PALEOCLIMATE PATTERNS

Chatters (1991; Chatters et. al., 1991; Chatters and Hoover, 1992) summarizes existing proxy evidence for climatic and ecological change in the Columbia Basin of northwestern America for the last 10,000 to 13,000 yr. Inferences based on glacial and fluvial geomorphology, cave and lacustrine sedimentation patterns, palynology, fossil timberlines, subfossil paleontology, and stable isotope study of inland lakes are integrated into a detailed record of environmental change. Findings identify a terminal late Wisconsin environment kept cool and dry by masses of ice and glacial meltwater, supporting a mosaic of now-allopatric plant and animal communities. This was followed in the early Holocene (between 10,000 and 8500 yr ago) by a period of warmer-than-modern summers, colder-than-modern winters, and low but spring-dominant precipitation that supported extensive grasslands and the faunas associated with them. By 8000 yr ago, summers and winters were both relatively warm, and precipitation was at least 33% below modern levels. This climate pattern resulted in reduced stream flows with late spring flow maxima and extensive development of shrub-steppe vegetation throughout most of the region.

Following are the effects the climate of 6000 and 8000 yr ago had on streams and rivers of the region, according to Chatters et al. (1991):

- Stream flows were more than 30% below modern amounts.
- Many small, low-elevation, perennial streams became intermittent (dry during part of the year).
- Streams had finer bed loads (greater sedimentation).
- Water temperatures were higher.
- The spring peak flow (freshet) ended 3 to 4 weeks earlier than it does today.

Chatters (1991) indicates that the climate cooled but remained dry between 5400 and 4500 yr ago; it then underwent a period of transition to wetter, cooler conditions between 4500 and 3900 yr ago. Rivers flooded frequently, and forests expanded into steppe zones. From 3900 until 2400 yr ago, the climate was cool in summer and cold in winter, with winter-dominant precipitation at least 30% above modern amounts. Alpine glaciers underwent the first synchronous advance of the Holocene. Rivers stabilized, and their peak flows were delayed into midsummer. Steppe vegetation increased in density as subalpine and probably montane forests expanded beyond their modern margins.

Warmer, drier conditions returned between 2400 and 2000 yr ago, reducing vegetation density and bringing a renewal of severe flooding. Climate then improved somewhat, and modern climates were established. The Little Ice Age (dating between about A.D. 1250 and about 1850), although evident in alpine cirques, had little apparent effect on regional environments except for a slight drying around 550 years ago. The Carp Lake Pollen Coring Project (contracted to Golder Associates, Inc., Redmond, Washington, with a subcontract to Dr. Cathy Whitlock of the University of Oregon) is an ongoing task directed at coring and analyzing the pollen from Carp Lake, near Goldendale, Washington, southeast of the Hanford Site. Fossil pollen that has been accumulating over the years in the lake-bottom sediments are being analyzed to determine the type of vegetation that once grew around the lake and the climate that produced it. The Carp Lake site completes a northeast to southwest transect of pollen sites that cuts through the Pasco Basin (Figure 4). The lake coring operation recovered nearly 20.19 m of lake sediment—providing a record of past climatic changes dating back 75,000 yr or more.

The age assignment of the core is base on 12 radiocarbon ages of bulk sediment and tephrochronology. An age-depth model was selected that best fit a curve to a plot of age-against-depth to assign ages to samples between dates. However, two other age-depth models are possible. Depending on which age-depth model is applied, the core could range

Figure 4. Carp Lake, located at extreme southeast end of northwest-to-southeast transect of pollen sites that bracket Hanford Site, is being studied by researchers from the University of Oregon.

in age from as little as 66,000 to as much as 104,450 yr. Further work is expected to increase confidence in the model selected. Today, Carp Lake is located in a very climatically sensitive area near the lower limit of the ponderosa pine forest, which is not too far from the shrub-steppe boundary. Its position on the flanks of the Cascade Range means that the wet montane conifer forest is not too far above it. Any changes in either the upper or lower boundary of the ponderosa pine forest should readily be distinguished at Carp Lake by the resultant distinctive pollen of each of the vegetation types. Past movements of these two vegetation boundaries are clearly evident in the pollen record, which is divided into nine assemblage zones that describe alterations between periods of montane conifer forest, pine forest and shrub-steppe.

The climate record at Carp Lake indicates that the Columbia Basin was under the influence of the Ice Age climate 75,000 yr ago. While continental glaciers expanded out of the area that is now Canada toward the present-day state of Washington, the Columbia Basin was much colder than it is at present. These cold conditions lasted until about 10,000 yr ago with little interruption. Over the last 10,000 yr, the climate in the Columbia Basin warmed significantly as the earth cycled out of the Ice Age and into climates more like the present. At Carp Lake, modern vegetation was established only in the last 3900 yr; the middle Holocene was characterized by a more open ponderosa pine forest than today; and the early Holocene was characterized by a moderate expansion of shrub-steppe, whereas most of the Wisconsin contained in the record was characterized by a vast expansion of shrub-steppe. Throughout the record, mean annual precipitation ranged from 25-50% below modern levels (in the early Holocene and early Wisconsin) to 28% above (in the mid-Wisconsin interval, when the site was characterized by montane forest for a short time). At no time did precipitation levels reach three times that of present day. (Three times modern precipitation has been taken as an upward bounding condition of precipitation to be used in barrier performance assessment, to be discussed in the next section.)

PERFORMANCE ASSESSMENT AND LONG-TERM FUTURE CLIMATE CHANGE

Regional studies are still developing about the potential impact of the so called "greenhouse effect," but good preliminary work is provided by Leovy and Sarachik (1991) and Franklin et al. (1991). A handful of very complex computer models (called Global Circulation Models, or GCM) have been developed that can model the earth's climatic system and test ideas about future climatic change. Most of the GCM used to assess the impact of increasing amounts of atmospheric carbon dioxide and other important gases on global climate predict increased temperature (Houghton et al., 1990; Houghton et al., 1992). The paleoclimatological studies for eastern Washington, discussed above, provide a valuable analog for barrier studies. This is because, 6000 to 8000 yr ago, temperatures were judged to be approximately 1°C to 2°C warmer and slightly drier than current conditions. This provides a change similar to that predicted due to the increases in atmospheric concentrations of greenhouse gases and is expected to result in a global temperature increase of 1.5 to 4.5°C over the next 100 yr (Houghton et al., 1992).

As for changes in precipitation under greenhouse warming, the GCM are not as consistent in their predictions for the next 100 years (Houghton et al., 1990; Crowley and North, 1991). The warmer atmosphere will have a greater capacity to hold more water. Globally, this would mean that the climate would be wetter, but some regions will likely be effectively drier because of the greater evaporation potential.

In spite of the potential future warming due to the greenhouse effect, ultimately this will be overwhelmed as we plunge into the next Ice Age. It has been over 10,000 yr since earth was locked in the grips of an Ice Age, and in view of the patterns that have been in place for at least the last 800,000 yr (Berger et al., 1984; Hays et al., 1976), the earth is due to move back into a glacial climate. How rapidly that will occur is a matter of debate, and the impact of the anthropogenic trace gases that increase the greenhouse effect is just now beginning to be evaluated. A number of recent studies suggest that the movement into and out of a glacial climate may be quite abrupt (Allen and Anderson, 1993; Berger and Labeyrie, 1987; Dansgaard et al., 1993; GRIP members, 1993; Phillips et al., 1990a,b). Such studies suggest that although global climate change may be driven by slow changes in orbital parameters, the response may be nonlinear in nature, with feedback mechanisms that push the system toward one state or the other (either glacial or nonglacial) and that transition from one state to the other may be quite rapid—possibly as short as a few years or decades or centuries. All these observations have important implications for bounding potential future climate change and for long-term performance assessment.

In an important paper, Berger et al. (1991) review seven models of different complexity developed to predict the global climate for then next 10,000 to 100,000 yr (i.e., Calder, 1974; Petersen and Larsen, 1978; Melice and Berger, 1989; Imbrie and Imbrie, 1980; Kukla, 1981; Berger, 1981; Berger et al., 1990). All models are in relatively good agreement with the computation of astronomical forcing of climate change. In the absence of anthropogenic disturbances, the long-term cooling trend that began some 6000 yr ago will continue for the next 5000 yr; this will be followed by a stabilization at about 15,000 yr, by a cold interval centered at approximately 25,000 yr, and, finally, by a major glaciation at about 55,000 yr into the future. If, however, the Greenland ice sheet were to disappear in the next few centuries (due to greenhouse warming, for example), the northern hemisphere ice sheets would not reappear before 15,000 yr from now; the climate would then be different from its natural evolution up to 65,000 yr into the future, with the maximum ice volume amounting to only two-thirds of its naturally expected size at that time.

Winograd et al. (1992) has questioned the regular cyclic timing, used in part to date the pace of the ice ages (e.g., Hays et al., 1976), by using a well-dated and uninterrupted 500,000 yr paleotemperature record contained in ^{18}O concentrations in calcite deposits below the water table in a Nevada crevasse called Devils Hole. Rather than a steady beat of interglacial-and-ice-age lasting a combined 110,000 yr as predicted by celestial mechanics, Winograd et al. (1992) found an aperiodic swing to the climate cycle. The implication Winograd et al. (1992) sees for waste isolation is that the present interglacial may last another 10,000 yr, and continental glacial ice reaching its maximum extent may occur 100,000 yr or more into the future. This is important for this report because of its potential for contributing to catastrophic flooding.

On the other hand, isotope and chemical analyses of the Greenland Ice-core Project (GRIP, 1993) ice core from Summit, central Greenland, reveal that climatic changes in Greenland that preceded the last interglacial period came as a series of severe cold periods (changes up to 10°C) that began extremely rapidly (possibly in as little as a couple of decades) and lasted from decades to centuries. The GRIP authors note that because the last interglacial seems sightly warmer than the present one, its unstable climates raise questions about the effect of future global warming.

SUMMARY

The object of the Hanford climate studies is to characterize potential extremes to help bound the range. The Carp Lake pollen record suggests that if glacial climates were to return to the Pasco Basin in the next few thousand years, then a colder and drier climate would again return to the Columbia Basin. Because the sediment record at Carp Lake stopped short of the last interglacial/glacial transition, as recommended by the technical review panel, Golder Associates Inc. is being directed by Westinghouse Hanford Company to go back to the lake with a commercial drilling rig mounted on a barge to take samples for a longer record. This should provide specific information on the local climate manifestations for the Hanford Site region and, possibly, provide information on how rapidly the transition took place.

REFERENCES

Adams, MR and NR Wing. **1986**. *Protective Barrier and Warning Marker System Development Plan*, RHO-RE-PL-35 P. Rockwell Hanford Operations, Richland, WA.

Allen, BD and RY Anderson. **1993**. Evidence from western North America for rapid shifts in climate during the last glacial maximum. Science 260:1920-1923.

Berger, AL. **1981**. The Astronomical Theory of Paleoclimates, pp. 501-525. In: *Climatic Variations and Variability: Facts and Theories*, A Berger, (ed.). D. Reidel Publishing Company, Dordrecht, The Netherlands.

Berger, WH and LD Labeyrie (eds.). **1987**. *Abrupt Climatic Change: Evidence and Implications*. D. Reidel Publishers, Dordrecht, The Netherlands.

Berger, A, J Imbrie, J Hays, G Kukla, and B Saltzman (eds.). **1984**. *Milankovich and Climate*. D. Reidel Publishing Company, Dordrecht, The Netherlands.

Berger, A, H Gallee, T Fichefet, I Marsiat, and C Tricot. **1990**. Testing the astronomical theory with a coupled climate-ice sheet model. Palaeogeogr. Palaeoclimatol. Palaeoecol. (Global and Planetary Change Suppl.) 89:125-141.

Berger, A, H Gallee, and JL Melice. **1991**. The Earth's Future Climate at the Astronomical Timescale, pp. 148-165. In: *Future Climate Change and Radioactive Waste Disposal: Proceedings of International Workshop*, GM Goodess and JP Palutikof (eds.). NSS/R257. U.K. Nirex Radioactive Waste Disposal, Ltd., Harwell, UK.

Cadwell, LL (ed). **1991**. *Hanford Site Protective Barrier Development Program: Fiscal Year 1990 Highlights*, PNL-7831. Pacific Northwest Laboratory, Richland, WA.

Calder, N. **1974**. Arithmetic of ice ages. Nature 252:216-218.

Chatters, JC. **1991**. Long-Term Climate Change Effects. In: *Hanford Site Protective Barrier Development Program: Fiscal Year 1990 Highlights*, PNL-7831, LL Cadwell (ed.). Pacific Northwest Laboratory, Richland, WA.

Chatters, JC and KA Hoover. **1992**. Response of the Columbia River fluvial system to holocene climatic change. Quat. Res. 37:42-59.

Chatters, JC, DA Neitzel, MJ Scott, and SA Shankle. **1991**. Potential impacts of global climate change on Pacific Northwest spring Chinook salmon (*Oncorhynchus tshawytscha*): An exploratory case study. Northwest Environ. J. 7:71-92.

Crowley, TJ and GR North. **1991**. Paleoclimatology. In: *Oxford Monographs on Geology and Geophysics*, No. 16. Oxford University Press, New York.

Dansgaard, W, SJ Johnsen, HB Clausen, D Dahl-Jensen, NS Gunderstrup, CU Hammer, CS Hvidberg, JP Steffensen, AE Sveinbjornsdottir, J Jouzel, and G Bond. **1993**. Evidence for general instability of past climate form a 250-kyr ice-core record. Nature 364:218-220.

Franklin, JF, FJ Swanson, ME Harmon, DA Perry, TA Spies, VH Dale, A McKee, WK Ferrell, JE Means, SV Gregory, JD Lattin, TD Schowalter, and D Larsen. **1991**. Effects of global climatic change on forests in northwestern North America. Northwest Environ. J. 7:233-254.

GRIP (Greenland Ice-core Project) Members. **1993**. Climate instability during the last interglacial period recorded in the GRIP ice core. Nature 364:203-207.

Hays, JD, J Imbrie, and NJ Shackleton. **1976**. Variations in the earth's orbit: Pacemakers of the Ice Ages. Science 194:1121-1132.

Houghton, JT, GJ Jenkins, and JJ Ephraums. **1990**. *Climate Change: The IPCC Scientific Assessment*. Cambridge University Press, Cambridge, UK.

Houghton, JT, BA Callander, and SK Varney. **1992**. *Climate Change 1992: The Supplementary Report to the IPCC Scientific Assessment*. Cambridge University Press, Cambridge, UK.

Imbrie, J and JZ Imbrie. **1980**. Modeling the climatic response to orbital variations. Science 207:943-953.

Kukla, GJ. **1981**. Pleistocene Climates on Land, pp. 207-232. In: *Climate Variations and Variability: Facts and Theories*, A Berger (ed). D. Reidel Publishing Company, Dordrecht, The Netherlands.

Leovy, C and ES Sarachik. **1991**. Predicting climate change for the Pacific Northwest. Northwest Environ. J. 7:169-201.

Melice, JL and A Berger. **1989**. Modèle de prévision statistique du type Box-Jenkins pour la prévision du climat des 10,000 prochaines années, Scientific Report 1989/6. Institut d'Astronomie et de Géophysique, G. Lemaitre (ed.). Université Catholique de Louvain-la-Neuve, Belgium.

Petersen, EL and SE Larsen. **1978**. A statistical study of a composite isotopic paleotemperature series from the last 700,000 years. Tellus 30:193-200.

Petersen, KL, JC Chatters, and WJ Waugh. **1993**. *Long-Term Climate Change Assessment Study Plan for the Hanford Site Permanent Isolation Barrier Development Program*, WHC-EP-0569 REV 1. Westinghouse Hanford Company, Richland, WA.

Phillips, FM, AC Campbell, R Roberts, and CV Kruger. **1990a**. Abrupt Interglacial/Glacial Transition in the 1.4 Ma Water Balance Record from Searles Lake, California, pp. 28. In: *American Quaternary Association Program and Abstracts*, June 4-6, 1990. University of Waterloo, Waterloo, Ontario, Canada.

Phillips, FM, MG Zreda, SS Smith, D Elmore, PW Kubik, and P Sharma. **1990b**. Cosmogenic chlorine-36 chronology for glacial deposits at Bloody Canyon, eastern Sierra Nevada. Science 248:1529-1532.

Stone, WA, JM Thorp, OP Gifford, and DJ Hoitink. **1983**. *Climatological Summary for the Hanford Area*, PNL-4622. Pacific Northwest Laboratory, Richland, WA.

Wing, NR and GW Gee (eds.). **1990**. *Hanford Site Protective Barrier Development Program: Fiscal Year 1989 Highlights*, WHC-EP-0318. Westinghouse Hanford Company, Richland, WA.

Winograd, IJ, TB Coplen, JM Landwehr, AC Riggs, KR Ludwig, BJ Szabo, TP Kolesar, and M Revesz. **1992**. Continuous 500,000-year climate record from vein calcite in Devils Hole, Nevada. Science 258:255-260.

INDEX TO AUTHORS